The Anthropocene: Politik—Economics—Society—Science

Volume 10

Series editor

Hans Günter Brauch, Mosbach, Germany

More information about this series at http://www.springer.com/series/15232
http://www.afes-press-books.de/html/APESS.htm
http://www.afes-press-books.de/html/APESS_10.htm

Michael Durant Thomas

The Securitization of Climate Change: Australian and United States' Military Responses (2003–2013)

UNSW
THE UNIVERSITY OF NEW SOUTH WALES

Springer

Michael Durant Thomas
School of Humanities and Social Science
Australian Defence Force Academy,
University of New South Wales
Canberra
Australia

More on this book is at: http://afes-press-books.de/html/APESS_10.htm

ISSN 2367-4024 ISSN 2367-4032 (electronic)
The Anthropocene: Politik—Economics—Society—Science
ISBN 978-3-319-49657-3 ISBN 978-3-319-49658-0 (eBook)
DOI 10.1007/978-3-319-49658-0

Library of Congress Control Number: 2017937526

Cover photo: Sunset in the Australian Outback close to Alice Springs, July 2010. © Hans Güner Brauch, who granted permission.
The photo on page iii shows an Australian surf rowing boat, south of Cape Helles, Turkey on 23 April 2015. Behind the rowers in the distance is a large storm coming across the Mediterranean. The author kayaked from Istanbul to Anzac Cove with Jason Wilson and Turkish paddlers in memory of those soldiers, from both sides, who fought in the Gallipoli Peninsula campaign of 1915. © The author.

Copy-editing: PD Dr. Hans Günter Brauch, AFES-PRESS e.V., Mosbach, Germany
Style and English Language Editor: Dr. Vanessa Greatorex, Chester, England, UK

Printed on acid-free paper

This Springer imprint is published by Springer Nature
The registered company is Springer International Publishing AG
The registered company address is: Gewerbestrasse 11, 6330 Cham, Switzerland

This book is dedicated to those people who have worked tirelessly throughout their lives to bring us greater clarity and understanding of the dangers posed by climate change. In an era of 'fake news' and 'alternative facts', may they continue to have the courage to speak truth to power.

Foreword

Nowadays I am often asked: "What could the military possibly have to do with climate change?" So, it is with great pleasure that I introduce this book by Michael Thomas because it addresses this question precisely.

With an analyst's eye for the detail, but tempered with a military practitioner's outlook, Michael has written a narrative account of U.S. and Australian military responses to the challenges of climate change, highlighting the various climate initiatives and programmes undertaken by these organisations since 2003. In reading his account I am particularly drawn to the conclusion that Australia has trailed well behind its major alliance partner in taking up the challenge (although, future US military leadership on climate action is not guaranteed under President Trump).

Around the globe military forces are significantly exposed to climate risks. They use vast natural land holdings, possess large and dispersed infrastructure assets and consume significant amounts of energy of all natures, but particularly fossil fuels, as well as being major national employers. The world's military forces have a huge carbon footprint!

Yet, the same military forces also play a unique role in contributing to global peace and security through upholding and defending our nation states and, in the west, promoting the importance of the rule of law. For these reasons governments have a responsibility to ensure that military forces are adequately prepared for the range of regional and international climate security risks—including a substantial increase in the demand for humanitarian assistance and disaster relief operations. Given the challenges from climate change consequences that lie ahead there is a fair question to ask about whether we are doing enough, or not.

At the end of this book Michael poses a serious question for consideration. How will our militaries in the future look back on their efforts to head off a potential climate crisis in this critical decade? For many senior officers in our military forces this is a highly-charged question because the demands of acknowledging the primacy of the civil authority sometimes creates a political straight jacket which is blind to the reality of the demands flowing from climate change consequences.

Further, if we assess that insufficient has been done so far in this critical decade, is there time to recover the situation?

The answer lies, in part, through our militaries showing leadership. To paraphrase Georges Clemenceau, "climate change is too important to be left to the politicians". Solving the climate change challenge will need everyone—military included—to bring their efforts to bear on the problem using unique skill-sets and other contributions for building sustainable, climate friendly assets and having the willpower to tackle the key issues head on. Few people are better placed to understand this demand than our military personnel, who live in a world of uncertainties all the time.

As the former commander of the Australian Defence Force and current Australian representative on the Global Military Advisory Council on Climate Change I view climate change as a major, and urgent, threat to global peace and security because of the instabilities it has the potential to create. The military has an important role to play in taking the agenda for change forward. But, military forces alone cannot be a substitute for governments, businesses, and communities taking assertive, positive action now as time is running out in the critical decade.

March 2017
Canberra, Australia

Admiral Chris Barrie
(Former Chief of the Australian Defence Force)
Current Australian representative on the Global Military
Advisory Council on Climate Change

Preface

It is my great pleasure to contribute a preface to this book—an important contribution to the emerging climate security discourse. Like Michael, as a military practitioner I have witnessed the ravages of conflict up-close and personal. In 1992 I served in Somalia as a junior US Marine and have conducted missions all over the world since.

Frequently though, I return to thinking about Somalia.

Walking through the streets of Mogadishu was like entering the pages of Robert D. Kaplan's *Coming Anarchy*. For me at least, it seemed clear that the convergence of environmental degradation and political dysfunction had rendered an entire population of people to the fate of a famine which claimed over 220,000 lives. The juxtaposition of modern militarised hovercraft laden with relief supplies gliding smoothly but noisily onto the coastline of a community that in many aspects was stuck in the eighteenth century is one that is hard to forget—a different kind of clash of civilisations that was a harbinger of what lay ahead.

As I progressed in my military career I became acutely aware that a common underlying thread amongst many such combat missions was the element of environmental degradation leading to turmoil and conflict. Increasingly though the military also became first-responders in the event of natural disaster, providing humanitarian aid, reconstruction and leadership as the security situation beckoned. In recent times this appears to have accelerated—climate change emerging as a twenty-first century mega-trend that is re-writing the rule book: 'Mega Typhoons', 'super storms', 'once-in-500-year' droughts, floods, wildfires and 'extreme weather' are now commonplace in the Anthropocene. The military is at the forefront in responding to these events. As the US Navy asks: 'If not us, then who?'

Dr. Thomas carefully researched book not only does a fine job zeroing in and explaining the geo-security and political impacts of the world-wide changes in the environment but he also expertly frames and elucidates how the phenomenon has already begun to impact armed forces' missions sets, facilities and operational readiness. His findings such as the "potential for increased tension over changes to resource availability, access and use may flare into low-level and wide-scale conflict involving national, regional and international military intervention" meshes squarely with my own conclusions as well as those of other climate security researchers and is a warning that should be heeded. The science driven findings found in his work connects dots that create an outline of the challenge that lie ahead for military organisations, most specifically, the new forces that will continue to not only degrade operational readiness, but that will also create entirely new mission sets. Beyond this, Dr. Thomas also poses a deeper question: what role does the military have in hastening societal and political action on climate change?

Lastly, Michael's message is a clarion call for action directed at Armed Forces across the world; alerting them to prepare for a time where the non-traditional joins the traditional at the centre of national security concern and action. We would all be wise to heed the call in the face of the slow moving emergency—*climate change.*

Oliver Leighton Barrett
U.S. Navy (Retired)
Foreign Policy Association Contributor

Acknowledgements

This book owes itself entirely to those that have supported me along the way.

Dr. Gavin Mount and Associate Professor Stuart Pearson, from the University of New South Wales, have provided continuous support and encouragement. Gavin, thank you for your unbounded enthusiasm and shaping my understanding of securitisation theory. To Stuart, thank you for being a great mentor. Without your experience, steady-hand and precise interventions in moments of crisis, this research would simply not have been possible. You nudged it in the right direction at crucial junctures.

The Australian Defence Force has also been instrumental. On the one hand, while this book adopts a somewhat critical perspective of ADF climate policy, it remains an institution open to strong debate and passionate ideas. The team from the ADF Global Change and Energy Sustainability Initiative are firmly in this tradition and have been important in injecting policy relevant discourse on the subject within the ADF. Likewise, I've been fortunate to have had timely encouragement, informal discussions and inspiration on the subject with Bruce Thom and David Karoly from the Wentworth Group of Concerned Scientists, Anthony Bergin of the Australian Strategic Policy Institute, Admiral Chris Barrie (retired), Associate Professor Matt McDonald from The University of Queensland and Alix Pearce of the Climate Council.

Many thanks also to the series editor, Hans Günter Brauch. Your early writing on the subject, in conjunction with your colleagues, has influenced my thinking and shaped my outlook in untold ways. Thank you for your guidance, support and advice in putting this book together. I would also like to thank Vanessa Greatorex for her dedicated language editing and the Springer team in Heidelberg (Germany) and Chennai (India) who made this book happen.

And lastly to my family, Jodie, Patrick and Elsie-Mae, my parents and to my essential 'support crew', Denis and Judy. Thank you for your tireless support and encouragement throughout the development of this book.

Canberra, Australia — Michael Durant Thomas
January 2017

Contents

Abbreviations

ADF	Australian Defence Force
ADM	Admiral
ALP	Australian Labor Party
ARRA	American Recovery and Reinvestment Act
AU	Australia
BAU	Business-As-Usual
BMU	German Ministry for Environment, Nature Conservation and Nuclear Safety
CDF	Chief of Defence Force (Australia)
CJCS	Chairman of the Joint Chiefs of Staff
CNO	Chief of Naval Operations (United States)
COCOM	Combatant Command
COIN	Counter Insurgency
COP	Conference of the Parties
CPRS	Carbon Pollution Reduction Scheme
CSP	A Cooperative Strategy for the twenty-first Century Seapower
DoD	Department of Defense (US) or Department of Defence (Australia)
DoDI	Department of Defense Instruction
DoE	Department of Environment
DSG	Defence Support Group
DSTO	Defence Science Technology Organisation
DWP	Defence White Paper
EEZ	Exclusive Economic Zone
EIT	Economies-In-Transition
ELF	Enhanced Land Force
ENSO	El Niño Southern Oscillation
EO	Executive Order
EPA	Environmental Protection Agency
ESSP	Earth System Science Partnership

EU	European Union
FMOC	Future Maritime Operating Concept
FOI	Freedom of Information
FPR	Force Posture Review
FY	Financial Year
GDP	Gross Domestic Product
GEHSHA	Global Environmental and Human Security Handbook for the Anthropocene
GFC	Global Financial Crisis
GHG	Greenhouse Gas(es)
GCESI	Global Change and Energy Sustainability Initiative
GMACCC	Global Military Advisory Council on Climate Change
HADR	Humanitarian Aid and Disaster Relief
HDI	Human Development Index
HMAS	Her Majesty's Australian Ship
HNA	Hardened and Networked Army
ICAO	International Civil Aviation Organization
ICSU	International Council for Science
IGBP	International Geosphere-Biosphere Programme
IHDP	Human Dimensions Programme on Global Environmental Change
INDC	Intended Nationally Determined Contributions
IPCC	Intergovernmental Panel on Climate Change
IPCC AR4	Intergovernmental Panel on Climate Change Fourth Assessment Report
IPCC AR5	Intergovernmental Panel on Climate Change Fifth Assessment Report
IT	Information Technology
MDG	Millennium Development Goals
MAB	Military Advisory Board
MINDEF	Minister of Defence
MoD	Ministry of Defence (UK)
NCCO	Naval Climate Change Coordination Office
NDAA	National Defense Authorization Act
NDS	National Defense Strategy
NGO	Non-Government Organisation
NMS	National Military Strategy
NSS	National Security Strategy
OPEC	Organisation of Petroleum Exporting Countries
PPBE	Planning-Programming-Budgeting-Execution
POM	Programme Objective Memorandum
QDR	Quadrennial Defense Review
RAAF	Royal Australian Air Force
RADM	Rear Admiral
RAN	Royal Australian Navy

RAND	(US) Research and Development Corporation
SECDEF	Secretary of Defence
SSPP	Strategic Sustainability Performance Plan
TFCC	Task Force Climate Change
UFC	United Facilities Criteria
UN	United Nations
UNDP	United Nations Development Programme
UNFCCC	United Nations Framework Convention on Climate Change
UNSC	United Nations Secretary General
US	United States of America
USAF	United States Air Force
USCENTCOM	United States Central Command
USEUCOM	United States European Command
USNORTHCOM	United States Northern Command
USPACOM	United States Pacific Command
USMC	United States Marine Corp
USN	United States Navy
WCRP	World Climate Research Programme
WBGU	German Advisory Council on Global Change
WMD	Weapons of Mass Destruction
WWF	World Wildlife Fund

List of Figures

List of Tables

Part I
Introduction

Chapter 1
The Strategic Dissonance of Australia's Climate Security Response

Abstract This chapter outlines how the Australian Department of Defence, and the Australian Defence Force (ADF) in particular, developed a strategic apathy in its early response to addressing the challenges posed by climate change. It sketches the wider sectoral, political and allied-military responses to climate change and highlights existing risks and knowledge gaps in ADF climate response. Four main reasons are identified for ADF climate apathy: incompatible timeframes, institutional reluctance, more appropriate use of resources, and the challenge of an avowedly non-partisan institution grappling with an ostensibly political issue. The chapter concludes by describing the purpose and structure of this book.

Keywords Climate change · Climate politics · Climate security · Climate military · ADF

1.1 Climate Change in Australian Military and National Security Discourse

Climate change is an issue that hardly requires an introduction. In Australia, it has been prominent in federal politics since at least 2006, when it was the number one foreign policy concern and (in 2007) the most important domestic policy priority (Wesley 2012: 632). In 2004, Australian Prime Minister John Howard (2004) declared that climate change was 'one of the major challenges confronting the world this century', and in 2008 his successor, Kevin Rudd, (in)famously labelled it the 'greatest moral, economic and social challenge of our time' (Rudd 2008d). Although Tony Abbott (Australian Prime Minister 2013–2015) had once labelled climate change as 'absolute crap' (quoted in Rintoul 2009) and was widely seen as a 'sceptic', his successor and current Prime Minister, Malcolm Turnbull stated to the 2016 Pacific Island Forum that '[f]or Australia there is no more pressing need for regional action than on climate change' (Turnbull 2016). Scholars and the political commentariat alike have variously identified climate change (and its

M.D. Thomas, *The Securitization of Climate Change: Australian and United States' Military Responses (2003–2013)*, The Anthropocene: Politik—Economics—Society—Science 10, DOI 10.1007/978-3-319-49658-0_1

derivative issues) as a major factor at the 2007, 2010 and 2013 Australian federal elections (Macintosh et al. 2010; Hartcher 2014).

Today, climate change remains a 'serious and pressing problem' for most Australians (Oliver 2015: 1). There are good reasons for this. Since the late 1970s the global scientific community has warned with increasing confidence that Business-As-Usual (BAU) anthropogenic greenhouse gas (GHG) emissions will increase planetary warming 2°C above pre-industrial levels by 2050. The United Nations (UN) Intergovernmental Panel on Climate Change (IPCC) Fifth Assessment Report stated, '[h]uman influence on the climate system is clear … Warming of the climate system is unequivocal' (IPCC 2015: 2) and its consequences have been covered by globally prominent sectors *beyond* the scientific community (Stern 2007; Garnaut 2008; Asian Development Bank 2010; World Bank 2013; ICAO 2015; World Vision 2015; International Federation of Red Cross 2015; Pope Francis 2015). Various global leaders have repeatedly identified climate change as *the* strategic issue set to define this century (UN 2008; UNSC 2011; UNSG 2015). Indeed, for all the time, attention and resources directed towards issues such as the South China Sea and the global fight against terrorism, former US president Barack Obama often placed climate change at the centre of global strategic concerns:

> For all the immediate challenges that we gather to address this week – terrorism, instability, inequality, disease – there's one issue that will define the contours of this century more dramatically than any other, and that is the urgent and growing threat of a changing climate (Obama 2014).

Given the ubiquity of climate change, Australia's political, social, economic and even military sectors cannot be considered exceptional to the effects of climate change. As a major agricultural exporter reliant on stable climatic conditions, with 85% of its population and infrastructure clustered along coastal margins, and as one of the world's highest per capita GHG emitters situated in a region of under-developed small island states exposed to climatic extremes magnified by El Niño Southern Oscillation (ENSO) events, Australia has been repeatedly cited as being vulnerable to climate change (Pittock 2009). Although many of these risks had previously been documented in a piecemeal fashion, the 2008 *Garnaut Climate Change Review* consolidated and raised them to national prominence for the first time (Garnaut 2008).

The release of the *Garnaut Review* triggered a litany of national, sub-national and sectoral reports that have detailed the impacts of climate change across Australian society and its political economy. A 2011 report by the then Department of Climate Change titled *Climate Change Risks to Coastal Buildings and Infrastructure*, for instance, identified $226 billion in replacement costs for commercial, industrial, road, rail and residential assets in the event of a 1.1 m sea level rise (Commonwealth of Australia 2011: 3). A follow-on report in 2014 by the Australian Climate Council, *Counting the Costs: Climate Change and Coastal Flooding*, noted the Australian coastline was 'highly vulnerable' to climate change and emphasised major risks to habitats, tourism, infrastructure, health, wetlands and other ecosystems (Steffen et al. 2014: iv). Despite some unevenness, planning for climate change has been undertaken by a wide range of federal, state and local government agencies, including the police forces and civil emergency services (Commonwealth of Australia 2007b, 2010; Gibbs/Hill 2011; Baker/McKenzie 2011; NCCARF 2013; Victorian Government 2011). Climate change has also increasingly featured across Australia's private sector, including the domestic insurance industry (Insurance Council of Australia 2008), the banking sector (Commonwealth Bank 2008; NAB 2015), health professionals (CAHA 2010), biodiversity (Steffen et al. 2009; Williams et al. 2009), the tourism sector (Tremblay et al. 2008) and a myriad of others, even including a climate change vulnerability assessment of the East Coast Tasmanian Rock Lobster Fishery (Pecl et al. 2009).[1] Fundamentally, few sectors in the Australian economy either have been, or are anticipated to be, untouched by the physical or regulatory impacts of climate change.

Notwithstanding all this attention, there is at least one area that—until very recently—has escaped scrutiny. Namely, Australia's national security sector and, more specifically, the Australian Defence Force (ADF). Although works exist in the Australian scholarly community on broader climate security matters (Edwards 1999; Barnett 2003, 2009; Barnett et al. 2010; Chaturvedi/Doyle 2010; McDonald 2012; Christoff/Eckersley 2014), Australian military journals (Lawson 2007; Thomas 2011, 2013) and think-tank reports (Dupont/Pearman 2006; Bergin/Townsend 2007; Press et al. 2013; Sturrock/Ferguson 2015; Barrie et al. 2015), analysis by the ADF on how it has responded, and plans to respond, to climate change has been limited. Indeed, while Defence White Papers (Defence 2009b, 2013b, 2016c) raised some aspects, the ADF itself has not published a *single* publicly available document that focuses exclusively on how it is strategically responding to climate change (Thomas 2015: 98). Missing from this discussion are explanations about how climate change may impact critical homeland security-infrastructure, deployable military capabilities, international and regional security, workforce health and well-being or how climate change regulation will affect Defence industry and capability acquisition costs.[2] This lack of published output has not just been limited to formal policy documents. Looking more broadly, Australian military leaders have not outlined a climate change strategy or policy approach for the ADF in any major speeches, articles or media interviews. As a self-proclaimed 'strategy-led organisation'—responsible for vast manpower,

[1]For a comprehensive list of Australian initiatives relating to climate change, see Department of Environment at: http://www.environment.gov.au/climate-change/adaptation/publications (12 June 2016).

[2]While the ADF has not progressed any significant climate policy, public analysis of its response has increased in the lead-up to the release of the next Defence White Paper. Two important reports were the Centre for Policy Development, *The Longest Conflict: Australia's Climate Security Challenge* and the Climate Council's *Be Prepared: Climate Change, Security and Australia's Defence Force.*

infrastructure and GHG emissions—the lack of ADF climate policy stands in contrast to that progressed by the civil sector (Defence 2010c: i).[3,4]

To trace the roots of this emerging dissonance and set the context for this book, it is necessary to overview Australia's climate security policy settings. The first observation is that climate change remained absent from Australia's strategic policy agenda at precisely the time it was identified elsewhere as becoming important. This was first described in 2006 by Alan Dupont and Graeme Pearman in *Heating Up The Planet*, when they argued 'that the wider security implications of climate change have been largely ignored [in Australia] and seriously underestimated in public, policy, academia and the media' (Dupont/Pearman 2006: viii). In late 2006 the then Secretary of the Department of Defence, Mr. Ric Smith, conceded that Defence had done no internal analysis of the security implications of climate change (Commonwealth of Australia 2006: 70).

The second observation is that even once climate change was identified as a public policy priority around 2007, it remained at the margins of Australia's national security agenda. In 2008, then Prime Minister Kevin Rudd stated that 'less attention has been given to the security implications' and that he would initiate its 'formal incorporation … [into] Australia's national security policy and analysis process' (Rudd 2008g). But the outcome was decidedly mixed. In *Heavy Weather: Climate and the Australian Defence Force*, the think-tank Australian Strategic Policy Institute (ASPI) stated that the 2009 Defence White Paper (*Force 2030*) had dismissed climate change as an issue for future generations (Press et al. 2013). Likewise, the 2013 *National Security Strategy* stated Australia should [only] 'monitor' the issue and cast it as a 'broader global challenge' that '*may* contribute to instability and tension' at some point in the future (Commonwealth of Australia 2013d: 31; emphasis added). Despite subtle changes to this message in the 2016 Defence White Paper, various think-tank reports and commentators decried a lack of substance and action on the issue (Sturrock/Ferguson 2015; Barrie et al. 2015).

The absence of climate change in national security circles, particularly the military, was also evident. In one of his first major speeches as Chief of [Australian] Defence Force (CDF), General David Hurley (CDF in tenure 2011–2014) identified four fundamental strategic shifts occurring in the world: climate change or any form of global environmental change was not listed amongst them. On his appointment as CDF, General Hurley also issued an 'All Points Bulletin' where he listed his

[3]The ADF has a 'strategy' for literally dozens of issues. Examples include: *ADF Indigenous Employment Strategy*; *Defence Air Quality Strategy*; *Energy Management Strategy*; *Waste Minimisation Strategy*; *Defence Pollution Management Strategy; Defence Environmental Education Strategy; Defence Noise, Vibration and Electromagnetic Radiation Generation Strategy,* to name a few.

[4]The Australian Department of Defence ('Defence') has an annual budget of $AUD31.9 billion (13th largest) in 2015 according to SIPRI (Perlo-Freeman et al. 2016) and a full-time uniformed and public service workforce of around 75,000 people (Defence 2016b). Defence also manages an estate and infrastructure portfolio of more than 3 million hectares, 300 managed properties and 25,000 buildings (Defence 2016e).

priorities as CDF. Again, climate change was not mentioned (Hurley 2011). In 2015 Hurley's replacement, Air Chief Marshal Mark Binskin, stated that, while the ADF plans for the consequences of a changing climate, it does 'not specifically plan for climate change' per se (Binskin cited in Commonwealth of Australia 2015b: 42). Climate change has also been missed by major ADF reports and reviews. The 2012 *Force Posture Review* commissioned by the then Minister of Defence to assess whether the ADF was positioned to meet Australia's future strategic and security challenges did not mention, discuss or analyse any matters relating to climate change (Hawke/Smith 2012). The *2012–2017 Defence Corporate Plan* that outlined future directions and key priorities including sustainment and development of the organisation as well as the key risks confronting it, did not mention climate change. Similar accounts are to be found in service-level strategy documents, including the Australian Army's annual assessment of the land operating environment out to 2035, the *Future Land Warfare Report* (Australian Army 2014).

This thinking has also been manifest in lower-level ADF policy documents that might have been expected to address climate change. The ADF's lead environmental policy statement, the *Defence Environmental Strategic Plan 2010–2014* used the term 'climate change' only twice in thirty-four pages and otherwise used the description 'climate variability' (Defence 2010b). Although this document focused on ADF compliance with legislative obligations, no discussion existed on how climate change might impact the vast Defence estate, including the most obvious of risks such as predicted sea-level rise to its sizable maritime infrastructure or other low-lying Defence bases. At two pages in length, the 2007 *Defence Energy Policy* also suggested the issue had been taken lightly. While describing Defence as one of the highest consumers of energy within the Australian Government, the policy targeted small, low impact energy reductions in areas such as lighting, heating, hot water services, cooking appliances and swimming pools (Defence 2007a). The ADF's 2008 *Combat Climate Change* 'initiative' was striking if only in conflating the exact same policies from the 2007 *Energy Policy*, *Defence Environmental Strategic Plan (2010–2014)* and the 2006 *Energy in Government Operations*. Its adoption of limited energy efficiency targets for individuals and its somewhat underwhelming presence (unpublished in hard-copy, available only online at the Defence Support Group webpage and removed from the public ADF website completely in 2017) made this less a departmental-wide strategy than a narrowly focused sub-set of its environmental policy.[5]

Similar accounts can be found in the 2006–07 *Energy Use in Australian Government Operations*, where the government championed the ADF as demonstrating 'leadership' on energy efficiency through low-level approaches. This pamphlet described how a regional Defence office had developed an energy strategy that, after meeting with a 'great deal of concern' from information technology

[5]Since this author first visited the Defence Support Group *Combat Climate Change* web-page in 2011 not one single change to the website was recorded. The website, and presumably the policy, was removed by the ADF at the start of 2017.

(IT) staff, was eventually successful in decreasing Brindabella Park ('Office 3') electricity consumption by '23.4%' (Commonwealth of Australia 2008b: 43). The punch-line to this anecdote—and the idea that Defence failed to adopt a strategic outlook—was delivered when it was recognised that the strategy would not have succeeded if not for the implementation of a rewards based system of 'minties and chocolates … double movie passes [and] Boris, the Energy Hog' who was particularly 'embraced by most personnel and contributed to the overall success of the project' (Commonwealth of Australia 2008b: 43).

Even in places where serious discussion would be expected, for example in Defence journals, papers and seminars, there appeared little interest or conviction. A content search performed in 2015 of the ADF's leading journal, the *Journal of the Australian Profession of Arms,* under the search category 'environmental issues' yielded five articles: one on ADF cultural heritage issues, two on 'environmental security', another on environmental law and the fifth, a dubiously titled 1983 piece regarding 'The Military Potential of the Feral Horse and Camel in North West Australia'. In thirty-five years of publishing history the *Journal of the Australian Profession of Arms* has published just *two* articles exclusively relating to the impact of climate change on Defence (both written by this author in 2011 and 2013).

In sum, none of these documents detailed how the physical or regulatory effects of climate change would impact national security, the wider Department of Defence or the Australian Defence industry more broadly. As a case in point, the so-called carbon tax as part of the *Clean Energy Act 2011* was forecast by Treasury to cost the ADF $81.9 million for FY 2012–13, yet it was missed as a strategic issue (Press et al. 2013: 25).

Although more recent Australian Department of Defence publications have begun to address climate change (Defence 2013a, 2016d) and its leaders are beginning to address the subject (Campbell 2016), there still exists a number of knowledge gaps. A special report in 2013 by the Australian Strategic Policy Institute (ASPI), *Heavy Weather: Climate and the Australian Defence Force,* identified six such knowledge gaps. These included a need to develop better understanding on the impact of climate change on: (1) its missions; (2) capability development and acquisition; (3) adaptation and mitigation of the Defence estate; (4) supply chain (energy) security and fuel interoperability; (5) preparedness; and (6) personnel and training (Press et al. 2013). Follow-up reports by the Centre for Policy Development, *The Longest Conflict: Australia's Climate Security Challenge* (2015), and the Climate Council's *Be Prepared: Climate Change, Security and Australia's Defence Force* (2015) made similar findings and suggest they continued to be left unaddressed as recently as 2015.

Placed in a broader context, two key points became evident. The first was the emergence of a civil-military divide on the scale and urgency required to address climate change. Fundamentally, the civil sector appeared to be well ahead of the Australian military, even though the risks to the ADF were becoming increasingly established. The second point was an emerging concern in some quarters that the apathetic ADF response might give rise to potentially avoidable operational, capability and institutional risks. On consideration of these aspects, the question may be asked: Why did the ADF fail to develop a clear strategic response to climate change?

Four initial reasons are presented here, but are further developed in the ADF case study in Chap. 6 and comparative analysis in Chap. 8. First, climate change has been framed in some quarters as a long-term threat and in timescales incompatible with Defence budgeting cycles and ADF threat-horizons. The Australian Defence White Paper (DWP) 2009 stated as much when it declared the 'strategic consequences of climate change are … not likely to be felt before 2030' (Defence 2009b: 31). With more pressing security concerns to deal with, that assessment (from 2009) effectively registered climate change as a low policy priority that left the ADF 'without a guiding compass' (Sturrock/Ferguson 2015: 34).

Secondly, even though climate change increasingly featured as part of the ADF's Humanitarian Aid and Disaster Relief (HADR) considerations, there remained a perception that it had little to do with the core business of war-fighting. Rob Sturrock and Peter Ferguson (Centre for Policy Development) in *The Longest Conflict: Australia's Climate Security Challenge* labelled this 'institutional reluctance' (Sturrock/Ferguson 2015: 34). Likewise was the perception in Defence that other government departments (such as Departments of Agriculture, Health, Industry, Infrastructure, Environment, and Foreign Affairs), other government agencies (including CSIRO, Australian Energy Regulator or Renewable Energy Agency) or the national intelligence community (Office of National Assessments) remained far better placed to respond on matters of climate change. Furthermore, the period 1999–2013 was one of the most operationally demanding for the ADF, with simultaneous and multiple peace-keeping deployments in the Asia-Pacific and wars in Iraq and Afghanistan. With soldiers on the front line, the ADF was more motivated to focus on defeating contemporary enemies than worry about incremental rises in global sea-level.

The third reason concerned best use of limited resources. In a 2013 speech, the then Chief of the Australian Army, Lieutenant-General David Morrison, forcibly argued this point when he stated that building Defence capability meant ensuring 'real value is extracted from limited resources' (Morrison 2013: 9). To do this, Morrison warned of succumbing to the 'latest [strategic] fad', which can leave 'confused policy makers and force planners in their wake' (2013: 7). According to Morrison, 'we inhabit the world of Hobbes not Fukuyama—less solitary, but vastly more brutal' in which the ADF must focus on core issues, urging 'we exist to defeat the nation's enemies' rather than 'semantic and philosophical discourse' which tended to regard the ADF as 'experimental organisations for each transient intellectual speculation' (Morrison 2013: 6–8). Thus, for Morrison and the other service chiefs, climate change might simply have represented an unwanted philosophical distraction that did little to enhance the immediate war-fighting skills required to defeat peer-enemies and uphold Australia's national security. In this view, the allocation of resources to adapt or mitigate for future climate threats may not have been a mere distraction, but might have been perceived by the military as a risk to national security on the grounds it could have diverted scarce budget otherwise meant for war-fighting. The seeming re-emergence of geo-strategic issues—in the form of a more muscular Russian military; the on-going crisis in Syria (since 2011); of the rise of the political 'right' across Europe in response to border security; terrorism; and of the sustained growth of the Chinese military and its encroachment in the South China Sea—would all support such a focus on 'hard' security.

In line with General Morrison's thinking, Lieutenant-General Peter Leahy (a former Chief of the Australian Army) proffered that militaries are also conservative institutions that can be slow to adapt (Leahy 2010: 9). Perhaps this was true of climate change, and that the ADF remained cautious until the risks became clearer, more urgent and the science more granular? Another possibility pointed towards the idea that the ADF might not be the strategic organisation it purports to be. In this context, climate change may simply be too complex and overwhelming even for vast resources of the military-bureaucracy to adapt and mitigate. As the Professor of Strategic Studies at the Australian National University and former Departmental Deputy Secretary for Strategy, Hugh White, has suggested, 'The ADF is not a strategic organisation. It is very much focused at the tactical level' (quoted in Morris 2015). This position has been reflected elsewhere, notably by the strategist Peter Layton, who has argued 'the next DWP [Defence White Paper] is likely to be a non-strategy of risk management' (Layton 2014).

A final consideration related to sensitivities surrounding the *politics* of climate change—in particular, the divergent climate policies of Australia's two major political parties, the Liberal-National Party and Australian Labor Party (ALP), and how climate change had become a stark political differentiator or 'wedge issue' in domestic politics. Viewed this way, the ADF's reluctance to engage on climate change may be seen as an effort to retain its non-partisan persona, thereby actively avoiding political debate. Moreover, the political dimension of climate change discourse and its influence on the Australian military cannot be dismissed (this aspect forms a crucial finding of this book).

Suffice to say, the limited offerings by the Department of Defence on the subject of climate change has made it difficult to discern why it has been neglected. Furthermore, few contributions have been provided by the (Australian) scholarly academy, where research on the subject of the military and climate change has been sparse. To gain further insight into the ADF's climate dissonance, it is worthwhile to briefly compare how the international security community—including Australia's major allies the United States (US) and United Kingdom (UK)—have responded to climate change.

The Securitisation of Climate Change. Climate change first emerged as an important global governance issue in the 1980s when the UN established the IPCC in 1988 and then the Framework Convention on Climate Change (UNFCCC) in 1992. Widespread international interest was piqued in 2007 by the release of the IPCC Fourth Assessment Report (AR4) that declared 'warming of the climate system is unequivocal' (IPCC 2007: 30). The heightened sense that 'something must be done' witnessed unprecedented interest and was described by one leading securitisation scholar as a 'turning point' (Brauch 2009: 62). From 2007, climate change appeared as a mainstream issue and was prioritised on the global agenda at multiple G8/G20 summits, the European Union, the OECD, the Africa Union as well as debates on security consequences in the UN General Assembly (2008 and 2011) and UN Security Council (2007, 2011, 2013 and 2015).

As the perceived urgency of climate change increased, international interest and concern about its security implications grew. Scholars (Brauch 2012; Brzoska 2012b; Oels 2012) contend that the period between 2007 and 2011 was marked by high-level efforts to escalate the importance of climate change by framing it as a *securitised* issue, presenting it as a threat to human, national and international security. As climate change was placed on the global political (security) agenda, it was increasingly being considered by traditional securitising actors, namely the national defence ministries, military establishments and intelligence communities (IPCC 2015: 7). Michael Brzoska analysed the national security strategies of 31 countries for climate change as a security issue and concluded that 'climate change has become a widely, though not universally, accepted security issue among national security elites' (Brzoska 2012b: 175).[6] In short, high-level international efforts to securitise climate change resulted in the military sector becoming an important stakeholder in the field of 'climate security'. Indeed, by 2009 some scholars went so far as to conclude that 'recent evidence of abrupt climate change and threatening tipping points has bought the challenge of climate change into the urgent timescale of military contingency planning' (Spencer et al. 2009: 1).

This point was not lost on the militaries of Australia's two most historically important allies, the US and UK. Abetted by almost a decade of think-tank, scholarly and military literature on the security implications of climate change, the US and UK militaries appeared vastly more engaged on the subject than the ADF.[7]

[6]The international community, more broadly, is divided on the securitisation of climate change. Two main camps are evident. In one camp are countries working to ensure climate change remains a *non-securitised* issue addressed outside the logic of security politics. Although exceptions exist, this position is mainly advanced by Brazil, Russia, India and China (BRICS), developing countries and the Third World. The main argument held by the BRICS+ is that whilst climate change may present certain kinds of threats to international peace and security, it primarily remains a sustainable development issue where focus must be on delivering mitigation and adaptation strategies as well as finance mobilisation, technology transfer and capacity–building in developing countries. In contrast to these positions are 'pro-securitizing' countries that have sought to frame climate change as a securitised issue to be dealt through multi-lateral, non-securitised *and* securitised *forums*. Countries subscribing to this 'two-track' approach mainly consist of developed countries such as those within the EU, as well as the US, UK, Australia and a collection of other small island nations plus some developing countries most affected by climate change. These countries argue that securitising climate change raises its profile, engenders a sense of purpose and urgency into international mitigation and adaptation strategies and should be viewed as part of a wider strategy of 'preventative diplomacy'. See Thomas, Michael. 2013. 'The Securitisation of Climate Change: A Military Perspective.' *Journal of the Australian Profession of Arms*, no. 192: 7–18.

[7]A list of the climate-security literature within the US and UK is extensive. Prominent examples (to name a few) include: *An Abrupt Climate Change Scenario and Its Implications for US National Security* (Schwartz and Randall, 2003), *National Security and the Threat of Climate Change* (Centre for Naval Analyses, 2007), *The Age of Consequences: The Foreign Policy and National Security Implications of Global Climate Change* (Centre for Strategic and International Studies, 2007), *Broadening Horizons: Climate Change and the US Armed Forces* (Centre for New

A COMPARISON OF MILITARY ACTION ON CLIMATE CHANGE					
	Potential Effects of Climate Change On:	Military Climate Action Checklist	US DoD	UK MoD	ADF
1	Military Planning and Operations	Incorporated and mainstreamed climate change into National Strategic (Military) planning.	[green]	[green]	[orange]
		Shown leadership by appointing a Senior Military authority as Climate Change planning officer/envoy.	[green]	[green]	[red]
		Published a Climate Change Adaptation Strategy.	[green]	[green]	[red]
		Participating member of Interdepartmental / Interagency Climate Change Working Group.	[green]	[orange]	[orange]
		Analysis of climate change impacts on military base locations and base capacity (force posture).	[green]	[orange]	[red]
		Analysis of climate change impacts on how military is organised for combat missions, stabilisation operations and disaster relief (force structure).	[green]	[orange]	[red]
2	Military Training and Testing	Analysis of climate change risks to military training.	[green]	[orange]	[orange]
		Analysis of climate change impacts on individual soldier, sailor and airmen readiness.	[orange]	[red]	[red]
		Analysis of climate change impacts on the preparedness of the military to respond to operations and emergencies.	[green]	[orange]	[orange]
		Analysis / inclusion of climate change in military doctrine (eg., disaster relief doctrine or aid to civil community).	[orange]	[orange]	[orange]
		Analysis of health impacts of climate change on military forces and operational areas.	[orange]	[red]	[red]
3	Military Estate (Built and Natural Infrastructure)	Mandated renewable energy targets for military bases.	[green]	[orange]	[red]
		Conducted Risk Assessment of sea-level rise and inundation on military bases.	[green]	[green]	[green]
		Conducted a Risk Assessment of climate affected extreme weather events on military bases.	[green]	[green]	[red]
		Implemented 'Green' building codes and energy programs across military bases.	[green]	[green]	[green]
4	Military Acquisition and Supply Chain	Mandated fuel and energy efficiency goals in purchase of major military hardware and platforms (eg., use of bio-fuels and hybrids).	[orange]	[orange]	[orange]
		Analysed climate change risks to critical civilian infrastructure and civilian workforce and resultant impacts on military infrastructure, operations and training.	[orange]	[orange]	[red]
		Sustainable Procurement practices implemented (e.g. energy efficient civilian vehicle fleets, energy efficient lighting, heating, waste reduction strategies etc).	[green]	[green]	[green]

Key	
[green]	Implemented / Underway
[orange]	Partial Response
[red]	No Action or Evidence Found

Fig. 1.1 Comparison of ADF, US and UK military climate security response. *Source* Barrie et al. (2015). Reprinted with permission.

Both militaries, for instance, appointed senior military officials to oversee a strategic military response to address climate change.[8] Both had well-developed major policy papers detailing climate change adaptation strategies and, to varying degrees,

(Footnote 7 continued)

American Security, 2010), *MoD Climate Change Delivery Plan* (Ministry of Defence, 2010), *US Navy Climate Change Roadmap* (US Navy, 2010), *National Security Implications of Climate Change for US Naval Forces* (National Research Council, 2011).

[8]For the UK, see Rear Admiral Neil Morisetti at: https://www.gov.uk/government/people/neil-morisetti (11 May 2015). For the US, see Rear Admiral David W. Titley at: http://www.navy.mil/navydata/bios/navybio.asp?bioID=438 (11 May 2015).

declared positions to reduce energy consumption and limit emissions. Various scholarly works have also highlighted the US and UK militaries as important securitising actors responding to climate change (Brzoska 2012b; Hayes/Knox-Hayes 2014; Youngs 2015).

Although imperfect and not immune from criticism, the measures undertaken by the US and UK militaries in relation to climate change appeared in stark contrast to those adopted by the ADF (see Fig. 1.1). Thus, the ADF's strategic dissonance towards climate change appeared not just limited to the domestic horizon. Preliminary indications during research for this book revealed that such a dissonance extended to the international realm, where allied militaries with long histories of mutual cooperation, interoperability and policy alignment seemingly undertook proportionally larger efforts and showed far greater interest in responding to climate change.

1.2 General Purpose of This Book

At the broadest level, this book examines—and to some extent, documents—how armed forces have responded to the strategic challenges posed by climate change since 2003. While the opening chapters provide a broad overview of the science of climate change and general climate securitisation discourses, the main undertaking is the two case studies that focus on climate securitisation within select parts of the Australian and US militaries. Although this scope may be considered somewhat narrow (i.e., the prioritisation of national security concepts exclusively limited to military institutions), it is intended that by just focusing on two militaries the book is able to provide a deeper understanding than might otherwise be possible through an examination of multiple militaries or wider security perspectives (e.g., human security, environmental security or even analysis of national foreign policy objectives). In any case, the findings in this study may have wider applicability to other countries that possess similarly-sized armed forces and comparable political structures. Given that so little research has been done on how militaries are responding to climate change, it is hoped that this will at least stimulate discussion and provide a baseline for future research.

Distinctly, this book also examines the *intersection* between the political and the military spheres. This should be considered an important aspect, since the issue of climate change presents as both a general public-policy issue but also as a security issue. This can be problematic for some armed forces and military institutions that —for good reason—are often reluctant to be seen as entering, or weighing-in, on issues that are manifestly of a political nature. A close examination of the key political actors—their intent, outlook and political mandate for broader climate action—is therefore crucial to understanding the policy freedom and constraints that military leaders operate within. In this regard, Australia represents a particularly useful case study given the divisive split on climate policy between the major political parties that has existed since at least 2009, if not earlier. Pointedly, it shows how the Australian military curtailed its climate response to avoid being seen as overtly political; and herein lies the greater importance and applicability of this book. By demonstrating that the armed forces and wider security institutions were prone to downplaying, or publicly avoiding, the security threats posed by climate change on account of keeping their non-partisan status, it suggests something of a dilemma for societies attempting to genuinely and transparently grasp and deal with the true nature of the threats posed by climate change. As the US moves from the Obama to the Trump Administration, the findings of this book may act as a useful guide, if not a warning, to what might happen in terms of US climate security discourse.

1.3 Guide to Reading

The book is presented in three parts. Part I outlines the research origins and introduces climate change as a security issue in the context of recent Australian national security discourse. Part II is a literature survey of climate change as a security issue. It commences by outlining the physical basis of climate change, the broad security risks and its attendant military impacts before turning to a narrative history of policy-oriented literature on climate securitisation. This book does not seek to add to the literature but rather to provide context for the case studies. Part III consists of three chapters that contain the climate security case studies. The first chapter, Chap. 5, describes the aim, method and analytical framework applied throughout the case studies. The case studies themselves, which form the core research, are in Chap. 6 (Australia) and Chap. 7 (US). Case study discussion and analysis is at Chap. 8.

This book has been written for both generalists and specialists with an interest in the subject of climate security. Where possible, the book elevates and places emphasis on the *military* aspects within climate security discourses. In this manner, the scope of this book is rather narrow; it does not consider the wider security implications of climate change addressed from the perspective of other actors, sectors or even government agencies (for example, the Department of Foreign Affairs, State Department and so on). At appropriate junctures, the chapters pose a number of questions that attempt to highlight where research gaps exist. For specialists already well-versed in climate change science and climate securitisation literature, it is recommended that Part III, containing the case studies and comparison chapter, be your main focus. For generalists who are seeking a broader contextual understanding of climate science and also wider climate security debates in both policy and scholarly literature, Part II is worth considering. Finally, while the original research focused on an examination of military responses to climate change between the years 2003 and 2013, attempts have been made to update the literature and policy developments through until December 2016.

References

Asian Development Bank. 2010. *Responding to Climate Change in the Pacific: Moving from Strategy to Action*. Pacific Studies Series: Manila: Asian Development Bank. Accessed on 18 July 2012, https://www.army.gov.au/sites/g/files/net1846/f/flwr_web_b5_final.pdf.

Australian Army. 2014. *Future Land Warfare*. Department of Defence. Canberra: Directorate of Future Land Warfare. Accessed on 5 May 2017, https://www.army.gov.au/sites/g/files/net1846/f/flwr_web_b5_final.pdf.

Baker & McKenzie. 2011. *Local Council Risk of Liability in the Face of Climate Change—Resolving Uncertainties A Report for the Australian Local Government Association*. Sydney: Australian Local Government Association. Accessed on 10 October 2015, http://www.environment.gov.au/system/files/resources/d9b2f9cf-d7ab-4fa0-ab0e-483036079dc7/files/alga-report.pdf.

Barnett, Jon. 2003. "Security and climate change." *Global Environmental Change*, vol. 13, no. 1:7–17. doi:10.1016/S0959-3780(02)00080-8.

Barnett, Jon. 2009. "Environmental Security in the Asia-Pacific Region: Contrasting Problems, Places, and Prospects." In *Facing Global Environmental Change Environmental, Human, Energy, Food, Health and Water Security Concepts*, edited by Hans Günter Brauch, Úrsula Oswald Spring, John Grin, Czeslaw Mesjasz, Patricia Kameri-Mbote, Navnita Chadha Behera, Béchir Chourou and Heinz Krummenacher, vol. 4, pp. 939–950. Berlin: Springer.

Barnett, Jon, Richard A. Matthew, and Karen L. O'Brien. 2010. "Global Environmental Change and Human Security: An Introduction." In *Global Environmental Change and Human Security*, edited by Richard A. Matthew, Jon Barnett, Bryan McDonald and Karen L. O'Brien, pp. 3–32. Massachusetts: The MIT Press.

Barrie, Chris, Will Steffen, Alix Pearce, and Michael Thomas. 2015. *Be Prepared: Climate Change, Security and Australia's Defence Force*. Sydney: Climate Council. Accessed on 20 September 2015, https://www.climatecouncil.org.au/uploads/fa8b3c7d4c6477720434d6d1089 7af18.pdf.

Bergin, Anthony, and Jacob Townsend. 2007. *A change in climate for the Australian Defence Force*. Special Report: 7. Canberra: Australian Strategic Policy Institute. Accessed on 12 February 2013, https://www.aspi.org.au/publications/special-report-issue-7-a-change-in-climate-for-the-australian-defence-force/SR7_climagechange_ADF.pdf.

Brauch, Hans Günter. 2009. "Securitizing Global Environmental Change." In *Facing Global Environmental Change: Environmental, Human, Energy, Food, Health and Water Security Concepts*, edited by Hans Günter Brauch, Úrsula Oswald Spring, John Grin, Czeslaw Mesjasz, Patricia Kameri-Mbote, Navnita Chadha Behera, Bechir Chourou and Heinz Krummenacher, vol. 4, pp. 65–102. Berlin: Springer.

Brzoska, Michael. 2012b. "Climate Change as a Driver of Security Policy." In *Climate Change, Human Security and Violent Conflict: Challenges for Societal Stability*, edited by Jurgen Scheffran, Michael Brzoska, Hans Gunter Brauch, Peter Michael Link and Janpeter Schilling. Berlin: Springer.

CAHA. 2010. "Climate change is a health issue." Published online: Climate and Health Alliance. Accessed on 10 March 2015, http://caha.org.au/wp-content/uploads/2011/03/CAHA-Briefing-Paper-1_Climate-change-is-a-health-issue1.pdf.

Campbell, Angus. 2016. "Chief of Army opening address to the 2016 Chief of Army's Exercise." In *Chief of Army: Speeches and Transcripts*. Canberra: Commonwealth of Australia. Accessed on 5 May 2017, https://www.army.gov.au/our-work/speeches-and-transcripts/chief-of-army-opening-address-to-the-2016-chief-of-armys-exercise.

Chaturvedi, Sanjay, and Timothy Doyle. 2010. "Geopolitics of Climate Change and Australia's 'Re-engagement' with Asia: Discourses of Fear and Cartographic Anxieties." *Australian Journal of Political Science*, vol. 45, no. 1:95–115. doi:10.1080/10361140903517734.

Christoff, Peter, and Robyn Eckersley. 2014. "No island is an island: Security in a Four Degree World." In *Four Degrees of Global Warming: Australia in a hot world*, edited by Peter Christoff, pp. 190–204. London: Routledge.

Commonwealth Bank. 2008. *Commonwealth Bank of Australia Environment Policy*. Published online: Commonwealth Bank. Accessed on 12 October 2015, https://www.commbank.com.au/about-us/download-printed-forms/EnvironmentalPolicy_0808.pdf.

Commonwealth of Australia. 2006. "Supplementary Budget Estimates." In *Standing Committee on Foreign Affairs, Defence and Trade Official Committee Hansard*, 1 November 2006. Canberra: Commonwealth of Australia. Accessed on 19 March 2012, http://www.aph.gov.au/Parliamentary_Business/Senate_Estimates/fadtctte/estimates/sup0607/index.

Commonwealth of Australia. 2007b. *National Climate Change Adaptation Framework*. Plan of Collaborative Action on Climate Change: Canberra: Commonwealth of Australia. Accessed on 2 October 2012, http://www.environment.gov.au/system/files/resources/eaaf0350-9781-4006-957c-a5801fadc466/files/nccaf.pdf.

Commonwealth of Australia. 2008b. *Energy Use in the Australian Government's Operations 2006–2007*. Water Department of the Environment, Heritage and the Arts. Canberra: Department of the Environment, Water, Heritage and the Arts. Accessed on 4 June 2011, http://industry.gov.au/Energy/EnergyEfficiency/Non-residentialBuildings/GovernmentBuildings/Documents/report-2006.pdf.

Commonwealth of Australia. 2010. *Adapting to Climate Change in Australia An Australian Government Position Paper*. Department of Climate Change. Canberra: Commonwealth of Australia. Accessed on 5 October 2012, http://www.environment.gov.au/system/files/resources/93e900d7-3307-41c8-88e1-e0aa314c4cb3/files/adapt-climate-change-position-paper.pdf.

Commonwealth of Australia. 2011. *Climate Change Risks to Coastal Buildings and Infrastructure: A supplement to the First-Pass National Assessment*. Department of Climate Change and Energy Efficiency. Canberra: Commonwealth of Australia. Accessed on 12 July 2012, https://www.environment.gov.au/system/files/resources/0f56e5e6-e25e-4183-bbef-ca61e56777ef/files/risks-coastal-buildings.pdf.

Commonwealth of Australia. 2013d. *Strong and Secure: A Strategy for Australia's National Security* Canberra: Commonwealth of Australia. Accessed on 5 May 2017, http://apo.org.au/files/Resource/dpmc_nationalsecuritystrategy_jan2013.pdf.

Commonwealth of Australia. 2015b. "Supplementary Budget Estimates 2015–2016." In *Foreign Affairs, Defence and Trade Legislation Committee Hansard*. Canberra: Commonwealth of Australia. Accessed on 2 November 2015, http://www.aph.gov.au/Parliamentary_Business/Senate_Estimates/fadtctte/estimates/sup1516/index.

Defence, Department of. 2007a. *Defence Energy Policy*. Defence Support Group. Canberra: Commonwealth of Australia.

Defence, Department of. 2009b. *Defending Australia in the Asia Pacific Century: Force 2030*. Canberra: Commonwealth of Australia. Accessed on 5 May 2017, http://www.defence.gov.au/whitepaper/2009/docs/defence_white_paper_2009.pdf.

Defence, Department of. 2010b. *Defence Environmental Strategic Plan 2010–2014*. Canberra: Defence Support Group for Department of Defence. Accessed on 16 June 2011, http://www.defence.gov.au/environment/strat_plan.pdf. [Author note: Hyperlink inoperative, author retains original. Superseded reference accessible at: http://www.defence.gov.au/estatemanagement/governance/Policy/Environment/Policy/EnvironmentStrategy2016.PDF].

Defence, Department of. 2010c. *The Strategy Framework 2010*. Canberra: Department of Defence. Accessed on 14 March 2011, http://www.defence.gov.au/publications/docs/TheStrategyFramework2010.pdf.

Defence, Department of. 2013a. *Defence Energy Integration Framework*. Department of Defence. Canberra: Commonwealth of Australia. Accessed on 5 May 2017, http://www.natolibguides.info/ld.php?content_id=1675730.

Defence, Department of. 2013b. *Defence White Paper*. Canberra: Commonwealth of Australia. Accessed on 10 January 2014, http://www.defence.gov.au/whitepaper/2013/docs/WP_2013_web.pdf.

Defence, Department of. 2016b. *Defence Annual Report 2014–15*. Department of Defence. Canberra: Commonwealth of Australia. Accessed on 1 November 2016, http://www.defence.gov.au/annualreports/14-15/downloads/DAR_2014-15_Vol_1.pdf.

Defence, Department of. 2016c. *Defence White Paper*. Canberra: Commonwealth of Australia. Accessed on 31 October 2016, http://www.defence.gov.au/WhitePaper/Docs/2016-Defence-White-Paper.pdf.

Defence, Department of. 2016d. *Environmental Strategy 2016–2036*. Canberra: Defence Estate and Infrastructure Group (E&IG). Accessed on 16 October 2016, http://www.defence.gov.au/estatemanagement/governance/Policy/Environment/Policy/EnvironmentStrategy2016.PDF.

Defence, Department of. 2016e. "Estate and Infrastructure Group." Canberra: Commonwealth of Australia. Accessed on 1 November 2016, http://www.defence.gov.au/EIG/.

Dupont, Alan, and Graeme Pearman. 2006. *Heating up the Planet: Climate Change and Security*. Lowy Institute Paper 12. Sydney: Lowy Institute. Accessed on 10 March 2010, http://www.lowyinstitute.org/publications/heating-planet-climate-change-and-security.

Edwards, Michael J. 1999. "Security Implications of a Worst-case Scenario of Climate Change in the South-west Pacific." *Australian Geographer*, vol. 30, no. 3:311–330.

Garnaut, Ross. 2008. *The Garnaut Climate Change Review: Final Report*. Melbourne: Cambridge University Press.

Gibbs, Meredith, and Tony Hill. 2011. *Coastal Climate Change Risk: Legal and Policy Responses in Australia*. Department of Climate Change and Energy Efficiency. Canberra: Commonwealth of Australia. Accessed on 12 March 2012, https://www.environment.gov.au/climate-change/adaptation/publications/coastal-climate-change-risk.

Hartcher, Peter. 2014. "Tony Abbott must shift with the tide on climate change." *The Age*, Fairfax Media, 9 December 2014, http://www.theage.com.au/federal-politics/political-opinion/tony-abbott-must-shift-with-the-tide-on-climate-change-20141208-122ie7.html.

Hawke, Allan, and Ric Smith. 2012. *Australian Defence Force Posture Review*. Canberra: Commonwealth of Australia.

Hayes, Jarrod, and Janelle Knox-Hayes. 2014. "Security in Climate Change Discourse: Analysing the Divergence between US and EU Approaches to Policy." *Global Environmental Politics*, vol. 14, no. 2:82–101. doi:10.1162/GLEP_a_00230.

Howard, John W. 2004. "Address to National Press Club Canberra, Securing Australia's Energy Future." In *Transcripts from the Prime Ministers of Australia*, 15 June 2004. Canberra: Department of the Prime Minister and Cabinet. Accessed on 13 March 2012, http://pandora.nla.gov.au/pan/10052/20040821-0000/www.pm.gov.au/news/speeches/speech912.html.

Hurley, David J. 2011. "Chief of the Defence Force—Address to the Australian Strategic Policy Institute." In *Defence News and Media*, 18 August 2011. Canberra: Department of Defence. Accessed on 12 May 2013, http://news.defence.gov.au/2011/08/19/chief-of-the-defence-force-%E2%80%93-address-to-the-australian-strategic-policy-institute/. [Author note: Hyperlink inoperative, author retains original].

ICAO. 2015. "Uniting Aviation on Climate Change." International Civil Aviation Organization. Accessed on 23 September 2015, http://www.icao.int/environmental-protection/Pages/act-global.aspx.

Insurance Council of Australia. 2008. *Improving Community Resilience to Extreme Weather Events*. Published online: Insurance Council of Australia. Accessed on 10 October 2015, http://www.insurancecouncil.com.au/assets/files/community%20resilience%20policy%20150408.pdf.

International Federation of Red Cross. 2015. "Aggravating factors: climate change." Published online: International Federation of Red Cross and Red Crescent Societies. Accessed on 1 October 2015, http://www.ifrc.org/en/what-we-do/disaster-management/about-disasters/aggravating-factors/climate-change/.

IPCC. 2007. "Summary for Policymakers." In *Climate Change 2007: The Physical Science Basis. Contribution of Working Group I to the Fourth Assessment Report of the Intergovernmental Panel on Climate Change*, edited by S. Solomon, D. Qin, M. Manning, Z. Chen, M. Marquis, K. B. Averyt, M. Tignor and H. L. Miller. Cambridge: Cambridge University Press.

IPCC. 2015. "Climate Change 2014 Synthesis Report." In *Climate Change 2014: Synthesis Report. Contribution of Working Group I to the Fifth Assessment Report of the Intergovernmental Panel on Climate Change*, edited by Core Writing Team, Rajendra K. Pachauri and Leo Meyer. Geneva: Intergovernmental Panel on Climate Change. Accessed on 14 March 2015, http://www.ipcc.ch/pdf/assessment-report/ar5/syr/SYR_AR5_FINAL_full.pdf.

Lawson, Robert. 2007. "Climate Change in the Asia-Pacific Region: Security Implications for Australia." In *Centre for Defence and Strategic Studies Australian Defence College Shedden Papers*. Canberra: Department of Defence. Accessed on 5 May 2012, http://www.defence.gov.au/ADC/Publications/Shedden/2008/Publctns_ShedPaper_050310_ClimateChangeintheAsiaPacificRegion.pdf.

Layton, Peter. 2014. "What strategy for the new Defence White Paper?" In *The Strategist*, 13 March 2014. Canberra: Australian Strategic Policy Institute. Accessed on 23 March 2014, http://www.aspistrategist.org.au/long-live-the-force-structure/.

Leahy, Peter. 2010. "How do we know when we are at war?" In *Lowy Institute Perspectives: Occasional Papers and Speeches*, 22 July 2010. Sydney: Lowy Institute. Accessed on 12 May 2013, http://www.lowyinstitute.org/publications/how-do-we-know-when-we-are-war.

Macintosh, Andrew, Deb Wilkinson, and Richard Denniss. 2010. "Climate Change." In *The Rudd Government: Australian Commonwealth Administration 2007–2010*, edited by Chris Aulich and Mark Evans, pp. 596–652. Canberra: The Australian National University E Press. Accessed on 3 March 2014, http://press.anu.edu.au/apps/bookworm/view/The+Rudd+Government%3A+Australian+Commonwealth+Administration+%092007+-+2010/5091/ch11.xhtml#toc-anchor.

McDonald, Matt. 2012. "The Failed Securitization of Climate Change in Australia." *Australian Journal of Political Science*, vol. 47, no. 4:579–592. doi:10.1080/10361146.2012.731487.

Morris, Sophie. 2015. "Tony Abbott's affinity with ADF reflected in parliament." *The Saturday Paper*, Morry Schwartz. Accessed on 5 May 2017, https://www.thesaturdaypaper.com.au/news/politics/2015/03/14/tony-abbotts-affinity-with-adf-reflected-parliament/14262516001605.

Morrison, David. 2013. "Speech to Royal United Services Institute, Sydney." In *Chief of Army Speeches and Transcripts*, 26 March 2013. Canberra: Department of Defence. Accessed on 5 May 2017, http://103.11.78.168/Our-work/Speeches-and-transcripts/Royal-United-Services-Institute.

NAB. 2015. "Climate Change." Published online: National Australia Bank. Accessed on 12 October 2015, https://www.nab.com.au/about-us/corporate-responsibility/responsibility-management-of-our-business/environment/climate-change.

NCCARF. 2013. *Emergency management and climate change adaptation*. Policy Guidance Briefs: Brisbane: National Climate Change Adaptation Research Facility. Accessed on 5 October 2015, https://www.nccarf.edu.au/sites/default/files/attached_files_publications/EM-A4-Webview.pdf.

Obama, Barack H. 2014. "Remarks by the President at U.N. Climate Change Summit." In *White House Briefing Room*, 23 September 2014. Washington, DC: White House. Accessed on 12 January 2015, http://www.whitehouse.gov/the-press-office/2014/09/23/remarks-president-un-climate-change-summit.

Oliver, Alex. 2015. *Lowy Institute Poll 2015*. Sydney: The Lowy Institute. Accessed on 12 September 2015, http://www.lowyinstitute.org/files/final_2015_lowy_institute_poll.pdf.

Pecl, Gretta, Stewart Frusher, Caleb Gardner, Marcus Haward, Alistair Hobday, Sarah Jennings, Melissa Nursey-Bray, André Punt, Hilary Revill, and Ingrid van Putten. 2009. *The east coast Tasmanian rock lobster fishery—vulnerability to climate change impacts and adaptation response options*. Department of Climate Change. Canberra: Commonwealth of Australia. Accessed on 3 October 2014, http://www.environment.gov.au/system/files/resources/48b0b14e-9234-480f-9a6b-602274396991/files/rock-lobser-report.pdf.

Perlo-Freeman, Sam, Aude Fleurant, Pieter Wezeman, and Siemon Wezeman. 2016. *Trends in World Military Expenditure, 2015*. SIPRI Military Expenditure Database: Stockholm: SIPRI. Accessed on 1 November 2016, http://books.sipri.org/files/FS/SIPRIFS1604.pdf.

Pittock, A. Barrie. 2009. *Climate Change: The Science, Impacts and Solutions*. Second ed. Melbourne: CSIRO Publishing.

Pope Francis. 2015. *On care for our common home, Encyclical Letter Laudato Si' Of the Holy Father Francis*. Vatican: Vatican Press.

Press, Anthony, Anthony Bergin, and Eliza Garnsey. 2013. *Heavy Weather: Climate and the Australian Defence Force*. Special Report 49. Canberra: Australian Strategic Policy Institute. Accessed on 12 March 2014, https://www.aspi.org.au/publications/special-report-issue-49-heavy-weather-climate-and-the-australian-defence-force.

Rintoul, Stuart. 2009. "Town of Beaufort changed Tony Abbott's view on climate change." The Australian. Accessed on 31 October 2016, http://www.theaustralian.com.au/archive/politics/the-town-that-turned-up-the-temperature/story-e6frgczf-1225809567009.

Rudd, Kevin M. 2008d. "Building a Better World Together, Speech at Kyoto University." In *PM Transcripts: Transcripts from the Prime Ministers of Australia*, 09 June 2008. Canberra: Department of Prime Minister and Cabinet. Accessed on 2 April 2013, http://pmtranscripts.pmc.gov.au/release/transcript-15955.

Rudd, Kevin M. 2008g. "The First National Security Statement to the Parliament Address by the Prime Minister of Australia The Hon. Kevin Rudd MP." In *PM Transcripts: Transcripts of the Prime Ministers of Australia*, 4 December 2008. Canberra: Department of Prime Minister and Cabinet. Accessed on 5 March 2013, http://pmtranscripts.pmc.gov.au/release/transcript-16289.

Spencer, Tom, Nick Mabey, Chad Briggs, Elena Belluci, and Geraud de Ville. 2009. *Climate Change & The Military: The State of the Debate*. Institute for Environmental Security Military Advisory Council: Belgium: Institute for Environmental Security. Accessed on 4 April 2013, http://www.tomspencer.info/articles/StateoftheDebate.pdf.

Steffen, Will, Andrew A. Burbidge, Lesley Hughes, Roger Kitching, David Lindenmayer, Warren Musgrave, Mark Stafford Smith, and Patricia A. Werner. 2009. *Australia's Biodiversity and Climate Change: A Strategic Assessment of the vulnerability of Australia's biodiversity to climate change*. Canberra: Commonwealth of Australia. http://www.environment.gov.au/system/files/resources/eab369d6-76f9-46c8-beb4-aaae8ece112e/files/biodiversity-vulnerability-assessment.pdf.

Steffen, Will, John Hunter, and Lesley Hughes. 2014. *Counting the Coast: Climate Change and Coastal Flooding*. Sydney: Climate Council. Accessed on 5 April 2014, http://www.climatecouncil.org.au/uploads/56812f1261b168e02032126342619dad.pdf.

Stern, Nicholas. 2007. *Stern Review on the Economics of Climate Change*. Cambridge: Cambridge University Press.

Sturrock, Robert, and Peter Ferguson. 2015. *The Longest Conflict: Australia's Climate Security Challenge*. Melbourne: Centre for Policy Development.

Thomas, Michael. 2011. "Climate Change and the ADF." *Journal of the Australian Profession of Arms*, no. 185:34–43.

Thomas, Michael. 2013. "The Securitisation of Climate Change: A Military Perspective." *Journal of the Australian Profession of Arms* vol. Nov/Dec, no. 192:7–18.

Thomas, Michael. 2015. "Climate securitization in the Australian political-military establishment." *Global Change Peace and Security*, vol. 27, no. 1:97–118. doi:10.1080/14781158.2015.990879.

Tremblay, Pascal, Anna Boustead, Stephen Turton, Robyn Wilson, Tracey Dickson, Bruce Hayllar, Rosallie Chapple, Sue Uzabeaga, Kate Hammill, Beatrice Pegard, Mario Rimini, Phoenix Lawhon Isler, John Merson, Bradley Jorgensen, Maureen Rogers, Peter Hayman, Graham Brown, Jenny Davies, Amanda Lynch, Carolina Roman, Lee Tryhorn, Tien Duc Pham, Peter Forsyth, and Ray Spurr. 2008. *The Impact of Climate Change on Australian Tourism Destinations*. Published online: Commonwealth of Australia. Accessed on 10 October 2015, http://www.bmwhi.org.au/docs/The%20Impacts%20of%20Climate%20Change%20Summary.pdf.

Turnbull, Malcolm. 2016. "Remarks at the Pacific Island Forum." In *Media Transcripts*, 9 September 2016. Canberra: Commonwealth of Australia. Accessed on 31 October 2016, https://www.pm.gov.au/media/2016-09-09/remarks-pacific-island-forum.

Victorian Government. 2011. *Review of the Climate Change Act 2010*. Department of Sustainability and Environment. Melbourne: Department of Premier and Cabinet. (Website archived, document in possession of author and available on request. Second Independent Review of the Climate Change Act 2010 (2015) is available at: https://www.climatechange.vic.gov.au/__data/assets/pdf_file/0019/55306/Independent-Review-of-the-Climate-Change-Act-2010.pdf).

Wesley, Michael. 2012. "Climate views have moved on." *The Australian*, Lowy Institute for International Policy, 2 July 2012. Accessed on 23 March 2013, http://www.lowyinstitute.org/publications/climate-views-have-moved.

Williams, Richard J., Ross A. Bradstock, Geoffrey J. Cary, Neal J. Enright, A. Malcolm Gill, Adam C. Liedloff, Christopher J. Whelan, Alan N. Andersen, David J. M. S. Bowman, Peter J. Clarke, Garry D. Cook, Kevin J. Hennessy, and Alan York. 2009. *Interactions between climate change, fire regimes and biodiversity in Australia: A Preliminary Assessment*. Water Department of Climate Change and Department of the Environment, Heritage and the Arts. Canberra: Commonwealth of Australia. Accessed on 5 October 2014, http://www.environment.gov.au/system/files/resources/ab711274-004c-46cb-a9f1-09dba68d98c3/files/20100630-climate-fire-biodiversity-pdf.pdf.

World Bank. 2013. *Turn Down the Heat: Climate Extremes, Regional Impacts and the Case for Resilience*. Washington, DC: World Bank. Accessed on 5 May 2017, http://documents.worldbank.org/curated/en/975911468163736818/Turn-down-the-heat-climate-extremes-regional-impacts-and-the-case-for-resilience-full-report.

World Vision. 2015. "Climate change: The effects on children." Published online: World Vision. Accessed on 23 September 2015, http://www.worldvision.com.au/global-issues/work-we-do/climate-change/climate-change-the-effects-on-children.

Youngs, Richard. 2015. *Climate Change and European Security*. London: Taylor and Francis.

Part II
Climate Security: The Physical and Policy Basis

Chapter 2
Climate Security—The Physical Basis

Abstract This chapter provides an introduction to the physical science of climate change, current and projected levels of anthropogenic GHG emissions and a brief account of the political challenges associated with a coordinated and binding international policy response to limit global warming. This wider examination of climate change is also undertaken within a security and military context. Perforce, 'climate security' is defined upfront and the key physical (environmental) risks of climate change are mapped to potential security and military impacts. The chapter also examines military GHG emissions (estate and operations) for the Australian, US and UK militaries. It concludes that the military sector is a significant contributor to global emissions.

Keywords Climate science · Climate risk · Climate security · Military emissions · Military GHG emissions · Climate security definition

2.1 Defining Climate Security

Before reviewing the science of climate change it is important to have a working understanding of what is meant by the term 'climate security'. Reflecting earlier debates within the field of security studies, definitions of 'climate security' (like the term 'security' itself) remain imprecise, fluid and contested (Wolfers 1952; Buzan 1991; Dalby 1992; Youngs 2015). In the US, climate security has been defined as a 'threat multiplier' with a tendency to focus on national security implications; in contrast the UN has framed it as a 'threat minimiser' with a focus on its implications for human security. Others again have identified it as a 'nexus' issue that affects the 'broader determinants of stability' that encompass the economy, patterns of resource production and consumption as well as the direct physical impacts of global warming (Youngs 2015).

A common thread amongst the various definitions is the acceptance that climate change presents as a major security threat operating with varying effects across multiple time-frames, scales, locations and domains. Adopting this wide approach,

M.D. Thomas, *The Securitization of Climate Change: Australian and United States' Military Responses (2003–2013)*, The Anthropocene: Politik—Economics—Society—Science 10, DOI 10.1007/978-3-319-49658-0_2

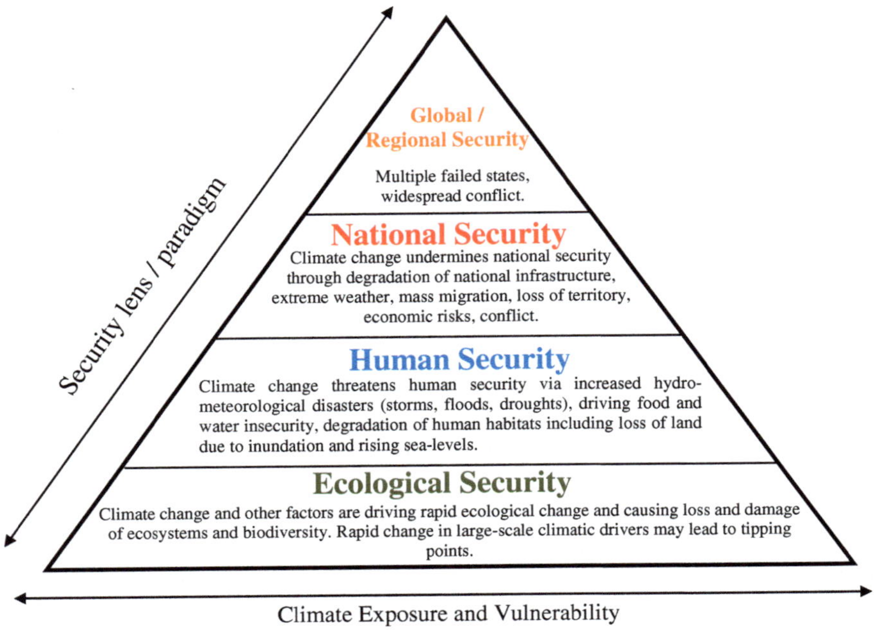

Fig. 2.1 Climate security framework. Security of ecosystems and biodiversity forms the basis of all other forms of security. Since climate change undermines the base, it degrades the pyramid. *Source* The author

'climate security' can be said first and foremost to concern the return to, and maintenance of, atmospheric GHG concentrations within the range of 280–350 ppm[1] for the purpose of stabilising global average surface temperatures below 1.5 °C in order to preserve and sustain planetary ecological systems. For the purpose of this book, 'climate security' can be defined more narrowly as: *the absence of threats to individuals, states, and the international system that arise from anthropogenic warming.*[2] Throughout this book, climate security is generally viewed through the prism of national security; that is to say, the nation-state is the referent object to be secured against an adversely changing climate. This is not intended to downplay or disregard other legitimate forms of security, but should rather be viewed simply as an effort to have a workable research boundary. The armed forces, as critical enablers of national security (but not the only enablers), are central actors in this discourse.

[1]Using CO_2 (carbon dioxide) as the GHG indicator.

[2]While much discussion exists on what target is appropriate, this definition uses the 1.5 °C target that was previously adopted by the Association of Small Island States (AOSIS) as well as many NGOs and scientists (Tschakert 2015).

Figure 2.1 is a simplified representation of the linkages of climate security across the security domains described above. This simplified representation shows that each security 'level' (or 'lens') is reliant upon, or sustained by, the ones below it; and that climate change has the potential to undermine the most fundamental basis (i.e., planetary ecological systems) from which all other forms of security (human, national and global) are made possible. The horizontal axis indicates 'climate exposure' and 'climate vulnerability', indicating that ecosystems (including biodiversity) are the most exposed and vulnerable to climate change.

2.2 Overview of Climate Science

This section overviews the fundamental science of climate change. While this will be familiar territory to many involved in disciplines and sectors addressing climate change, it is an often overlooked area and one that is arguably given less attention than it deserves by security sector professionals more generally (Sturrock/Ferguson 2015). For brevity, this literature survey does not attempt to build on the science of climate change per se, but rather aims to provide wider context on the scale of the challenges at hand and an update on the most recent data (as at the end of 2016).

The foundation science of climate change was laid down in the nineteenth century by naturalists, philosophers and part-timers investigating the 'riddle of the ice age' (Weart 2008: 5). Key discoveries during this period included the basic mechanism of global warming (1827, Joseph Fourier); the discovery of the main greenhouse gases (1859, John Tyndall); the quantification of those gases to global warming (1896, Svante Arrhenius); and the postulation in 1907 that industrial pollution was the main cause. Swedish scientist Svante Arrhenius is particularly noteworthy. Building on the experiments and observations of other scientists, he stated (in 1896): 'If the quantity of carbonic acid increases in geometric progression, the augmentation of the temperature will increase nearly in arithmetic progression' (Arrhenius 1896: 267). He surmised that a 50% increase of 'carbonic acid' (carbon dioxide, CO_2) would lead to temperature increases of more than 3 °C (Arrhenius 1896: 266).

Figure 2.2 shows the Earth's energy balance. Of the 341 watts per square meter (W m^{-2}) of solar radiation reaching the Earth's atmosphere, roughly 30% (102 W m^{-2}) is reflected back into space, while 70% (239 W m^{-2}) is either reflected or absorbed and then re-radiated as long-wave (infrared radiation) at the Earth's surface (Trenberth et al. 2009).[3] In the lower reaches of the Earth's atmosphere, GHGs

[3]The figures in the text differ slightly from those given in the diagram. The diagram is from NASA Earth Observatory while the figures in the text are from the citation. For an explanation, see NASA at: http://earthobservatory.nasa.gov/Features/EnergyBalance/page4.php (2 November 2016).

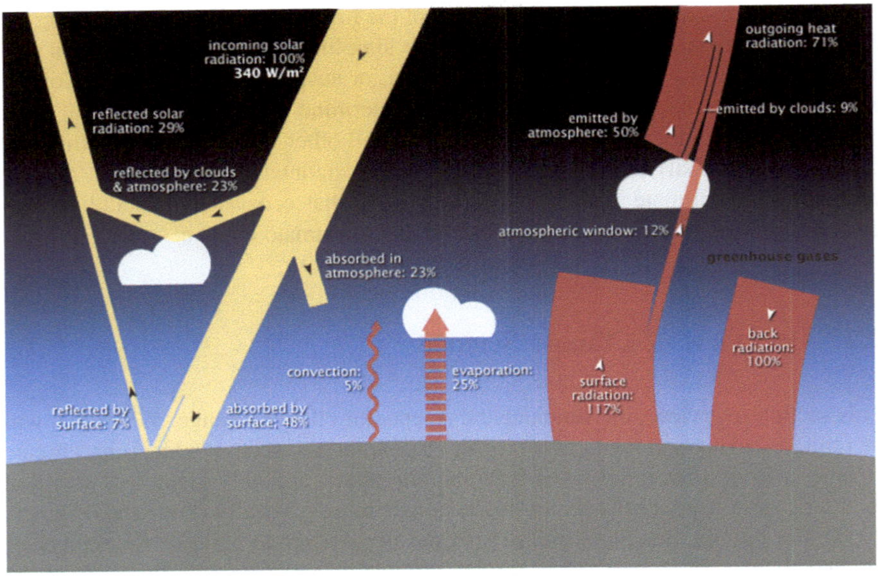

Fig. 2.2 Earth energy balance. *Source* NASA Earth Observatory (2016). Reprinted with permission.

absorb a small fraction of the re-radiated long-wave radiation (about 5%), scattering heat energy back to Earth ('back radiation') and warming the lower atmosphere. This is called the 'greenhouse effect' and, without it, global average temperatures would be about 30 °C colder than current conditions (Cotton/Peilke 2007: 121).

As John Tyndall discovered in 1859, the two most abundant gases in the atmosphere (nitrogen and oxygen) exert almost no greenhouse effect (LeTreut et al. 2007). Rather, the two main GHGs are water vapour and CO_2. Although water vapour is a critically important GHG, humans only have a very small direct influence on the actual amount in the atmosphere. CO_2, by contrast, is primarily emitted into the atmosphere through human industrial action via combustion of fossil fuels. Invariably, climate scientists pinpoint CO_2 as the most important GHG due to a combination of increasing anthropogenic emissions combined with its high 'radiative forcing', high atmospheric concentrations, long residence time (centuries) and global dispersion pattern (Blasing 2012). Methane (CH_4), nitrous oxide (N_2O) and a range of other minor gases including ozone gases, hydroflourocarbons (HFCs), perfluorocarbons (PFCs) and sulphur hexafluoride (SF_6) also contribute as GHGs.

Any change to the Earth's energy balance (a 'pertubation') requires what is known as a 'climate forcing'. A climate forcing can occur in one of three ways: (1) changing the amount of incoming solar radiation (e.g., orbital changes), (2) changing the amount of solar radiation that is reflected (e.g., albedo, aerosols), or (3) altering the long-wave radiation from Earth back towards space (e.g., by changing the level of GHG concentrations) (LeTreut et al. 2007: 96). In 1957 attention was drawn to the last of these when a research programme funded by the US Navy's Office of Naval Research at the Scripps Institution of Oceanography (notably led by Roger Revelle) published results indicating that sequestration of CO_2 by the world's oceans was not as rapid as originally thought. Further analysis by Charles Keeling and others showed an accumulation of atmospheric CO_2 and again led to the postulation that increasing anthropogenic emissions might produce a positive climate forcing (i.e., warming) by the end of the twentieth century (Weart 2008).

Like Svante Arrhenius and the nineteenth century scientists before them, those working in the field of 'climate science' in the 1950s did not consider this to be a problem. Indeed, it was viewed by some as a practical way to actually warm the planet, particularly across northern Europe! However, the scale of industrialisation

Fig. 2.3 Atmospheric CO_2 concentrations and growth rate, 1960–2016. *Source* NOAA ESRL (2016). Reprinted with permission.

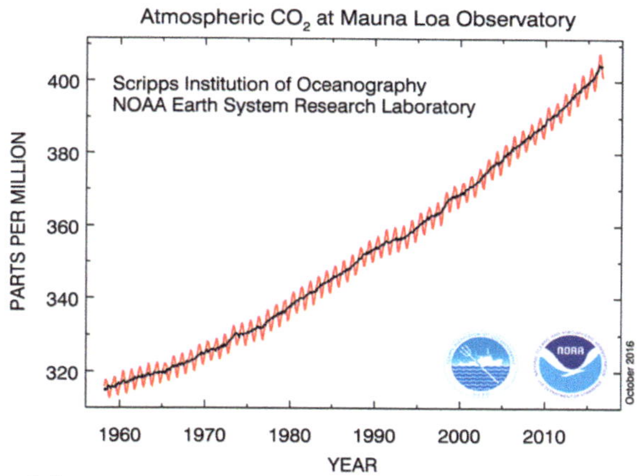

(a) Atmospheric CO_2 concentration, 1960–2016.

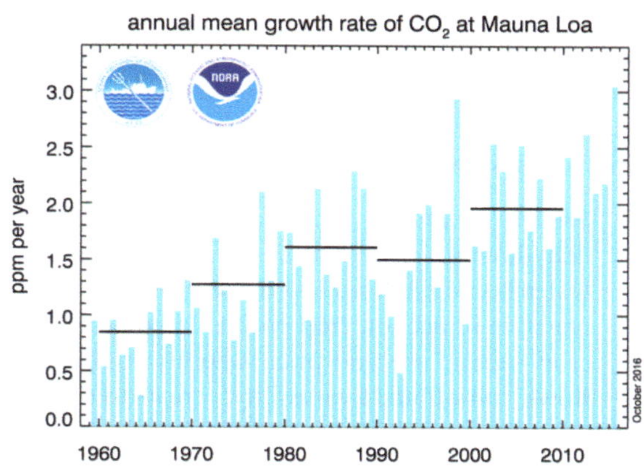

(b) Annual mean growth rate of atmospheric CO_2 1960–2016

through the rapid uptake of coal, oil and gas, combined with rapid population increases—particularly following World War II—witnessed an unforeseen growth in GHGs emissions and concomitant atmospheric concentration levels. Thus, for the first time in 800,000 years, atmospheric CO_2 now exceeds 400 ppm (higher when factoring in 'masking agents' such as aerosols) and has been accumulating in the atmosphere at increasing rates (see Fig. 2.3). Moreover, since industrialisation, humans have emitted more than 550 billion tonnes (GtC) of carbon into the atmosphere (with 40% remaining in the atmosphere and about 30% each in oceanic and terrestrial sinks) (IPCC 2015: 4). Scientists now understand atmospheric GHGs to be the dominant contributor to the Earth's energy imbalance, with the Goddard Institute for Space Studies (GISS) recording a globally averaged energy imbalance of 1.73 W m^{-2} (Hansen et al. 2011) and IPCC AR5 placing total anthropogenic radiative forcing over 1750–2011 at 2.3 W m^{-2} (IPCC 2015: 44). Despite the cooling effect of aerosols and other agents, the overall positive net forcing has resulted in a globally averaged surface temperature increase approaching 1 °C since 1850, with further warming 'locked-in' as a consequence of the climate's inertia (Hansen et al. 2011).

2.3 Climate Change and Implications for Security and Military Forces

Having overviewed the science of climate change, this section maps the observed changes in the climate system to possible impacts on security and on military forces. The observed changes are drawn from IPCC AR5 (IPCC 2015: 39–54), while accompanying security and military impacts have been synthesised from a variety of subject literature (CNA 2007; NIC 2008; UNSG 2009; Carmen et al. 2010; Thomas 2012; CNA 2014; Barrie et al. 2015):

- *Atmospheric warming and altered rainfall patterns.* Globally averaged combined land and ocean surface temperatures show a warming of 0.85 °C over the period 1880–2012, with global scale changes in precipitation patterns, intensification of heavy precipitation, desertification and reductions in renewable surface and groundwater. Warming is not evenly distributed across the globe, impacting some countries more and others less.
  - *Security Impacts*: Increased risk of heat-related mortality (heat waves, dehydration, heat stroke, heat exhaustion); changes in distribution of water-borne illnesses and disease vectors; risks to food security as a result of negative impact on crop production (particularly wheat, maize, rice and soy), crop access, use and price stability; risks to water security through altered flows, run-off and availability; threats to the capacity of governments to meet

the basic needs of their people and risks to development from the combined and associated impacts listed above (and below).
  - *Military Impacts*: Potential for increased tension over changes to resource availability, access and use may flare into low-level and wide-scale conflict involving national, regional and international military intervention. Increased warming may also impact military training regimes (changes to soldier/sailor/airmen productivity due to heat, duration of exercises, seasonality), military training grounds (e.g., increased risk of wild-fire, changes in suitability of training grounds from longer fire-seasons or permafrost thaw), and military logistics (e.g., long-term storage of supplies requiring stable temperature ranges).

- *Extreme Weather*: Extreme weather events have become more common since the mid-twentieth century. Examples include heatwaves, heavy precipitation, coastal flooding, storm surge, flooding, landslides, wildfires, droughts and more intense cyclones/hurricanes/typhoons in certain regions.
  - *Security Impacts*: Increased extreme weather combined with population growth has the potential to cause increased destruction of infrastructure and agriculture, loss of life, displacement and loss of economic output that may place increased strain on fragile states and capacity of developed nations to respond. Also, risks to development from associated impacts listed above.
  - *Military Impacts*: Potential increase in domestic and offshore Humanitarian Aid and Disaster Relief (HADR) missions could influence military force structure, capability acquisitions, training regimes and strategic scenario planning. Extreme weather may also impact military bases and hardware through direct physical destruction and degradation.

- *Ocean warming and acidification.* IPCC AR5 has 'high confidence' that ocean warming dominates the increase in energy stored in the climate system (accounting for more than 90% of energy accumulated between 1971 and 2010). There is 'virtual certainty' that the upper ocean (0–700 m) has warmed since 1971 and 'high confidence' that the ocean has absorbed 155 GtC since 1750, resulting in pH decrease of 0.1 and a 26% increase in ocean acidity.
  - *Security Impacts*: Potential changes in the geographic range, regional abundance, seasonal activities and migration patterns of marine species could result in new/increased tensions over marine resources.
  - *Military Impacts*: Potential impacts on naval crews and coastguards from increased requirement to protect commercial fishing fleets and EEZs. Changes to ocean thermal stratification, warming and acidity could all impact naval surface and sub-surface operations and capability as well as ship maintenance and marine infrastructure.

- *Sea level rise.* IPCC AR5 has 'high confidence' that the rate of sea-level rise since the mid-nineteenth century has been larger than the mean rate during the previous two millennia. There is 'high confidence' that glacial mass loss and thermal expansion account for 75% of sea level rise. Over the period 1901–2010 global mean sea-level rose by 0.19 m and the *rate* of sea level rise is increasing (it is 'very likely' that the mean global sea level rise between 1971–2010 was 2.0 mm/year and between 1993–2010 was 3.2 mm/year). Sea level rise is not evenly distributed, impacting some countries more, others less.
  - *Security Impacts*: Destruction and loss of coastal crops due to inundation, intrusion and salinisation; damage to infrastructure along coastal communities, cities; gradual loss of sovereign territory (particularly low-lying islands) and displacement/migration.
  - *Military Impacts*: Physical loss and damage to low-lying military infrastructure, bases, training grounds, ports, airstrips, barracks. Other considerations include the potential loss and damage of critical civilian infrastructure that supports military mobilisation and basing (e.g., emergency services, utilities) (e.g., power, water, sewer, gas).
- *Cryosphere.* Melting ice and permafrost across the Arctic, Greenland and Antarctica are driving sea-level rise, increased coastal erosion, impacts on biodiversity, changes in thermokarst lakes, altered river runoff, impacts on coastal communities and cities; opening of new trade routes and resource fields. The Arctic is particularly vulnerable, with summer sea-ice decreasing at a rate of between 0.73 and 1.07 million km^2 per decade ('very likely').
  - *Security Impacts*: Changes to the hydrological system have potential to create competition (or cooperation) between countries over access to water. Threats to water security are especially relevant in glacial-fed rivers of Himalaya ranges (Indus, Ganges, Brahmaputra, Irrawaddy, Salween, Mekong, Yangtze and Yellow) that support more than one-quarter of the world's population through long-term change effecting river runoff and flows which impact irrigation, agriculture and drinking water. Rapid change in the Arctic has potential for new trade routes, tourism and resource exploitation (fishing grounds, oil and gas) that may generate competition and tensions.
  - *Military Impacts*: Rapid change in the Arctic may drive requirement for increased military presence conducting asset protection, assistance in clean-up operations and emergency response. Arctic presence may also see countries develop new military capabilities, installations/bases opening up or being revamped, and increased military exercises being conducted across the Arctic. Changes in river-flows across the key Asian river basins may see increased military focus on asset protection (e.g., dams, river alterations, hydro-electricity plants, irrigation schemes) and possible deployment for water-related conflict resolution.

2.4 Global and Military GHG Emissions

This section briefly reviews global GHG emissions, the primary cause of anthropogenic climate change and, by implication, a contributor to the security and military impacts just mentioned. It opens with an overview of the current global GHG emissions picture, identifying the major emitting nations, sources of GHGs and possible future emissions scenarios. In keeping with the focus for this book, attention then turns to an examination of military GHG contributions. While much information exists on national GHG tallies, including the various sectoral contributions (e.g., aviation, agriculture, transport and so on), there is less clarity, understanding and knowledge of the size and scale of global military GHG contributions (Michaelowa/Kock 2001). Some of the challenges in obtaining clear information on military GHG data include: (1) a lack of a clear guiding international or national regulatory or legal requirement for nations to account for and report military emissions; (2) difficulties in data collection relating to offshore/non-domestic consumption and usage of fossil fuels by military forces involved in permanent off-shore basing arrangements, exercises, routine patrols and operations, or war; (3) an unwillingness by nations to report military emissions for operational or security-related reasons; and (4) challenges in developing an appropriate set of emissions boundaries that attribute responsibility in an equitable manner (e.g., if nation A contributes military forces to nation B who are both operating within the boundaries of nation C, then which nation should emissions be attributed to? Alternatively, if multiple nations contribute military assets to supply-line protection or anti-piracy operations for the greater regional or global good, and are operating in international waters, then which nations should emissions be attributed to?). The section examines GHG emissions for the Australian, US and UK militaries and uses a general approach to estimating global emissions for the military sector.

2.4.1 Current Global GHG Emissions and Projections

Since industrialisation (1750), the primary source of anthropogenic GHG emissions has been the extraction and combustion of fossil fuels and land-use change (e.g., deforestation). Total cumulative anthropogenic GHG emissions since industrialisation is 470–640 GtC (IPCC 2015: 4).[4] On an annual basis, fossil fuel and land use change emissions in 2016 are estimated at 36.4 $GtCO_2$ or around 9.9 GtC (LeQuéré et al. 2016: 630).[5] This is 63% above 1990 levels (ibid). Key sources of emissions

---

[4]The most recent report by the Carbon Dioxide Information Analysis Center (CDIAC) places total cumulative emissions 1870–2016 at 565 $\pm$ 55 GtC (LeQuéré et al. 2016: 632).

[5]1 Pg = 1 Petagram = 1×10^{15} g = 1 billion metric tonnes = 1 Gigatonne (Gt).
1 kg carbon (C) = 3.67 kg carbon dioxide (CO_2).
1 PgC = 3.67 billion tonnes CO_2 = 3.67 $GtCO_2$.

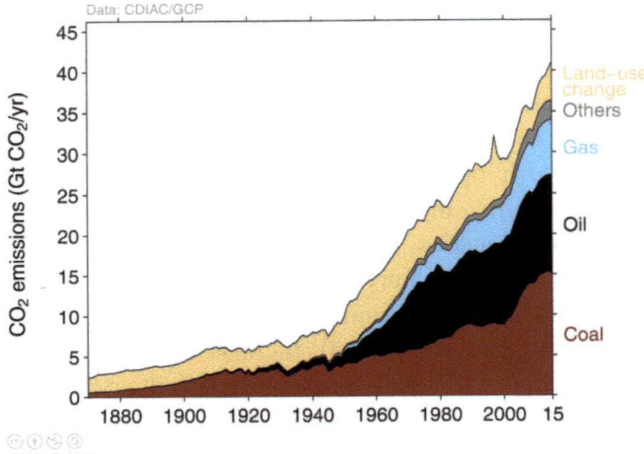

Fig. 2.4 Total global GHG emissions by source. *Source* LeQuéré et al. (2016). Reprinted with permission under CDIAC guidelines.

are at Fig. 2.4 and major emitting countries are at Fig. 2.5. To have a 66% change of keeping global warming below 2 °C to avoid dangerous climate change, then remaining quota is 856 Gt CO_2, or about 240 GtC (LeQuéré et al. 2015). Assuming current emission rates of about 10 GtC per year, then this quota will be consumed around 2040.

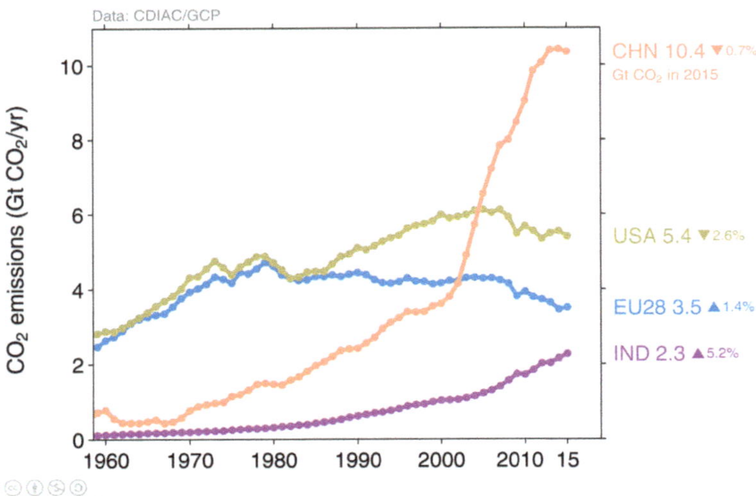

Fig. 2.5 Major fossil fuel emitters, 1960–2015. *Source* LeQuéré et al. (2016). Reprinted with permission under CDIAC guidelines.

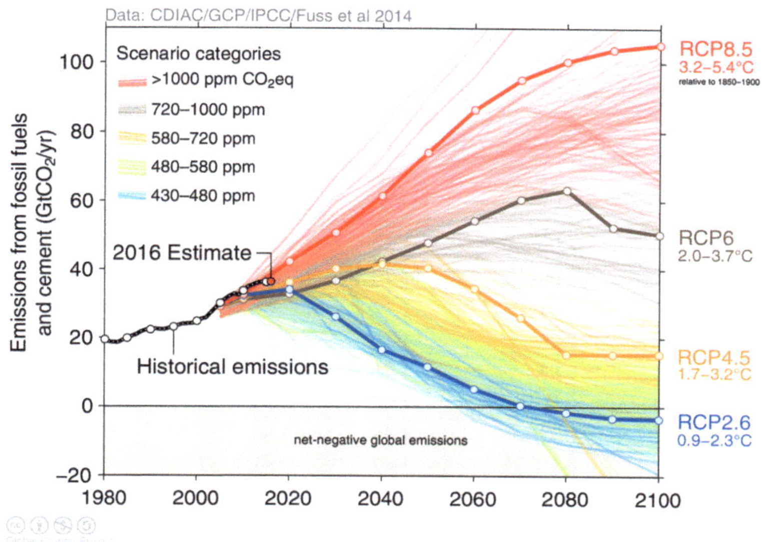

Fig. 2.6 Representative concentration pathways, 2014–2100. *Source* LeQuéré et al. (2016). Reprinted with permission under CDIAC guidelines.

Emissions scenarios, referred to as Representative Concentration Pathways (RCPs), were developed for IPCC AR5 and to provide policy-makers with plausible descriptions of how future climate scenarios may unfold. The RCP names are given after a possible range of radiative forcing values in the year 2100 (i.e., 2.6, 4.5, 6.0 and 8.5 W m^{-2}, respectively). Figure 2.6 shows the various emissions scenarios with the expected temperature increase at 2100, should a given RCP scenario unfold. These indicate that to have a 'likely chance' of keeping temperatures of below 2 °C, current emissions need to peak about 2020 and sustain a steady reduction until net negative emissions are achieved by around 2070 (RCP 2.6). Based on recent historical emissions, however, such a scenario appears unlikely, with studies indicating that anthropogenic emissions have tracked at the top end of all emissions scenarios developed since 1985 (ESSP 2012). On this basis, scientists estimate that, under RCP8.5, globally averaged surface temperatures are now more likely to be in the range 4.78–7.36 °C by 2100 (Friedrich et al. 2016).

2.4.2 *Military GHG Emissions*

Military GHG emissions continue to be a source of interest on the basis of their oft-stated sizeable contribution to global emissions and the lack of clear and transparent data (Savage 2015; Flounders 2014; Neslen 2015). This section shows

Table 2.1 Australian DoD GHG emissions 2001–2012[6]

AS DoD GHG emissions (Mt CO_2e)							
Year	**AS DoD estate[a]**	**AS DoD operations[b] (2)**	**AS DoD total[c] (b + c)**	**AS Gov't**	**% (d/e)**	**Total AS emissions**	**% (d/g)**
a	b	c	d	e	f	g	h
2001	0.76	1.34	2.10	3.18	66.18	491.44	0.43
2002	0.73	1.13	1.86	2.93	63.50	494.74	0.38
2003	0.73	1.09	1.82	2.86	63.43	495.15	0.37
2004	0.76	0.99	1.76	2.61	67.26	511.71	0.34
2005	0.76	1.01	1.77	2.61	67.73	518.85	0.34
2006	0.81	0.96	1.77	2.59	68.42	522.52	0.34
2007	0.82	0.93	1.75	2.70	64.80	529.84	0.33
2008	0.78	0.90	1.67	2.83	59.11	533.69	0.31
2009	0.76	0.72	1.48	2.66	55.70	537.89	0.28
2010	0.75	0.87	1.62	2.82	57.51	533.92	0.30
2011	0.74	0.96	1.71	2.89	59.09	534.09	0.32
2012	0.74	0.93	1.68	2.84	59.01	537.38	0.31

GHG emissions for the Australian Department of Defence (AS DoD), United States Department of Defense (US DoD) and UK Ministry of Defence. Efforts are made to show emissions resulting from respective Defence infrastructure, facilities and buildings (the 'Estate'), as well as those resulting from military operations ('Defence Operations' are emissions generated from the domestic use of military equipment including planes, ships, tanks and so on). Notable exclusions in these tallies are off-shore military exercises and operations, land use, land use change and forestry (LULUCF) and whole-of-life-cycle emissions resulting from military-industrial research, development and manufacturing (generally covered under other sectoral reporting requirements). This section concludes by estimating total global military GHG emissions which are further compared against individual country contributions.

---

[6]Million Metric Tonnes CO_2 equivalent (Mt CO_2e) (Emissions data from the individual EEGO reports differ slightly from those in the most recent Excel database provided by the AS Government (refer to 'Energy Use in the Australian Government Operations: Annex C'). This table has used data from the most recent agency reporting, and includes updates from each Agency expressing previous 'under-reporting' (refer to 'Department of Defence, Agency Comments' within cited source). Final EEGO report was published in 2013 (covering 2012), following election of the Abbott government.). *Sources* Commonwealth of Australia (2013b, c, 2016)

[a]Covers all buildings and facilities within established Defence bases (does not include office building and stores outside base perimeters)

[b]Covers emissions from domestic Defence operations for aircraft, tanks, ships and combat vehicles. Excludes offshore emissions used on operations, exercises or war

[c]Figures exclude Australian War Memorial and the Department of Veterans' Affairs

AS DoD GHG Emissions. Table 2.1 shows AS DoD GHG emissions between 2001–2012. Table 2.1, Column D shows that GHG emissions in this period varied from a minimum of 1.48 Million tonnes of CO_2 equivalent (Mt CO_2e) in 2009 to a maximum of 2.10 Mt CO_2e in 2001. To place this in context, this is roughly equivalent to the total national emissions of many small developing nations, such as Haiti (2.3 Mt CO_2), Gabon (1.9 Mt CO_2), Guinea (1.9 Mt CO_2) or Madagascar (1.9 Mt CO_2), or the combined total national emissions of several smaller countries, e.g., Timor-Leste (0.2 Mt CO_2), Solomon Islands (0.2 Mt CO_2), Tonga (0.1 Mt CO_2), Greenland (0.6 Mt CO_2), and Fiji (0.8 Mt CO_2). Globally, if the AS DoD were a country, it would be ranked roughly between 143rd and 157th largest in the world (CDIAC 2015; UNFCCC 2015). Table 2.1, Column F shows that the AS DoD contributes between 55 and 68% of all Australian government emissions. As a percentage of the overall Australian GHG emissions (i.e., for the entire economy), this represents less than one-half of 1% (i.e., ranging from 0.28% in 2009 to 0.43% in 2001).

US DoD GHG Emissions. Table 2.2 shows US DoD estate and operational emissions over the period 2008–2015. All US DoD emissions figures are obtained from the US Strategic Sustainability Performance Plans (SSPP), which have been published by US DoD since 2010, and US Department of Energy Comprehensive Annual Energy and Data Sustainability Performance publications. Table 2.2 shows that US DoD emissions (including US DoD international bunkers and Scope 3 emissions) vary from 75.62 CO_2e in 2014 to 93 Mt CO_2e in 2010. This represents roughly three-quarters of all US government emissions and about 1.2% of all US domestic emissions.[7] To place this in context, it is almost one-sixth of the entire Australian economy (based on 84 Mt CO_2e for US DoD and 500 Mt CO_2e for Australia). If the US DoD were a single country, it would rank around the 40th largest emitter in the world, larger than some 176 other countries (CDIAC 2015; UNFCCC 2015).[8]

UK DoD GHG Emissions. Table 2.3 shows UK Ministry of Defence (MoD) estate and operational emissions over the period 2010–2014. Although some

---

[6] Million Metric Tonnes CO_2 equivalent (Mt CO_2e) (Emissions data from the individual EEGO reports differ slightly from those in the most recent Excel database provided by the AS Government (refer to 'Energy Use in the Australian Government Operations: Annex C'). This table has used data from the most recent agency reporting, and includes updates from each Agency expressing previous 'under-reporting' (refer to 'Department of Defence, Agency Comments' within cited source). Final EEGO report was published in 2013 (covering 2012), following election of the Abbott government.). *Sources* Commonwealth of Australia (2013b, c, 2016)

[a]Covers all buildings and facilities within established Defence bases (does not include office building and stores outside base perimeters)

[b]Covers emissions from domestic Defence operations for aircraft, tanks, ships and combat vehicles. Excludes offshore emissions used on operations, exercises or war

[c]Figures exclude Australian War Memorial and the Department of Veterans' Affairs

[7]Noting that this does not include any emissions associated with defence industrial manufacturing and production.

[8]Based on 2014 emissions.

Table 2.2 US DoD emissions, 2008–2013

US DoD GHG emissions (Mt CO_2e)								
Year	**US DoD estate**[a]	**US DoD operations**[b]	**US DoD intern'l bunkers**[c]	**US DoD total (b + c + d)**	**US Gov't**[d]	**% (e/f)**	**Total US emissions**[e]	**% (e/h)**
a	**b**	**c**	**d**	**e**	**f**	**g**	**h**	**i**
2008	34.49	50.44	8.20	93.13	121.20	76.84	7216.00	1.29
2009	N/A	N/A	7.40	N/A	N/A	N/A	N/A	N/A
2010	34.37	49.59	8.20	92.16	117.76	78.26	6985.00	1.32
2011	33.76	48.83	7.00	89.59	114.71	78.10	6865.00	1.31
2012	31.99	44.99	7.00	83.99	114.71	73.22	6643.00	1.26
2013	31.23	39.57	6.00	76.80	106.66	72.00	6800.00	1.13
2014	31.00	38.12	6.50	75.62	97.21	77.79	6870.00	1.10
2015	30.82	38.91	6.25[f]	75.98	97.89	77.62	N/A	N/A

Million Metric Tonnes CO_2 equivalent (Mt CO_2e). *Sources* US DoD (2012b, 2016a), US DoE (2016), US EPA (2016)

[a]Covers installations, facilities, buildings etc. Does not include US Army Corps of Engineers and other military related agencies. Includes Scope 1, 2, and 3 emissions. Data available at: http://ctsedwweb.ee.doe.gov/Annual/Report/ComprehensiveGreenhouseGasGHGInventoriesByAgencyAndFiscalYear.aspx (12 November 2016)

[b]Covers domestic military emissions of tactical ships, planes, vehicles (includes Scope 1, 2 and 3 emissions). Data available at: http://ctsedwweb.ee.doe.gov/Annual/Report/ComprehensiveGreenhouseGasGHGInventoriesByAgencyAndFiscalYear.aspx (12 November 2016)

[c]For US military emissions related to International Bunkers, see Table A-168. Page A-235, data available at: https://www.epa.gov/sites/production/files/2016-04/documents/us-ghg-inventory-2016-annex-3-additional-source-or-sink-categories-part-a.pdf (12 November 2016)

[d]For US Government agency emissions, see Department of Energy website, data available at: http://ctsedwweb.ee.doe.gov/Annual/Report/ComprehensiveGreenhouseGasGHGInventoriesByAgencyAndFiscalYear.aspx (12 November 2016)

[e]For US emissions see US EPA Inventory of US Greenhouse Gas Emissions and Sinks 1990–2014, specifically see Figure ES-1: US Greenhouse Gas Emissions by Gas, page ES-4, data available at: https://www.epa.gov/sites/production/files/2016-04/documents/us-ghg-inventory-2016-main-text.pdf (12 November 2016)

[f]This is an estimate based on the mean of the previous two years' data—i.e., (6.0 + 6.5/2)

data was unavailable, it is worth highlighting that, as a percentage of the overall national economy, the UK MoD contributes between 0.74–0.85% of all UK emissions. This is approximately mid-point between the Australian and US militaries.[9]

Estimating Global Military GHG Emissions. Based on total global GHG emissions of 52,763.43 Mt CO_2e (World Bank 2016), and assuming that any given

[9]Major exclusions from all data sets are emissions produced by offshore (global) operations (including from offshore military installations and tactical platforms operating internationally), emissions from land use, land use change and forestry (LULUCF) on military training grounds and bases, and emissions produced by the civilian military-industrial manufacturing sector during its manufacturing processes.

Table 2.3 UK MoD GHG emissions, 2010–2015

UK MoD GHG emissions (Mt CO_2e)								
Year	**UK MoD estate**[a]	**UK MoD operations**[b]	**N/A**	**UK MoD total (b + c + d)**	**UK Gov't**	**% (e/f)**	**Total UK emissions**[c]	**% (e/h)**
a	b	c	d	e	f	g	h	i
2010	1.43	3.33	N/A	4.77	N/A	N/A	602.10	0.79
2011	1.44	3.25	N/A	4.70	N/A	N/A	553.40	0.85
2012	1.29	3.12	N/A	4.41	N/A	N/A	570.50	0.77
2013	1.24	2.86	N/A	4.11	N/A	N/A	557.30	0.74
2014	1.22	3.06	N/A	4.28	N/A	N/A	514.40	0.83
2015	1.17	1.83	N/A	3.00	N/A	N/A	N/A	N/A

Sources UK MoD (2013, 2015), UK DECC (2016)
[a]UK MoD, Sustainable MOD Annual Report 2014/15, Table 3, Greenhouse Gas GGC target performance (p. 3), at: https://www.gov.uk/government/uploads/system/uploads/attachment_data/file/447951/20150723-Sustainable_MOD_Annual_Report-internet-ver.pdf (12 November 2016)
[b]Figures for MoD Operations can be deduced by subtracting estate emissions with gross emissions data at Annex A (refer to above citation). Earlier UK MoD reports also provide a breakdown of gross MoD GHG emissions data and can be used to check accuracy. For example, see 'graphical table' at Annex D to MoD Annual Sustainability Report and Accounts 2011–2012, at: https://www.gov.uk/government/uploads/system/uploads/attachment_data/file/35010/20121205_mod_annualreport_Annex_D.pdf (12 November 2016)
[c]For UK emissions see Department of Energy and Climate Change 2014 UK greenhouse gas emissions: final figures—data tables, Table 1, row 12, at: https://www.gov.uk/government/statistics/final-uk-greenhouse-gas-emissions-national-statistics-1990-2014 (11 November 2016)

military represents between 0.3–1.1% of all national emissions, total global military emissions may be estimated at between 158.29 and 580.39 Mt CO_2e for estate and domestic military operations. The higher-end estimate would place the global military sector between the 10th and 15th largest global emitter (depending on what year it is compared). A mid-way point has the military sector at nearly 370 Mt CO_2e per year—still within the top 20 emitting nations. Clearly, these are very broad estimates and further analysis, including a full set of assumptions, would be required before higher levels of confidence could be achieved. Major omissions from these estimates include offshore operations and wars, LULUCF, and the emissions from the military industrial-manufacturing base (although these are accounted for separately in national inventories according to UNFCCC rules). Despite this, as a rough-order-magnitude estimate, the results suggest what has long been suspected: that the military sector is a significant, but not overwhelming, contributor to overall global GHG emissions. As major global emitters, however, military forces are known contributors to the very security challenges that they themselves may ultimately be called upon to resolve.

2.5 International Political Response to Climate Change

Having now reviewed the fundamental science of climate change as well as the emissions and security aspects, this section provides a brief summary of the main political attempts to address climate change. Scientific and political recognition of the need to reduce GHG emissions (and limit global warming) first arose in the 1970s. This early momentum culminated in the establishment of the IPCC (1988) and UNFCCC (1992). These institutions are regarded as the main forums for collective international action to reduce emissions and address the global consequences of climate change. The UNFCCC, in particular, was mandated to establish internationally binding quantitative emissions reductions via international agreements ('protocols').[10] Parties to the UNFCCC are classified as either 'Annex I Parties' that mainly include developed countries and those defined as 'economies in transition' (EIT), as 'Annex II Parties' consisting of developed countries from Annex I (but not EITs), or as 'Non-Annex I' countries mostly made up of developing countries. The main forum of the UNFCCC is an annual meeting between country representatives known as the Conference of the Parties (COP) that has been held every year since 1995.[11]

Until the Paris Agreement in 2015/16, the most prominent outcome from UNFCCC was the Kyoto Protocol that sought a binding 'top-down' international agreement to reduce GHG emissions according to underlying national differences in emissions, wealth, capacity and historical responsibility (enshrined in the principal of 'common but differentiated responsibilities' (UNFCCC 1997: Article 10).[12] Though the IPCC has previously trumpeted the Kyoto Protocol (Barker et al. 2007: 89), criticism was widespread and variously decried as a 'narrow, thin, and in most of the world, [an] ultimately symbolic' measure (Keohane/Victor 2011: 10). In response to the failure of the Kyoto Protocol, the UNFCCC adopted a more flexible approach in the lead-up to the 2015 COP 21 (Paris), particularly with respect to the

[10]The UNFCCC came into force in March 1994 and, as of 2014, had 196 signatories. The IPCC remains the primary international authority responsible for synthesising global climate change science from peer reviewed published scientific and technical literature. Its mission is to publish the latest scientific updates, identify uncertainties in climate science that require further research and to prepare response policies. Since 1990, the IPCC has reported via a series of Assessment Reports (1990, 1995, 2001, 2007 and 2013) that represent the latest global scientific research on climate change. Each Assessment Report has noted with increasing conviction that climate change is occurring, that humans are responsible and that its consequences will be serious and far-reaching.

[11]A Meeting of the Parties (MOP) has also been held annually in conjunction with COPs. The MOP is a meeting of those countries who have emissions responsibilities under the Kyoto Protocol.

[12]In total, there are 191 parties that are signatories to the Kyoto Protocol (from which some nations have since withdrawn), which was adopted in 1997 and came progressively into force from February 2005. Of the countries that ratified the Kyoto Protocol, 37 undertook legally binding emissions reductions of at least 5% below 1990 levels for the first commitment period (2005–2012), with the second commitment period (2013–2020) increasing commitments to at least 18% below 1990 levels (UNFCCC 2012).

concept of Intended Nationally Determined Contributions (INDCs). Under this process, countries openly declared their contributions prior to the conference (a 'bottom-up' process).

Irrespective of the *type* of mechanism in a post-Kyoto world, any future success of a global agreement must overcome perceptions of unfairness (real or imagined) as well the ability to account for the rapidly evolving structural changes in the international system that has made UN-styled consensus-based agreement far more difficult to achieve (Leal-Arcas 2011). Some scholars have framed the issue as an ethical dilemma (Baer et al. 2010; Gardiner 2010b; Jamieson 2010; Caney 2010) which has been used to promote the idea that developing nations should bear responsibility since 'given the historical responsibility for cumulative emissions of GHG in the developed world [and resultant high levels of prosperity], inaction implies a neglect of overall responsibility and ethical considerations' (Rajendra K. Pachauri quoted in Gardiner 2010a: 7). Others, however, look to future projections and consider the relative decline of developed nations against developing ones and a requirement for structural change in how to address the costs of adaptation and mitigation. Rafael Leal-Arcas has argued that approximately 50 non-Annex I countries now have higher per capita incomes than the poorest of the Annex I countries with Kyoto commitments; 40 non-Annex I countries rank higher on the Human Development Index in 2007 than the lowest ranked Annex I country; and that developing economies are out-performing many industrialised countries who have remained in economic stagnation or recession since the 2007 Financial Crisis (Leal-Arcas 2011). These points are reinforced by the IEA forecast that in the next 25 years the non-Annex I countries will account for 75% of all global emissions (IEA 2011). Fundamentally, shifts occurring in international relations between the major powers have made consensus-based agreement more difficult to achieve:

> In the UN machinery, consensus among the parties is required … The turn today towards a multipolar world indicates that approaches based on consensus are unlikely to produce results. No country, or group of countries, today is in a position to forge a global deal … of environmental goals (Leal-Arcas 2011: 14).

Leal-Arcas then goes to the crux of the issue by pinpointing the enduring nature of state self-interest:

> When negotiating trade agreements, parties have an interest to negotiate as they believe they will benefit from the agreement. By contrast, in climate change negotiations, most parties see mitigation as a burden to their economies, i.e., *they negotiate to ensure that they do not have to do more than other parties in the negotiation* (Leal-Arcas 2011: 43; emphasis added)

Australian political-economist Ross Garnaut has previously likened this situation to a 'prisoner's dilemma', whereby the international community remains split into contending blocs, refusing to cooperate even though (in the medium to long-term) it remains in everyone's best interest to do so (Garnaut 2008). These insights reinforce what Harriet Bulkeley and Peter Newell have described as the 'intractable problem' that climate change presents for global governance (Bulkeley/Newell

2010). These authors note that climate change is a problem of unique complexity, where multiple scales of political decision-making, the fragmented nature and blurred roles of states and non-state actors, and the deeply embedded nature of emissions production and consumption make many aspects of climate change an issue beyond the capacity of traditional governance processes. It is a theme highlighted by other scholars (Knieling/Filho 2013), and in a related manner has been identified as a 'governance trap' in which climate change is simply 'too big a problem' resulting in a situation where 'the governing and the governed seek action from the other but where none is forthcoming' (Newell et al. 2015: 536). Moreover, climate governance continues to remain in a 'state of flux' (Bulkeley/Newell 2010: 108 and 110), with recent assessments noting that a 'resurgence of climate scepticism, and pessimism about the possibility of collective action' has produced a 'crisis of climate politics' (Newell et al. 2015: 536). While COP 21 gave the appearance of assuaging some of these concerns, Garnaut's assessment that climate change remains 'harder than any other issue of high importance that has come before our polity in living memory' remains a potent descriptor (Garnaut 2008: xvii). Underscoring this point was the election of Donald Trump who, on a platform of climate-denial, presents the very real prospect of the US walking away from the Paris agreement and advances in emissions reductions made during the Obama Administration.

The fundamental point to be made for this book is that since the realisation of the problem some forty years ago, the collective global political order has systematically failed in its efforts to reduce emissions and thereby limit global warming and its attendant consequences. The failure of *normal* politics has led some to press the political establishment to adopt a war-footing response: widening the scope and raising the profile and urgency of the issue such that emergency measures may become justified (UNGA 2008; Spratt 2012).

2.6 Chapter Conclusion

Anthropogenic emissions of greenhouse gases since 1750 have been the dominant contributing factor of present-day global warming. Global warming is already wreaking change to the earth's climate systems through increased global surface temperatures, sea-ice and glacial loss, sea level rise, ocean acidification, increased precipitation, extreme weather and a range of other environmental impacts. These changes are projected to take place at increasing rates and intensity across the twenty-first century as global emissions increase. Currently, emissions are tracking on a BAU for a worse-case scenario such that it is highly likely that global average temperatures will exceed 4 °C by 2100 on pre-industrial levels, if not more. International policy response to reduce emissions has been stymied by self-interest. Developing nations, seeking to lift their people out of poverty, have sought access to cheap forms of electricity, predominately supplied through large-scale uptake of fossil fuels. Developed nations, who are responsible for the majority of historical emissions, have had a mixed record in shifting their economies away from fossil fuels. All told, despite the knowledge that rising emissions will cause continued global warming, global political leaders have been unable to stem increasing fossil fuel consumption. Attempts at top-down global agreements via the UNFCCC have so far failed to bring about meaningful reductions in emissions.

What relevance does this conclusion have for this book? First, it identifies climate change as a major—if not *the*—strategic challenge for the twenty-first century. By *strategic* it is meant that there have been no previous challenges on the pervasiveness, penetration or scale presented by climate change. Or, as Shannon O'Lear and Simon Dalby have stated, that climate change and its implications have 'had no obvious analogy in human affairs' (O'Lear/Dalby 2016: 3). Statements from the global scientific community that planetary temperatures may exceed 4 °C by the end of this century are *unprecedented* in civilisational history. This point serves to emphasise not just the scale of the problem but also the relative urgency if humankind is to reverse the trajectory of current emissions pathways. These are important concepts for this book, as it seeks to examine how the US and Australian militaries have approached climate change and to understand whether the two militaries have placed greater or less emphasis on both the scale and urgency of the threat than their political masters.

The second point is that climate change will present (if it is not already doing so) wide-ranging security challenges. In this respect, military forces are not exceptional to debates surrounding climate change. There is sufficient evidence to suggest militaries have a requirement to understand not just current climate impacts but also how future climate change will remake existing socio-political orders. In essence, militaries exist primarily as institutions designed to anticipate and respond to strategic risks that threaten the sovereign interest of their nation-states and the security of peoples living within their jurisdictions. In this respect, greater understanding of how militaries are responding to climate change is an important aspect in gauging whether the militaries are doing too much, just enough, or too little in warning the public and influencing political elite on the importance of the strategic risks of climate change.

References

Arrhenius, Svante. 1896. "On the Influence of Carbonic Acid in the Air upon the Temperature on the Ground." *Philosophical Magazine and Journal of Science*, vol. 41, no. 5:237–276.

Baer, Paul, Tom Athanasiou, Sivan Kartha, and Eric Kemp-Benedict. 2010. "Greenhouse Development Rights: A Framework for Climate Protection That Is "More Fair" Than Equal Per Capita Emissions Rights." In *Climate Ethics: Essential Readings*, edited by Stephen M. Gardiner, Simon Caney, Dale Jamieson and Henry Shue. Oxford: Oxford University Press.

Barker, T., I. Bashmakov, L. Bernstein, J. E. Bogner, P. R. Bosch, R. Dave, O. R. Davidson, B. S. Fisher, S. Gupta, K. Halsnæs, G. J. Heij, S. Kahn Ribeiro, S. Kobayashi, M. D. Levine, D. L. Martino, O. Masera, B. Metz, L. A. Meyer, G. J. Nabuurs, A. Najam, N. Nakicenovic, H. H. Rogner, J. Roy, J. Sathaye, R. Schock, P. Shukla, R. E. H. Sims, P. Smith, D. A. Tirpak, D.

Urge-Vorsatz, and D. Zhou. 2007. "Technical Summary." In *Contribution of Working Group III to the Fourth Assessment Report of the Intergovernmental Panel on Climate Change*, edited by B. Metz, O. R. Davidson, P. R. Bosch, R. Dave and L. A. Meyer. Cambridge, United Kingdom: Cambridge University Press.

Barrie, Chris, Will Steffen, Alix Pearce, and Michael Thomas. 2015. *Be Prepared: Climate Change, Security and Australia's Defence Force*. Sydney: Climate Council. Accessed on 20 September 2015, https://www.climatecouncil.org.au/uploads/fa8b3c7d4c6477720434d6d10897af18.pdf.

Blasing, T. J. 2012. "Recent Greenhouse Gas Concentrations." Carbon Dioxide Information Analysis Centre (CDIAC). Accessed on 28 January 2013, http://cdiac.ornl.gov/pns/current_ghg.html.

Bulkeley, Harriet, and Peter Newell. 2010. *Governing Climate Change*, edited by Thomas G. Weiss, *Global Institutions Series*. London: Routledge.

Buzan, Barry. 1991. *People, states and fear: an agenda for international security studies in the post-Cold War era*. London: Harvester Wheatsheaf.

Caney, Simon. 2010. "Cosmopolitan Justice, Responsibility, and Global Climate Change." In *Climate Ethics: Essential Readings*, edited by Stephen M. Gardiner, Simon Caney, Dale Jamieson and Henry Shue. Oxford: Oxford University Press.

Carmen, Herbert E., Christine Parthemore, and Will Rogers. 2010. *Broadening Horizons: Climate Change and the U.S. Armed Forces*. Washington, DC: Center for a New American Century. Accessed on 15 May 2017, https://s3.amazonaws.com/files.cnas.org/documents/CNAS_Broadening-Horizons_Carmen-Parthemore-Rogers.pdf.

CDIAC. 2015. "Global, Regional, and National Fossil-Fuel CO_2 Emissions." T. A. Boden, G. Marland and R. J. Andres (Eds.). Oak Ridge National Laboratory, Tenn., USA: U.S. Department of Energy, Carbon Dioxide Information Analysis Center. Accessed on 9 November 2016, http://cdiac.ornl.gov/trends/emis/meth_reg.html.

CNA. 2007. *National Security and The Threat of Climate Change*. Virginia: The CNA Corporation. Accessed on 12 April 2013, https://www.cna.org/cna_files/pdf/National%20Security%20and%20the%20Threat%20of%20Climate%20Change.pdf.

CNA. 2014. *National Security and the Accelerating Risks of Climate Change*. Alexandria: CNA. file:///C:/01.%20Michael%20Thomas%20Primary%20Folder/2.%20PhD_Primary%20Files_2%20Apr%202014/4.%20General%20Literature/Climate%20Change%20and%20Security/1.%20Global/CNA%20MAB_2014.pdf.

Commonwealth of Australia. 2013b. "Energy Use in the Australian Government's Operations 2011–12: Annex C." Energy and Tourism Department of Resources. Canberra: Commonwealth of Australia. Accessed on 8 November 2016, http://webarchive.nla.gov.au/gov/20160615065700/http://www.industry.gov.au/Energy/EnergyEfficiency/Non-residentialBuildings/GovernmentBuildings/EnergyUseOperations/Pages/default.aspx.

Commonwealth of Australia. 2013c. "Energy Use in the Australian Government's Operations (Reports 2001–2011)." Energy and Tourism Department of Resources. Canberra: Commonwealth of Australia. Accessed on 8 November 2016, http://webarchive.nla.gov.au/gov/20160615065736/http://www.industry.gov.au/Energy/EnergyEfficiency/Non-residentialBuildings/GovernmentBuildings/EnergyUseOperations/Pages/PreviousEditions.aspx.

Commonwealth of Australia. 2016. "National Greenhouse Gas Inventory—Kyoto Protocol classifications." Department of the Environment and Energy. Canberra: Commonwealth of Australia. Accessed on 9 November 2016, http://ageis.climatechange.gov.au/.

Cotton, William R., and Roger A. Peilke. 2007. *Human Impacts on Weather*. Fort Collins: Cambridge University Press.

Dalby, Simon. 1992. "Security, modernity, ecology: the dilemmas of post-Cold War security discourse." *Alternatives*, vol. 17, no.:95–134.

ESSP. 2012. "Global Carbon Budget 2012." Corinne Le Quéré, Glen Peters, Robbie Andrew, Bob Andres, Tom Boden, Thomas Conway, Skee Houghton, Jo House, Gregg Marland, Guido van der Werf, Anders Ahlström, Laurent Bopp, Pep Canadell, Philippe Ciais, Scott Doney, Clare Enright, Pierre Friedlingstein, Chris Huntingford, Atul Jain, Charlotte Jourdain, Etsushi Kato,

Ralph Keeling, Kees Klein Goldewijk, Samuel Levis, Peter Levy, Mark Lomas, Ben Poulter, Mike Raupach, Jörg Schwinger, Stephen Sitch, Benjamin Stocker, Nicolas Viovy, Charlie Wilson, Soenke Zaehle and Ning Zeng: Earth System Science Partnership. Accessed on 15 May 2017, http://www.globalcarbonproject.org/carbonbudget/archive/2012/CarbonBudget_2012.pdf.

Flounders, Sara. 2014. "The Pentagon, The Climate Elephant. The US Military Machine is the World's Worst Polluter of Greenhouse Gas Emissions." Online: Global Research: Centre for Research on Globalization. Accessed on 10 November 2016, http://www.globalresearch.ca/the-pentagon-the-climate-elephant-2/5402505.

Friedrich, Tobias, Axel Timmermann, Michelle Tigchelaar, Oliver Elison Timm, and Andrey Ganopolski. 2016. "Nonlinear climate sensitivity and its implications for future greenhouse warming." *Science Advances*, vol. 2, no. 11. doi:10.1126/sciadv.1501923.

Gardiner, Stephen M. 2010a. *Climate Ethics: Essential Readings*. Oxford: Oxford University Press.

Gardiner, Stephen M. 2010b. "Ethics and Global Climate Change." In *Climate Ethics: Essential Readings*, edited by Stephen M. Gardiner, Simon Caney, Dale Jamieson and Henry Shue. Oxford: Oxford University Press.

Garnaut, Ross. 2008. *The Garnaut Climate Change Review: Final Report*. Melbourne: Cambridge University Press.

Hansen, James, M. Sato, P. Kharecha, and K. von Scuckmann. 2011. "Earth's energy imbalance and implications." *Atmospheric Chemistry and Physics*, no. 11:13421–13449. doi:10.5194/acp-11-13421-2011.

IEA. 2011. "Executive Summary." In *2011 World Energy Outlook*. Paris: International Energy Agency. Accessed on 10 September 2012, http://www.iea.org/publications/freepublications/publication/English.pdf.

IPCC. 2015. "Climate Change 2014 Synthesis Report." In *Climate Change 2014: Synthesis Report. Contribution of Working Group I to the Fifth Assessment Report of the Intergovernmental Panel on Climate Change*, edited by Core Writing Team, Rajendra K. Pachauri and Leo Meyer. Geneva: Intergovernmental Panel on Climate Change. Accessed on 14 March 2015, http://www.ipcc.ch/pdf/assessment-report/ar5/syr/SYR_AR5_FINAL_full.pdf.

Jamieson, Dale. 2010. "Ethics, Public Policy, and Global Warming." In *Climate Ethics: Essential Readings*, edited by Stephen M. Gardiner, Simon Caney, Dale Jamieson and Henry Shue. Oxford: Oxford University Press.

Keohane, Robert O., and David G. Victor. 2011. "The Regime Complex for Climate Change." *Perspectives on Politics*, vol. 9, no. 1:7–23. doi:10.1017/S1537592710004068.

Knieling, Jörg, and Walter Leal Filho. 2013. "Climate Change Governance: The Challenge for Politics and Public Administration, Enterprises and Civil Society." In *Climate Change Governance*. Heidelberg: Springer.

Leal-Arcas, Rafael. 2011. "Top-Down Versus Bottom-Up Approaches for Climate Change Negotiations: An Analysis." *The IUP Journal of Governance and Public Policy* vol. 6, no. 4:7–52.

LeQuéré, C., R. Moriarty, R. M. Andrew, J. G. Canadell, S. Sitch, J. I. Korsbakken, P. Friedlingstein, G. P. Peters, R. J. Andres, T. A. Boden, R. A. Houghton, J. I. House, R. F. Keeling, P. Tans, A. Arneth, D. C. E. Bakker, L. Barbero, L. Bopp, J. Chang, F. Chevallier, L. P. Chini, P. Ciais, M. Fader, R. Feely, T. Gkritzalis, I. Harris, J. Hauck, T. Ilyina, A. K. Jain, E. Kato, V. Kitidis, K. Klein Goldewijk, C. Koven, P. Landschützer, S. K. Lauvset, N. Lefèvre, A. Lenton, I. D Lima, N. Metzl, F. Millero, D. R. Munro, A. Murata, J. E. M. S. Nabel, S. Nakaoka, Y. Nojiri, K. O'Brien, A. Olsen, T. Ono, F. F. Pérez, B. Pfeil, D. Pierrot, B. Poulter, G. Rehder, C. Rödenbeck, S. Saito, U. Schuster, J. Schwinger, R. Séférian, T. Steinhoff, B. D. Stocker, A. J. Sutton, T. Takahashi, B. Tilbrook, I. T. van der Laan-Luijkx, G. R. van der Werf, S. van Heuven, D. Vandemark, N. Viovy, A. Wiltshire, S. Zaehle, and N. Zeng. 2015. "Global Carbon Budget 2015." *Earth System Science Data Discussions*, vol. 7, no. 2:349–396. doi:10.5194/essd-7-349-2015.

LeQuéré, Corinne, Robbie M. Andrew, Josep G. Canadell, Stephen Sitch, Jan Ivar Korsbakken, Glen P. Peters, Andrew C. Manning, Thomas A. Boden, Pieter P. Tans, Richard A. Houghton, Ralph F. Keeling, Simone Alin, Oliver D. Andrews, Peter Anthoni, Leticia Barbero, Laurent Bopp, Frédéric Chevallier, Louise P. Chini, Philippe Ciais, Kim Currie, Christine Delire, Scott C. Doney, Pierre Friedlingstein, Thanos Gkritzalis, Ian Harris, Judith Hauck, Vanessa Haverd, Mario Hoppema, Kees Klein Goldewijk, Atul K. Jain, Etsushi Kato, Arne Körtzinger, Peter Landschützer, Nathalie Lefèvre, Andrew Lenton, Sebastian Lienert, Danica Lombardozzi, Joe R. Melton, Nicolas Metzl, Frank Millero, Pedro M. S. Monteiro, David R. Munro, Julia E. M. S. Nabel, Shin-ichiro Nakaoka, Kevin O'Brien, Are Olsen, Abdirahman M. Omar, Tsuneo Ono, Denis Pierrot, Benjamin Poulter, Christian Rödenbeck, Joe Salisbury, Ute Schuster, Jörg Schwinger, Roland Séférian, Ingunn Skjelvan, Benjamin D. Stocker, Adrienne J. Sutton, Taro Takahashi, Hanqin Tian, Bronte Tilbrook, Ingrid T. van der Laan-Luijkx, Guido R. van der Werf, Nicolas Viovy, Anthony P. Walker, Andrew J. Wiltshire, and Sönke Zaehle. 2016. "Global Carbon Budget 2016." *Earth System Science Data*, vol. 8, no. 2:605–649. doi:10.5194/essd-8-605-2016.

LeTreut, H., R. Somerville, U. Cubasch, Y. Ding, C. Mauritzen, A. Mokssit, T. Peterson, and M. Prather. 2007. "Historical Overview of Climate Change." In *Climate Change 2007: The Physical Science Basis. Contribution of Working Group I to the Fourth Assessment Report of the Intergovernmental Panel on Climate Change*, edited by S. Solomon, D. Qin, M. Manning, Z. Chen, M. Marquis, K. B. Averyt, M. Tignor and H. L. Miller, pp. 93–128. Cambridge: Cambridge University Press. Accessed on 15 June 2012, http://www.ipcc.ch/pdf/assessment-report/ar4/wg1/ar4-wg1-chapter1.pdf.

Michaelowa, Axel, and Tobias Kock. 2001. "Military Emissions, Armed Conflicts, Border Changes and the Kyoto Protocol." *Climatic Change*, vol. 50, no. 4:383–394.

NASA Earth Observatory. 2016. "The Atmosphere's Energy Budget." Goddard Space Flight Center, Washington DC: NASA Earth Observatory. Accessed on 2 November 2016, http://earthobservatory.nasa.gov/Features/EnergyBalance/page6.php.

Neslen, Arthur. 2015. "Pentagon to lose emissions exemption under Paris climate deal." Online: The Guardian. Accessed on 10 November 2016, https://www.theguardian.com/environment/2015/dec/14/pentagon-to-lose-emissions-exemption-under-paris-climate-deal.

Newell, Peter, Harriet Bulkeley, Karen Turner, Christopher Shaw, Simon Caney, Elizabeth Shove, and Nicholas Pidgeon. 2015. "Governance traps in climate change politics: re-framing the debate in terms of responsibilities and rights." *WIREs Climate Change*, vol. 6, no. Nov/Dec:535–540. doi:10.1002/wcc.356.

NIC. 2008. "National Intelligence Assessment on the National Security Implications of Global Climate Change to 2030", Thomas Fingar. Washington D.C.: National Intelligence Council. Accessed on 17 February 2013, http://online.wsj.com/public/resources/documents/FingarSFR.GCC06252008.pdf.

NOAA ESRL. 2016. "Full Mauna Loa CO_2 record." P. Tans and R. Keeling (Eds.). Boulder: National Oceanic and Atmospheric Administration Earth System Research Laboratory. Accessed on 2 November 2016, http://www.esrl.noaa.gov/gmd/ccgg/trends/.

O'Lear, Shannon, and Simon Dalby. 2016. *Reframing Climate Change: Constructing Ecological Geopolitics*. Edited by Shannon O'Lear and Simon Dalby. London: Routledge.

Savage, Lisa. 2015. "Elephant in the Room: The Pentagon's Massive Carbon Footprint." Online: Counterpunch. Accessed on 10 November 2016, http://www.counterpunch.org/2015/07/23/72279/.

Spratt, David. 2012. "Scientists call for war on climate change, but who on earth is listening." Published online: Climate Code Red. Accessed on 6 October 2015, http://www.climatecodered.org/2012/12/scientists-call-for-war-on-climate.html.

Sturrock, Robert, and Peter Ferguson. 2015. *The Longest Conflict: Australia's Climate Security Challenge*. Melbourne: Centre for Policy Development.

Thomas, Michael. 2012. "Assessing the Risk of Global Climate Change on the Australian Defence Force." e-IR. Accessed on 10 October 2012, http://www.e-ir.info/author/michael-thomas/.

Trenberth, Kevin E., J. T. Fasullo, and J. Kiehl. 2009. "Earth's Global Energy Budget." *Bulletin of American Meteorological Society*, vol. 90, no. 3:311–324. doi:10.1175/2008BAMS2634.1.

Tschakert, Petra. 2015. "1.5 °C or 2 °C: a conduit's view from the science-policy interface at COP20 in Lima, Peru." *Climate Change Responses*, vol. 2, no. 3. doi:10.1186/s40665-015-0010-z.

UK DECC. 2016. *Final UK Greenhouse gas emissions national statistics: 1990–2014 (2014 UK ghg: final figures—data tables)*. Department of Energy and Climate Change. London: UK Government. Accessed on 12 November 2016, https://www.gov.uk/government/statistics/final-uk-greenhouse-gas-emissions-national-statistics-1990-2014.

UK MoD. 2013. *Annex D to Sustainability Report—MOD Annual Report and Accounts 2011–2012*. Ministry of Defence. London: UK Ministry of Defence. Accessed on 12 November 2016, https://www.gov.uk/government/uploads/system/uploads/attachment_data/file/35010/20121205_mod_annualreport_Annex_D.pdf.

UK MoD. 2015. *Sustainable MOD: Annual Report 2014/15: Sustainability in the Ministry of Defence*. UK Ministry of Defence. London: UK Government. Accessed on 12 November 2016, https://www.gov.uk/government/uploads/system/uploads/attachment_data/file/447951/20150723-Sustainable_MOD_Annual_Report-internet-ver.pdf.

UNFCCC. 1997. *Kyoto Protocol to the United Nations Framework Convention on Climate Change*. Kyoto: United Nations. Accessed on 12 June 2013, http://unfccc.int/essential_background/kyoto_protocol/items/1678.php.

UNFCCC. 2012. "Kyoto Protocol: Doha Amendment." New York: United Nations. Accessed on 6 February 2013, http://unfccc.int/kyoto_protocol/items/2830.php.

UNFCCC. 2015. "National Inventory Submissions." New York: United Nations. Accessed on 9 November 2016, http://unfccc.int/national_reports/annex_i_ghg_inventories/national_inventories_submissions/items/8108.php.

UNGA. 2008. "Battle against climate change calls for 'war footing', Deputy Secretary-general says as General Assembly holds follow-up to February thematic debate." In *General Assembly Plenary Press Release*, GA/10725, 8 July 2008. New York: United Nations. Accessed on 6 October 2015, http://www.un.org/press/en/2008/ga10725.doc.htm.

UNSG. 2009. "Report of the Secretary-General: Climate change and its possible security implications (A/64/350)." In *United Nations General Assembly 64th session*, Item 114, A/64/350, 11 September 2009. New York: United Nations. Accessed on 15 May 2017, http://www.un.org/ga/search/view_doc.asp?symbol=A/64/350.

US DoD. 2012b. *Department of Defense Strategic Sustainability Performance Plan FY-2012*. Washington, DC: Department of Defense. Accessed on 9 November 2016, http://www.denix.osd.mil/sustainability/dod-sspp/unassigned/dod-sspp-fy-2012/.

US DoD. 2016a. *Department of Defense Strategic Sustainability Performance Plan FY 2016*. Washington, DC: Department of Defense. Accessed on 9 November 2016, http://www.denix.osd.mil/sustainability/dod-sspp/unassigned/department-of-defense-strategic-sustainability-performance-plan-fy-2016/ and http://www.denix.osd.mil/sustainability/dod-sspp/.

US DoE. 2016. "Greenhouse Gas Emissions." Washington, DC: US Department of Energy. Accessed on 10 November 2016, http://catalog.data.gov/dataset/federal-greenhouse-gas-inventories-and-performance/resource/347d9ac8-0388-4b41-b064-41404b9915ec.

US EPA. 2016. "Annex 3 (Part A) Methodological Descriptions for Additional Source or Sink Categories." United States: US Environmental Protection Agency. Accessed on 10 November 2016, https://www.epa.gov/ghgemissions/inventory-us-greenhouse-gas-emissions-and-sinks-1990-2014 and https://www.epa.gov/sites/production/files/2016-04/documents/us-ghg-inventory-2016-annex-3-additional-source-or-sink-categories-part-a.pdf.

Weart, Spencer R. 2008. *The Discovery of Global Warming*. Second ed. Cambridge: Harvard University Press.

Wolfers, Arnold. 1952. "'National Security' as an Ambiguous Symbol." *Political Science Quarterly*, vol. 67, no. 4:481–502.

World Bank. 2016. "Total greenhouse gas emissions". Online: World Bank. Accessed on 12 November 2016, http://data.worldbank.org/indicator/EN.ATM.GHGT.KT.CE.

Youngs, Richard. 2015. *Climate Change and European Security*. London: Taylor and Francis.

Chapter 3
Climate Security in Europe, the US and the UN Security Council

Abstract This chapter provides a narrative history of the policy-orientated literature on climate security, with a focus on highlighting climate discourse within US and European militaries, and the UN. With the fundamental science of climate change in place by the end of the 19th century, it was not until major environmental movements of the 1960s and '70s that rising GHG concentrations were tentatively, then more boldly, linked to adverse consequences for human, national and international security. By the 1980s climate change as a security issue had been elevated at major global meetings, the US Congress and the UN. Although wider debates on environmental security became prominent in the 1990s, it was not until the mid-2000s that climate change as a security issue was significantly addressed. Robust attempts were made by many Western nations between 2007 and 2010 to frame climate change as a security threat, culminating in it being debated by the UNSC. A counter bloc of countries, dominated by the so-called BRICS, argued against climate change being debated by the UNSC and for it to be addressed exclusively as a sustainable development issue in non-securitised, multi-lateral forums. Climate security in the US was extensively addressed by the US Congress, the National Intelligence Community and by US think-tanks from 2007, and it emerged as an issue of increasing significance for the US military from 2009. In Europe, although there existed moves by various leaders and institutions to frame climate change as a security issue, it was never fully translated into near or long term military planning.

Keywords Climate security · Climate securitisation · Climate politics · History of climate change · UN climate security

3.1 A General History of Climate Security

Climate change first emerged as a security concern out of the environmental movement of the 1960s. It was during this period that increased public, scientific and political awareness of environmental issues—particularly in the United States

M.D. Thomas, *The Securitization of Climate Change: Australian and United States' Military Responses (2003–2013)*, The Anthropocene: Politik—Economics—Society—Science 10, DOI 10.1007/978-3-319-49658-0_3

—came to prominence via government and non-government institutions, public campaigns and legislation.[1] As a security issue, climate change was partly addressed in Richard Falk's *This Endangered Planet* (Falk 1971) but more directly in US government research initiatives on energy and climate (Revelle et al. 1977) as well as the 1978 US *National Climate Program Act* (Public Law 95-367, 95th Congress). The last of these suggested 'climate change [can] affect food production, energy use, water resources and other factors vital to national security' (National Climate Program Act 1978). In 1981 Al Gore established the first congressional hearings on climate change (Corn 2006), while US scientist James Hansen anticipated that climate change might create 'drought-prone regions in North America and central Asia as part of a shifting climatic zones, erosion of the West Antarctic ice sheet with a consequence worldwide rise in sea level, and opening of the fabled Northwest Passage' (Hansen et al. 1981: 957).

Beginning in the 1970s, the United Nations and the international scientific community more generally, also began to examine the nexus between climate change and security. Although not a major focus, climate change as a security threat was indirectly addressed at major international conferences that included the UN Conference on the Human Environment (1972), a series of World Climate Conferences (1979, 1990 and 2009), the 'Villach meetings' of expert scientists (1985 and 1987) and the Brundtland Commission (1987). Political leaders on both sides of the Cold War divide voiced concern in international forums, among them Mikhail Gorbachev, who stated 'the relationship between man and the environment has become menacing … the threat from the sky is no longer missiles but global warming' (Brauch 2011a: 63–64). In 1985, the World Resources Institute (WRI) suggested that the impact of 'greenhouse warming could be catastrophic' (UN 1988), while security implications were addressed at the 1987 *World Conference on the Changing Atmosphere: Implications for Global Security,* that observed climate change was a major international security threat and represented an 'unintended, uncontrolled, globally pervasive experiment whose ultimate consequences could be second only to a global nuclear war' (WMO 1988: 292). International political momentum culminated with the establishment of the IPCC in 1988 and UNFCCC in 1992.

The end of the Cold War also marked a turning point in how security issues, including climate change, were conceived-of and addressed by the scholarly community (particularly from the fields of Security Studies and International Relations) as well as by the policy and intelligence community, and foreign policy establishment more widely. Whereas traditional security approaches (nurtured by concepts of American-centred Realism with a 'narrow' focus on hard-power and military capability) had dominated the security mind-set, the end of the Cold War

[1]For early and increased public awareness of environmental issues, see: Carson (1962), Hardin (1968), Ehrlich (1968), Meadows et al. (1972). Examples of organised institutions include the World Wildlife Fund (1961), Greenpeace (1969), Friends of the Earth (1969), US Environmental Protection Agency (1970), UN Environmental Programme (1972). Example of public campaigns and legislation include Earth Day (1970), Clean Air Act (1970), Clean Water Act (1972).

led to the emergence of critical perspectives and a *widening* and *deepening* of the security studies agenda (Buzan/Hansen 2009). In this context, 'widening' referred to a move away from an exclusive focus on military-security issues and towards other sectors, including economic, societal and environmental. Moreover, it gave rise to the idea that threats to national security were not exclusively defined by an opponent's military power, but could emerge from a complex milieu of circumstances; a gallimaufry of aggrieved non-state actors, rapid population growth (particularly in areas of poor governance), infectious disease, poverty, inequality, resource scarcity and environmental degradation. 'Deepening' of the security agenda referred to a move away from the state as the key referent object to be secured (Buzan et al. 1998). Thus, whereas the state was once the exclusive referent object, critical approaches located a variety of referent objects that led to 'new' concepts of 'human security' and 'environmental security'. Debates on whether to include these 'new' sources of threats and different referent objects within traditional national security agendas became a distinguishing feature of 1990s security studies discourse (Ullman 1983; Brown 1977, 1986; Paul Ehrlich/Anne Ehrlich 1988; Mathews 1989; Deudney 1990; Homer-Dixon 1991; Myers 1994; Levy 1995).

As part of this 'new' security outlook, scholars began to investigate the nexus between resource scarcity, population growth, environmental degradation and acute conflict. Significant empirical studies were undertaken throughout the 1990s with prominent contributions made by the Toronto Group (Homer-Dixon 1991, 1994, 1996, 1999) and the Environmental and Conflicts Project (ENCOP) (Klötzli 1994; Bächler et al. 1995; Lang 1995; Libiszewski 1995) as well the International Human Dimensions Programme on Global Environmental Change (IHDP), International Geosphere-Biosphere Programme (IGBP), Global Environmental Change and Human Security (GECHS) and German Advisory Council on Global Change (WGBU) (Tanzler et al. 2002: 48–49). The potential spill-over of conflict from human-induced environmental pressure into regional and international violence was highlighted by US and UN failures in Somalia (1992) and Rwanda (1994), both of which were well-captured by Robert Kaplan's *The Coming Anarchy* (1994).

Notably, the Toronto Group led by Thomas Homer-Dixon has been singled out by some as having particular influence on environmental security studies discourse, including the idea that it served to marginalise the issue of climate change by downplaying connections between localised impacts and global dynamics (Trombetta 2012: 154). As a consequence, climate change was arguably dismissed during the 1990s as a 'low priority issue' (Trombetta 2012: 154). Ultimately, the inconclusiveness of environmental security studies also led some to conclude that 'the bubble had burst', which resulted in the policy establishment moving to 'other theories … "the clash of civilizations" now claimed the spotlight' (Dabelko 1999: 14). September 11, followed by US-led wars in Afghanistan (2001) and Iraq (2003) as well as the ascendancy of neo-conservative politics in the US under President George W. Bush, reinforced this tendency (Trombetta 2008: 594; Buzan/Hansen 2009: 229–243).

Although terrorism and the war in Iraq dominated international politics and global security in the first half of the 2000s, climate change gained renewed prominence as a global issue between 2007 and 2010 as a result of the release of IPCC AR4 and the prospect of a global agreement in the lead-up to the Copenhagen Conference (Brauch 2009; Oels 2012; Trombetta 2012). Politically, climate change now registered throughout this period as a priority issue at the highest levels, including G8 summits (Gleneagles [2005], Heiligendamm [2007], Tokyo [2008]; G20 summits [2008, 2009]; the Conference of the Africa Union [2007]; the UNGA [discussed as a security issue in 2007 and 2008]; and the UNSC [first convened as a security issue in 2007]). In popular literature (Klare 2007; Dyer 2008), film (Guggenheim 2006) and mainstream media (Townsend/Harris 2004; Homer-Dixon 2007), climate change was portrayed as a 'new' threat that did not discriminate according to national wealth or status (Busby 2007: 1). In the scholarly community, climate change came to represent the new face of the old environmental-security debates of the 1990s, spawning new research opportunities, journals, as well as 'climate institutes', university courses and think-tanks dedicated to examining its every aspect.[2] It was during this period that climate change came to be framed increasingly as an urgent and existential security threat requiring emergency measures. In summary, between 2007 and 2010 concerted and systematic attempts were made by various nations, institutions and individual leaders to *securitise* the issue of climate change. In practice, securitisation was adopted primarily as a means to galvanise the international community towards a conclusive agreement on emission reductions (particularly at the Copenhagen conference) and as a fresh approach that sought a more comprehensive response to the security implications of climate change. The following sections trace this narrative in greater detail and—in keeping with the aims of this book—with a particular focus on climate securitisation in the security and military strategies of the EU, the US and the UN Security Council (UNSC).

3.2 Climate Security in Europe and the United States

3.2.1 Climate Security in European Policy

The European Union (EU), and its individual member countries more broadly, have long been regarded as an international exemplar in advancing strong action on climate change (Vis/Delbeke 2015). Examples abound, but prominent among these have been the strong EU commitments to tackle climate action at various international forums, the implementation of a progressive mix of climate policies at both the national and continental-wide levels, and the consistently assertive and public

[2]Examples abound; see, for example, the website of The Center for Climate and Security, at: https://climateandsecurity.org/ (23 December 2016).

pronouncements by European leaders on the urgency to act in regards to climate change. Major justifications given for strong European action on climate change have centred not solely on a concern for climate impacts but also on the opportunities presented by economic and global leadership (Hayes/Knox-Hayes 2014; Torney 2015). In this vein, a high-level 2007 German report urged the EU to become a leading player in global climate policy as a means to counteract emerging Great Power rivalries and convince both the US and rising Asian powers of the importance of avoid dangerous climate change (Schubert et al. 2008). To some extent, Europe has also been regarded as a leading actor in framing climate change as a security issue (Brauch 2009; ASP 2016). This has been most evident in its efforts to inject climate change as part of the UNSC agenda,[3] via public statements by its leaders declaring climate change a significant global security threat, and in its efforts to integrate climate change into its foreign and defence policies.

Notwithstanding, the overall impression of the EU climate security agenda is one of inconsistency. While climate change has been part of the formal European security agenda since at least 2003 (EU 2003) and witnessed significant uptake between 2007 and 2011 (Brauch 2009; Brzoska 2012b), it has subsequently drifted and stalled as the debt, migrant, terrorism and political crisis (capped by the 2016 Brexit referendum, continued spectre of Russian aggression and potential breakup of the EU) produced a re-prioritisation of the European security agenda (European Commission 2015; European Council 2016a, b). Hayes/Knox-Hayes (2014) have also suggested that climate security discourses were already underdeveloped on account of a European culture that has tended to place less emphasis on security affairs, as well as on governance structures that preclude the requirement for securitisation (Hayes/Knox-Hayes 2014). Others again have pinpointed to a less-developed climate security agenda than the US on the grounds that European militaries are simply not as exposed compared to the global disposition of US military forces (Stang 2014). Whatever the combination of reasons, while climate change entered into some European national security and military strategies (particularly in the UK), it largely failed to engender any longer-term presence or wider policy significance (Youngs 2015). In contrast to the US, climate change as a security threat failed to gain substantive traction in European militaries, where one analyst concluded that there existed 'little evidence of preparedness' (Youngs 2014: 13). Building on this theme, this section also poses the idea that, unlike in the US, many European leaders and policy statements actively and directly downplayed the role of the military in responding to climate change. Keenly aware to avoid the 'militarisation' of climate change, this particular framing arguably had the effect of signalling to military and strategy officials that climate security was not a priority matter for security forces and concomitant national security and military strategies.

[3]European nations have been instrumental in arranging debates on the security implications of climate change at the UNSC; the UK convened the 2007 debate, Germany the 2011 debate, the UK (and Pakistan) the 2013 debate, and Spain (and Malaysia) the 2015 debate.

Early indications of this outlook can be found by examining select statements from various European policy statements, articles and leaders. In 2002, for example, the German Ministry for the Environment (BMU) identified climate change as a 'major challenge to human security', but one that did not 'pose a military threat nor … [could be] solved with the traditional mind-sets … of military services' (Tanzler et al. 2002). Likewise, a 2007 German Advisory Council on Global Change (WBGU) report identified the steadying intensification of climate threats in the coming decades and placed priority on 'a well-functioning cooperation between development and security policy' since the ability of "classic' security policy's capacities to act are limited' (Schubert et al. 2008: 6). In 2007 the Swedish Defence Research Agency (FOI) commissioned a report that stated 'climate is in itself not a threat and certainly not one that can be countered by military means' (Halden 2007: 36). Similar themes have also been evident in the UK. Former UK Foreign Secretary Margaret Beckett, for example, identified climate change as a 'serious threat to international security' but one that 'the traditional tools of hard security—in simple terms bombs and bullets—are not going to be able to solve' (Beckett 2007). Former UK Special Representative for Climate Change John Ashton argued 'climate change is a security issue', but underscored the limited military role: 'there is no hard power solution … you cannot force your neighbour to change its carbon emissions at the barrel of a gun' (quoted in Brauch 2009: 89). In 2008, the EU High Representative and the European Commission released *Climate Change and International Security* which stated 'climate change … [is] a threat multiplier which exacerbates existing trends, tensions and instability' and 'threatens to overburden states and regions' (Council of the European Union 2008). Preceding the release of this report, however, the then British Foreign Secretary David Miliband and German Foreign Minister Frank-Walter Steinmeier stated:

> Ultimately, there is no hard power option for tackling the causes of the climate threat or for dealing with its direct impacts. You cannot use military force to build a low carbon global economy; no weapon system can halt the advance of a hurricane bearing down on a city, or hold back the rising sea (Miliband and Steinmeier quoted in Brauch 2009: 93; emphasis added).

Although somewhat dated, these early formulations offer an insight into the mind-set of European thinking on climate security. Conceptually, climate change was manifestly promoted and then established as a security issue, but it was not one to be solved by enlisting the military. Potentially, this was one of the factors that also contributed to an uneven level of involvement, engagement and penetration into European national security documents and military institutions as well as the security and intelligence agencies more broadly. The following examples illustrate this. In France, climate change was scarcely mentioned as a security or military concern in either the 2008 or 2013 Defence White Papers. The 2013 version, *Defence and National Security*, had a single mention of 'climate change' in 137 pages and vaguely noted that the 'precise regional consequences of global warming over the next few decades are still very uncertain' (Ministry of Defence 2013: 44). Perhaps unsurprisingly, but symptomatic of the lack of uptake, the French climate strategies of 2009 and 2011 failed to identify any security, let alone military, implications of climate change (Ministry of Ecology 2009, 2011). While Germany has been lauded for its efforts in arranging the second thematic debate of climate security at the UNSC in 2011 and in its wider climate efforts, the security implications and armed forces engagement on the issue remain underdeveloped (Youngs 2015: 75). The 2016 German Defence White Paper, for example, identified climate change as having 'significant and existential implications' and as 'increasingly relevant for security policy and indirectly for Germany' (German Government 2016: 42). Yet even here the deeper integration of climate change across German security institutions was flagged as a low priority, and only something to be achieved '[i]n the coming years' (ibid). Similar outlooks can be found across the continent in comparable European national security strategies[4] and NATO documents (Youngs 2014, 2015).[5] Moreover, although there appeared a lot of discussion, framing and speeches about climate change and its security implications by European political leaders, this did not necessarily appear to translate into anything meaningful across respective national security strategies or military documents (whether regulatory, administrative or operational).

The situation was somewhat different in the UK, however, where the Ministry of Defence actively set about integrating climate change into its planning considerations. These included the publication of a *Climate Change Strategy* (UK MoD 2010b), a *Climate Change Delivery Plan* (UK MoD 2010a), the appointment of Rear Admiral Neil Morisetti as the UK Climate and Energy Security Envoy (HM Government 2010: 93), and through various climate adaptation and mitigation initiatives such as the Climate Impact Risk Assessment Methodology (CIRAM) that assessed some 80 MoD bases at risk of climate change (UK MoD 2015). Climate change also received substantive treatment in the 2015 UK *National Security Strategy and Strategic Defence and Security Review* that described it as 'one of the biggest long-term challenges for the future to the planet' (HM Government 2015: 65). Despite this, the overall impression from UK and European climate security discourse was of one that had peaked around the time of the Copenhagen conference and, from 2010 and 2011 onwards, had come to adopt a somewhat lower profile within national security and military planning considerations. Arguably, this reduced trajectory reflected the fact that climate change had become a lower priority in European and UK politics where attention had shifted to more pressing priorities. Even in the UK MoD, climate publications that were once visibly promoted had, across the intervening years, been gradually compacted down into more discreet departmental pamphlets such as annual sustainability reports (UK MoD 2013, 2015). The UK MoD Climate and Security Envoy, appointed in 2009 at the height of UK climate securitisation to 'embed' climate strategy, was subtly discontinued

[4]For European national security papers, see: https://www.eda.europa.eu/procurement-gateway/information/national-directories/national-policies-and-strategies (22 November 2016).

[5]For NATO Strategic Concept, see: http://www.nato.int/nato_static_fl2014/assets/pdf/pdf_publications/20120214_strategic-concept-2010-eng.pdf (22 November 2016).

beyond 2012. In generalising the security situation, by 2016 European attention had shifted toward the 'arc of instability' surrounding the continent with a prioritisation on 'hard' security issues of mass migration, transnational terrorism, radicalisation, cyber threats and on grappling with a resurgent Russia (European Council 2016a: 1). By 2016, the UK government was also dealing with 'Brexit', somewhat limiting the capacity of officials to deal with the all the complexities of climate change.

3.2.2 *Climate Security in United States Policy*

Climate security discourse in the US has frequently been described as the most sustained and as having the greatest permeation into high level national security statements and military policy (Hayes/Knox-Hayes 2014; Youngs 2015; Barrie et al. 2015). Three broad phases of US climate security discourse can be identified. The first phase existed from the early identification of the problem in the 1970s through to the early 2000s. This phase was characterised by three aspects: (1) a dominance of scientific discussion and problem identification, (2) the gradual but circumspect rise of climate change as a US national security risk, and (3) a broad *lack* of engagement on the issue by US national security elite and the US DoD in particular. Prominent examples during this phase were the National Climate Program Act of 1978, efforts by various politicians to raise the concern of climate change as a security threat amongst US elite (e.g., Al Gore) and efforts by scientists who spoke publicly about their concerns of climate change as a security threat (e.g., see, for example, James Hansen quoted in Shabecoff (1988)).

The second phase of US climate security discourse commenced in the early 2000s and ended with the inauguration of Donald Trump on 20 January 2017. This phase was characterised by four important aspects: (1) the increased acceptance and consolidation of climate change as a legitimate security consideration within the US national security establishment; (2) the issuance of presidential executive orders as well as federal and state regulation that required climate security considerations to be addressed; (3) the issuance of US DoD strategies, reports and directives that resulted in the US military commencing adaptation and mitigation in response to climate security considerations; and (4) a general disengagement on the issue by the administration of President George W. Bush but a reversal of this outlook during the Obama Administration. To capture the extent of this, it is important to briefly outline a chronology of the key publications and events. A credible starting point remains a 2003 report by the US DoD Office of Net Assessments (Pentagon) that found climate change could lead to resource scarcity, mass migration and geo-political instability which might lead to 'skirmishes, battles and even war' (Schwartz/Randall 2003: 2). Often cited, the report also noted that 'the risk of abrupt climate change … should be elevated beyond a scientific debate to a US national security concern' (Schwartz/Randall 2003: 3). The inference here was that climate change had remained at the margins of US politics and that by securitising the issue it would enter the realm of 'high' politics and thereby grasp the attention of US policy makers and the military.

Between 2007 and 2010 climate security was increasingly addressed in a number of influential US think-tank reports, often with prominent contributions from a cadre of retired senior military officers.[6] In 2007 the Center for Naval Analysis (CNA), in conjunction with its Military Advisory Board (MAB) of former high-ranking officers, published *National Security and the Threat of Climate Change* that argued 'climate change poses a serious threat to America's national security' by acting as a 'threat multiplier for instability' (CNA 2007: 6 and 7).[7, 8] In the same year, the US Council on Foreign Relations (CFR) released *Climate Change and National Security* that—perhaps going further than other assessments—called for institutional reform in the US DoD, the US National Security Council and in the Office of the President to 'give voice to climate and security concerns' with the stated aim to integrate 'climate security concerns centrally into its National Security Strategy' (Busby 2007: 23). In 2010 the Center for a New American Security (CNAS) published *Broadening Horizons: Climate Change and the U.S. Armed Forces* (co-authored by a serving US Navy Commander) which stated that climate change would 'reshape the current and future security environment' (Carmen et al. 2010: 5). This report urged US military involvement in climate geo-engineering debates; the centralisation of an Arctic military command; the consideration of having nuclear reactors on military bases as a means to generate carbon-free energy; the promotion of US DoD renewable and energy efficiency programmes; and education for all US military services on how climate change will impact missions and capabilities. Other early influential US think-tank reports included combined works by the Center for Strategic and International Studies (CSIS) and CNAS titled *The Age of Consequences: The Foreign Policy and National Security Implications of Global Climate Change*. Echoing other reports, CSIS and CNAS also warned that climate change could have cascading security

---

[6]On several occasions since 2010 many retired military officers and national security experts have publicly urged action on climate change as a means of addressing the security consequences. See, for example: http://www.psaonline.org/2015/10/22/national-security-and-foreign-policy-leaders-warn-of-climate-change-threat-call-for-u-s-leadership/; and: https://climateandsecurity.files.wordpress.com/2016/09/climate-and-security-consensus-project-statement-2016_09.pdf (23 November 2016).

[7]This report was updated and re-released in 2014 as the *National Security and the Accelerating Risks of Climate Change* at: https://www.cna.org/cna_files/pdf/MAB_5-8-14.pdf (17 June 2016).

[8]The contribution of the Military Advisory Board under the auspices of CAN cannot be underestimated in galvanising US military climate response. Rear Admiral David Titley (who led the US Navy climate change programmes 2009–2011) reflected on this at the 2015 *The Australian Climate Security Summit*, stating: 'The 2007 Military Advisory Board (MAB) got the whole thing going in the US. It was a catalyst and prompt for the 2008 National Defense Authorization Act to get the US DoD to examine the issue in some detail. It was incredibly influential. The MAB was a cadre of respected and retired "grey-beards" operating outside the government framework. It was one of the key levers in our system to get congressional action' (Titley 2015).

consequences which could overwhelm traditional instruments of national security, including the military (Campbell et al. 2007). Other prominent US think-tank contributions on climate security (beyond 2010) have included Werz/Conley (2012), Huebert et al. (2012), Foley/Holland (2012), Slaughter et al. (2013), CNA (2014), CC&S (2015, 2016).

The second phase of US climate security discourse was also characterised by sustained interest, discussion and action by the US government. Although there were partisan attempts during the Bush Administration, including an unsuccessful Democrat-sponsored bill, 'Global Climate Change Security Oversight Act' (2007a), it was not until the Obama Administration that climate security came into its own. Granted the necessary political authority, this phase saw expansive action and discourse on climate security via Executive Order (Obama 2009a, 2013, 2015), Presidential Memorandum (Obama 2016), the US Congress (US House of Representatives 2007b, 2008a, b; US Senate 2006, 2007, 2009), the US intelligence community (NIC 2008; NRC 2013; Clapper 2013, 2015, 2016), the US Government Accountability Office (US GAO 2015), and the substantive entry of climate change into high-level national security planning documents such as the US National Security Strategy (NSS) and Quadrennial Defense Review (QDR) as well as into US Department of Defense regulatory, administrative and operational documents/strategies/plans/roadmaps/orders/directives/speeches and statements.

In terms of US DoD, the 2008 *National Defense Authorization Act* (NDAA 2008) directed the US military to examine the impacts of climate change and served as a catalyst for extensive coverage in the 2010 US NSS and QDR. Both of these documents were prominent strategic markers that, as will be shown in the US case study in Chap. 7, laid the foundation for DoD action. The 2010 US NSS framed climate change as a 'real, urgent and severe' danger that would 'lead to new conflicts over refugees and resources; new suffering from drought and famine; catastrophic natural disasters; and the degradation of land across the globe' (White House 2010: 47). The 2010 QDR stressed the need to adjust to the impacts of climate change on facilities and military capabilities and announced a 'complete a comprehensive assessment' of climate risks (US DoD 2010: 85). Similar statements and updated provisions were contained in the 2015 US NSS (White House 2015) and 2014 QDR (US DoD 2014c). The imperative to act on climate change from the highest levels of US government were also given prominence by several high-ranking serving US military officers who spoke on the risks of climate change (Mullen 2011; Admiral Locklear in Werrell/Femia 2014). Crucially, this phase also saw the US DoD and individual military service arms set about addressing climate risks through the publication of numerous climate change adaptation and mitigation strategies that included *US Navy Arctic Roadmap* (2009b), *US Navy Task Force Climate Change* (2009a), *US Navy Climate Change Roadmap* (2010b), *US DoD Strategic Sustainability Performance Plan* (2012b, 2013a, 2014b, 2016a), *US DoD Climate Change Adaptation Roadmap* (2012a, 2014a) and *US Army Corps of Engineers Adaptation Policy Statement* (USACE 2014). By 2015, efforts by the US DoD to address climate risks had filtered down to the tactical level Combatant Commanders (COCOMs) (US DoD 2015) and culminated in a 2016 DoD Directive, 4715.21 *Climate Change Adaptation and*

Resilience (2016b), that assigned responsibilities and resources to address climate change across the entire US DoD. Despite the abundance of high-level strategies, however, US DoD efforts on climate change have not been without problems. A 2014 US Government Accountability Office report, for example, highlighted that while climate change worked from a 'big picture' perspective, it had been difficult to translate, convert and implement into meaningful and practical action by front-line commanders responsible for managing large US military bases (US GAO 2014). For instance, many base officials had difficulty factoring climate change into their plans since climate data was insufficiently downscaled to be of any use. In essence, 'granularity' remained a problem. In other cases, base officials were uncertain about how to implement climate requirements efficiently since key information was missing: lack of definitions, unclear building codes and design standards, uncertainty on which climate scenarios and projections to use. As a result, climate change at the lower levels have struggled to be properly integrated into US DoD project management plans.

Notwithstanding these problems, from at least 2009 the US DoD had commenced a pro-active response to climate change that was far more engaged, sustained and of larger scope than was occurring in Europe and, in every likelihood, in other regions of the world, including Asia, Africa and South America. By 2016—certainly until the election of Donald Trump (considered by this author to be the commencement of a 'third phase' in US climate security discourse)—the US military appeared one of the most active militaries in the world in its response to climate change.

3.3 Climate Security in the United Nations Security Council

Since at least 2007, the United Nations has considered climate-security as part of its major deliberative and policy making organs, the UN General Assembly (UNGA) and the UN Security Council (UNSC). For its part, the UNGA has debated climate change on multiple occasions, and in 2009 released Resolution 63/281 requesting the Secretary-General to submit a comprehensive report on its possible security implications (UNGA 2009). More contentiously, the UNSC has also held two formal debates on the security implications of climate change (2007 and 2011) and two 'Arria-formula' meetings (informal sessions) in 2013 ('Security Dimensions of Climate Change') and 2015 ('The role of climate change as a threat multiplier for global security'). The other major UN securitising actor has been its former Secretary-General, Ban Ki-Moon. Since assuming the role in January 2007 and until his departure in December 2016, Ban Ki-Moon has regularly framed climate change as a major threat to international peace and security (UNSG 2011). Possibly, he has done more to elevate climate change as a security threat, and give precedence to the wider issue itself, than any other global leader (UN 2016). Significantly, the UN climate security discourse has been marked by disagreement between nations, and with the Secretary-General, on how to best approach the security consequences of climate change. The following section sketches the key UNSC debates, highlighting significant national statements on the matter, and identifying some possible implications for the military sector that flow from this.

Climate security remains an essentially contested subject in the international arena. Two broad approaches are evident. On one side are nations that reject the use of the UNSC as a forum to address climate change. In the broadest sense, while most nations recognise the potential security risks of climate change per se, many (but not all) have actively argued against climate change being framed and addressed by the UNSC. With notable exceptions, this position has largely been held by Russia, the G77 (including China, India, Brazil, Indonesia, South Africa and Pakistan) and the Non-Aligned Movement (NAM) of 120 nation-states (Sindico 2007; Brauch 2009; UNSC 2007, 2011). This bloc of countries (hereafter G77+) has argued that climate change should be dealt with primarily as a sustainable development issue in non-securitised, multi-lateral forums such as the UN General Assembly, UNFCCC, the World Meteorological Organisation (WMO), the UN Economic and Social Council and the UN Commission on Sustainable Development. The opposing position has argued for a more comprehensive approach to deal with climate change. First, as a sustainable development issue in non-securitised forums as advocated by the G77+, but—and this is the critical difference—also as a security issue in forums such as the UNSC and through respective regional and national security institutions and architectures. Nations advocating this position argue that climate change should also be treated as a security issue primarily because it galvanises political action and that the security threats resulting from climate change are imminent and sufficient such that they must now be addressed by security institutions. Nations adopting this position have variously included the US, Germany, UK, Australia, Canada and Japan but also a collection of other nations outside this bloc (e.g., particularly small island states threatened by sea level rise but also some 'dissenting' nations within the G77 and NAM). The UN Secretary-General has also repeatedly framed climate change as a security issue in this fashion (UNSG 2008, 2011, 2012).

As stated, the UNSC has convened two formal debates on climate change and its security implications. Heavily influenced by the UN 2005 World Summit Outcome (General Assembly Resolution 60/1), outcomes from a 2005 heads of government meeting that was subsequently endorsed by UNSC Resolution 1625, and by IPCC AR4, the UK convened the first UNSC debate on 17 April 2007 to explore 'the relationship between energy, security and climate' (UK 2007). This debate revealed sharp divisions in how to address climate security issues, with viewpoints generally falling along characteristic narratives of 'north' versus 'south', 'developed' versus 'undeveloped' and 'wealthy' versus 'poor'. Pakistan, for instance, argued that the 'ever increasing encroachment of the Security Council on the role and responsibilities of the other main organs' amounted to a 'distortion' of the UN principles that 'infringes on their authority' (UNSC 2007: 24). China added that discussion in the UNSC would do nothing to help developing countries respond to climate change and that the UNSC lacked the necessary 'expertise' (ibid: 13). Indonesia stated that the debate should not set a precedent, Russia appealed to 'avoid panicking and over dramatising the situation', while South Africa questioned the very legitimacy of the debate, 'climate change do[es] not as directly threaten international peace and security' (ibid: 14, 15, 17). Opposing this were the UN Secretary-General, Germany, France, the UK (and other European nations) and Japan, who all argued in favour of the UNSC fulfilling its mandate to prevent conflict and its causes; a 'duty to play its part' (ibid: 12). Underneath this overarching picture, however, various combinations and justifications revealed greater nuance. Notable was the subdued presence of the US representative who spoke briefly on energy security but failed to adopt a concrete position supporting the UNSC role on climate change or even discuss its security implications more broadly. As will be shown in the US case study in Chap. 7, this should not have come as a surprise, given that climate change during the Bush Administration was not a priority and certainly not one considered a security issue. The 2007 debate is also notable for those small countries who dissented from the larger and more powerful blocs that they are nominally aligned, Ghana, Congo, Panama, Peru and Papua New Guinea all supporting the debate and advocating UNSC involvement. Conversely, there were some surprises; the Maldives representative, while noting the grave security consequences for his nation, joined with the G77 and China in stressing the importance of addressing climate change as an issue of sustainable development (ibid: 23).[9]

Similar themes, complexities and nuances were evident in the subsequent UNSC debate that was convened by Germany and held on 20 July 2011. Brazil argued 'security tools are appropriate to deal with concrete threats to international peace and security, but they are inadequate to address complex and multidimensional issues such as climate change', besides which '[t]he possible security implications of climate change, however, are less obvious' (session 1 of UNSC 2011: 8). Russia noted their scepticism to climate securitisation by arguing it would 'bring no added value whatsoever and would merely lead to a further politicisation of the issue and increased disagreements' (ibid: 13). China repeated its 2007 position by declaring 'Climate change may affect security, but it is fundamentally a sustainable development issue', adding that the UNSC 'lacks expertise … means and resources …

[9]The interregnum between UNSC debates on climate security was filled by the publication of the 2009 UNSG report, *Climate Change and its possible Security Implications* (A/64/350). Promoted by some scholars as the most authoritative political declaration on the issue, this report identified five climate security implications: (1) *vulnerability* by threatening food security, human health and exposure to extreme events; (2) *development* by undermining State capacity; (3) *coping* and *security* by migration and resource competition; (4) *statelessness* by disappearance of territory; and (5) *international conflict* from undemarcated international resources (UNSG 2009). This report marked a continuation of the Secretary-General's advocacy for engaging the Security Council to address the climate change.

[and] is not a forum for decision-making with universal representation' (ibid: 9). Noteworthy was Bolivia's contribution which called for the UNSC to adopt a resolution to reduce global military spending by the establishment of a reduction of '10 or 20% in [global] defence and security expenditures' for the purpose of tackling the impacts of climate change in developing countries (session 2 of UNSC 2011: 26). In contrast, Nauru (on behalf of Small Island Developing States)[10] argued that the UNSC has a 'clear role in coordinating a response to the security implications of climate change' and called for the immediate appointment of a special representative to deal with it and for an 'assessment of the capacity of the UN system to respond' to the security dimension of climate change (ibid: 23). Although there were perceptible shifts in some country positions (Warren 2015), the major change in this debate, however, was the forceful contribution by the US representative, Ms Susan Rice, who urged the UNSC has 'an essential responsibility to address the clear-cut peace and security implications of climate change' (session 1 of UNSC 2011: 7). Countering the arguments of the G77+, the US lamented the ineffectiveness of the Security Council and the 'refusal of a few to accept our responsibility' as 'pathetic, short-sighted and, frankly, a dereliction of duty' (ibid). As will be shown in Chaps. 7 and 8, this forceful argument, mounted in support of framing climate change as a security issue and using security institutions to deal with the threats, reflected the change in political priorities of the Obama Administration. The 2011 debate closed with a unanimous vote to release a Presidential Statement that reaffirmed the UNFCCC as the key instrument to address climate change.

Similar country outlooks to those established in 2007 and 2011 were reflected in the 2013 and 2015 UNSC 'Arria-Formula' meetings on climate security. The 2013 debate convened by Pakistan and the UK was held on 15 February 2013, while the latter was convened by Spain and Malaysia and held on 30 June 2015. While country positions remained entrenched, the broad observation is that by 2015 the debate was much more subdued, with fewer countries participating, participants of lower seniority delivering shorter statements with less passion, and a resigned acquiescence by nations to accept a downgrading of the meeting agenda as reflected in the 'Arria-Formula' that is not recognised as a formal activity of the UNSC, is rather held in a side-bar conference room than the formal UNSC chamber, and the outcomes of which are not announced in the daily Journal of the UN (UN 2002). Moreover, a close reading of all country statements reveals that by 2015 there was a general sense that few new ideas existed on how to progress climate change from the angle of security and that momentum previously carried on the subject (of climate security) had dissipated.

What broad conclusions can be drawn from the UNSC debates and what, if any, relevance might they have for militaries? First, the securitisation of climate change in the UNSC reflected what had been recorded by scholars since 2012. Namely, that

[10]Nauru spoke on behalf of Fiji, the Marshall Islands, Micronesia, Palau, Papua New Guinea, Samoa, Solomon Islands, Tuvalu, Tonga, Vanuatu, Maldives, Seychelles and Timor-Leste.

the securitisation of climate change had failed to generate any new measures, let alone emergency ones, to deal with either the security implications of climate change or in galvanising nations to reduce their emissions (Oels 2012; McDonald 2012; Brzoska 2012b). Secondly, on matters of how to address the security implications of climate change, the international community was broadly divided between those who advocated for a comprehensive securitisation response to complement UNFCCC-led initiatives against those nations arguing that it be treated solely as a sustainable development issue in non-securitised forums. But operating beneath these ideological positions there appeared greater nuance. Some nations of the G77 and NAM, for example, dissented from their traditional blocs and aligned with Western initiatives in favour of adopting security measures. Furthermore, many Western nations also displayed a transient outlook that suggested the triumph of domestic politics over international solidarity around the issue of climate security. The US, which had a meek presence in the first debate, were, by 2011, under new political management and aggressively outlined the urgency to engage the UNSC and for nations to respond in a far more comprehensive fashion.[11]

What of the relevance for armed forces? Scholarly works—that have tended to be critical of the securitisation agenda, particularly concerning involvement of the military—provide an account of the main issues. Chris Penny, for example, has argued that normalisation in institutional mechanisms such as the UNSC might (under a worst case scenario) lead to application of Chap. 7 enforcement measures, thereby threatening state sovereignty (Penny 2007). Maria Trombetta has pursued this argument, writing that climate securitisation could result in 'confrontational politics, with states adopting politics to protect their territory … the Security Council adopting resolutions to impose emissions targets, and even military action against factories; and surveillance systems to monitor individual emissions' (Trombetta 2008: 599). Trombetta has further warned against militarising the environment and has questioned the credibility of climate security threats conjured up by 'think tanks that are close to the military' (Trombetta 2012: 159). This outlook has also been championed by Betsy Hartmann who opined that climate security is 'old wine in a new bottle' and argued against the potential colonisation of the climate change debate by the military with the risk it could impose authoritarianism, secrecy and illiberal practices and overtake the role traditionally performed by other institutions (Hartmann 2010: 234). Hartmann has lamented the lack of empirical data and a crisis narrative that goes uncontested and serves vested interests of private defence contractors, think tanks and the military (Hartmann 2010: 235). Thus far, however, adverse concerns about the militarisation of climate change appear unfounded (Youngs 2014). Military responses, such as in the US example, have appeared on the surface at least to have resulted in increased investment in humanitarian and disaster relief capacity, military emission reduction

[11]The same comment may be made of Australia, which, having arguing vigorously in the 2007 and 2011 debates, was by 2013 (and under a centre-right Conservative government) arguing in favour of removing climate change from the agenda of major international meetings (Scott/Vogler 2014; The New Daily 2014).

strategies and the enhancement of military bases against sea-level rise and extreme weather, and international cooperative engagement measures through military-to-military exercises focusing on disaster and climate relief. A more pointed critique of military forces addressing climate change, however, remains the practical challenge of aligning a decades-long strategic problem with the near-term outlook and real-time purpose and utility of armed forces (Brzoska 2012: 19; US GAO 2014: 1082).

The other key observation from these debates is that those nations arguing against addressing climate change as a security issue have had this outlook replicated downwards through their national security architectures and military institutions. As such, the touchstone national security documents and high-level military documents of these countries appear largely disengaged on climate security issues (Brzoska 2012a). As a case in point, the 2015 Chinese White Paper on Military Strategy, for example, failed to discuss or mention climate change (PLA 2015). Likewise, Russia's 2009 and 2016 national security strategies mention 'climate change' three times between them under 'environment management', but as an otherwise low-priority issue with no significance for the military (Russian Federation 2009, 2015). Current high-level security strategies for Brazil, *National Strategy of Defense* (Ministry of Defense 2008), and Pakistan, *National Internal Security Policy* (Ministry of Interior 2014), likewise fail to discuss climate change in any context.

In essence, the international climate securitisation agenda on display at the UNSC since 2007 has, in some respects, been representative of the extent to which individual nations have incorporated climate change into their national security institutions, thinking and processes. Currently, it remains an open question if this has been a positive or negative development.

3.4 Chapter Conclusion

This chapter has examined the securitisation of climate change from its historical beginnings in the 1960s through to its inclusion in high-level national security documents in the late 2000s. At international level, climate change became a contested security issue where nations were split on whether or not the UNSC was a suitable instrument to address the widely accepted security implications of climate change. While nations tended to align with their traditional blocs, the various debates progressively revealed a more complex picture. Some small island states and developing countries dissented from Russian and Chinese-led views, while the US flipped from a disengaged spectator to emphatic champion. In any case, the UNSC debates had petered out by 2015; devoid of fresh ideas, downgraded in status and, in every likelihood, overtaken by current (security) events and a UN-machinery focused on reaching agreement in Paris. Scholars had lamented the failed securitisation of climate change from 2012.

Notwithstanding, the climate security agenda has raised some important questions on the role and purpose of armed forces within the wider context of addressing climate change. Fundamentally it has been bifurcated between the spectre of militarisation, 'encroachment' and issues of sovereignty against the potential to galvanise political urgency to act, in lending particular skill-sets such as strategic scenario planning and in increasing military capacities as first-responders to climate related disasters. Pointedly, there remains little empirical or systemic analysis of how militaries have addressed the issue of climate change. The following chapters seek to address this gap.

References

ASP. 2016. "Global Security Defence Index on Climate Change." Washington DC: American Security Project. Accessed on 22 November 2016, https://www.americansecurityproject.org/wp-content/uploads/2014/09/Ref-0168-GSDI-Introduction-2014.pdf.

Bächler, Günther, David J. Eaton, Joseph W. Eaton, Malin Falkenmark, Partha S. Ghosh, Jad Isaac, Nurit Kliot, John Markakis, Robert Mayor, Katrina S. Rogers, Vaclav Smil, Kurt R. Spillmann, Ashok Swain, Peter Wallensteen, and Arthur H. Westing. 1995. *Environmental Crisis: Regional Conflicts and Ways of Cooperation*. CSS Environment and Conflicts Project: 14. Zurich.

Barrie, Chris, Will Steffen, Alix Pearce, and Michael Thomas. 2015. *Be Prepared: Climate Change, Security and Australia's Defence Force*. Sydney: Climate Council. Accessed on 20 September 2015, https://www.climatecouncil.org.au/uploads/fa8b3c7d4c6477720434d6d1089 7af18.pdf.

Beckett, Margaret. 2007. "The Urgernt Case for Climate Security." London: Royal United Services Institute. Published online 10 May 2007. Accessed on 15 May 2017, https://rusi.org/publication/urgent-case-climate-security.

Brauch, Hans Günter. 2009. "Securitizing Global Environmental Change." In *Facing Global Environmental Change: Environmental, Human, Energy, Food, Health and Water Security Concepts*, edited by Hans Günter Brauch, Úrsula Oswald Spring, John Grin, Czeslaw Mesjasz, Patricia Kameri-Mbote, Navnita Chadha Behera, Bechir Chourou and Heinz Krummenacher, vol. 4, pp. 65–102. Berlin: Springer.

Brauch, Hans Günter. 2011a. "Concepts of Security Threats, Challenges, Vulnerabilities and Risks." In *Coping with Global Environmental Change, Disasters and Security: Threats, Challenges, Vulnerabilities and Risks*, edited by Hans Gunter Brauch, Úrsula Oswald Spring, Czeslaw Mesjasz, John Grin, Patricia Kameri-Mbote, Bechir Chourou, Pal Dunay and Jorn Birkmann, vol. 5, pp. 61–106. Berlin: Springer.

Brown, Lester R. 1977. *Redefining National Security*. Washington, DC: Worldwatch Institute.

Brown, Lester R. 1986. "Redefining National Security." *Challenge*, vol. 29, no. 3:25–32.

Brzoska, Michael. 2012a. "Climate change and the military in China, Russia, and the United Kingdom, and the United States." *Bulletin of the Atomic Scientists*, vol. 68, no. 2:43–54. doi:10.1177/0096340212438384.

Brzoska, Michael. 2012b. "Climate Change as a Driver of Security Policy." In *Climate Change, Human Security and Violent Conflict: Challenges for Societal Stability*, edited by Jurgen Scheffran, Michael Brzoska, Hans Gunter Brauch, Peter Michael Link and Janpeter Schilling. Berlin: Springer.

Busby, Joshua W. 2007. *Climate Change and National Security: An Agenda for Action*. Council on Foreign Relations Special Report: 32. New York: Council on Foreign Relations Press.

Buzan, Barry, and Lene Hansen. 2009. *The Evolution of International Security Studies*. Cambridge: Cambridge University Press.

Buzan, Barry, Ole Wæver, and Jaap de-Wilde. 1998. *Security: A New Framework for Analysis*. Boulder: Lynne Rienner Publishers.

Campbell, Kurt M., Jay Gulledge, J. R. McNeill, John Podesta, Peter Ogden, Leon Fuerth, R. James Woolsey, Alexander T. J. Lennon, Julianne Smith, Richard Weitz, and Derek Mix. 2007. *The Age of Consequences: The Foreign Policy and National Security Implications of Global Climate Change*. Washington, DC: Center for a New American Century.

Carmen, Herbert E., Christine Parthemore, and Will Rogers. 2010. *Broadening Horizons: Climate Change and the U.S. Armed Forces*. Washington, DC: Center for a New American Century. Accessed on 15 May 2017, https://s3.amazonaws.com/files.cnas.org/documents/CNAS_Broadening-Horizons_Carmen-Parthemore-Rogers.pdf.

Carson, Rachel. 1962. *Silent Spring*. Boston: Mariner.

CC&S. 2015. *The U.S. Asia-Pacific Rebalance, National Security and Climate Change*. A Climate and Security Correlations Series: Washington, DC. Accessed on 10 November 2016, https://climateandsecurity.files.wordpress.com/2015/11/ccs_us_asia_pacific-rebalance_national-security-and-climate-change.pdf.

CC&S. 2016. *Briefing Book for a New Administration: Recommended Policies and Practices for Addressing the Security Risks of a Changing Climate*. Washington, DC. Accessed on 10 November 2016, https://climateandsecurity.files.wordpress.com/2016/09/climate-and-security-advisory-group_briefing-book-for-a-new-administration_2016_091.pdf.

Clapper, James. 2015. "Statement for the Record Worldwide Threat Assessment of the US Intelligence Community." In *Senate Armed Services Committee*. Accessed on 20 November 2016, https://www.dni.gov/files/documents/Unclassified_2015_ATA_SFR_-_SASC_FINAL.pdf.

Clapper, James. 2016. "U.S. Intelligence as a Pillar of Stability during Transition." In *DNI Clapper's As Delivered Remarks at the 2016 INSA & AFCEA Intelligence & National Security Summit*. Washington DC: Office of the Director of National Intelligence. Accessed on 15 May 2017, https://www.dni.gov/index.php/newsroom/speeches-interviews/speeches-interviews-2016/item/1627-dni-clapper-s-as-delivered-remarks-at-the-2016-insa-afcea-intelligence-national-security-summit.

Clapper, James R. 2013. "Worldwide Threat Assessment to the Senate Select Committee on Intelligence." Washington DC: US Public Affairs Office. Accessed on 14 November 2016, https://www.dni.gov/files/documents/Intelligence%20Reports/WWTA%20Remarks%20as%20delivered%2012%20Mar%202013.pdf.

CNA. 2007. *National Security and the Threat of Climate Change*. Virginia: The CNA Corporation. Accessed 17 May 2017, https://www.cna.org/cna_files/pdf/MAB_5-8-14.pdf.

CNA. 2014. *National Security and the Threat of Climate Change*. Virginia: The CNA Corporation. Accessed 17 May 2017, https://www.cna.org/cna_files/pdf/MAB_5-8-14.pdf

Corn, David. 2006. "Gore Warms Up." *The Nation*, Katrina vanden Heuvel, 25 May 2006. Accessed on 5 June 2014, http://www.thenation.com/article/gore-warms/.

Council of the European Union. 2008. *Climate Change and International Security: Paper from the High Representative and the European Commission to the European Council*. Brussels: Council of the European Union. Accessed on 12 March 2012, http://www.consilium.europa.eu/uedocs/cms_data/docs/pressdata/en/reports/99387.pdf.

Dabelko, Geoffrey. 1999. "The Environmental Factor." *Wilson Quarterly*, vol. 23, no. 4:14–19.

Deudney, Daniel. 1990. "The Case Against Linking Environmental Degradation and National Security." *Millennium: Journal of International Studies*, vol. 19, no. 3:461–476. doi:10.1177/03058298900190031001.

Dyer, Gwynne. 2008. *Climate Wars: The fight for survival as the world overheats*. Melbourne: Scribe Publications.

Ehrlich, Paul R. 1968. *The Population Bomb*. New York: Ballantine Books.

EU. 2003. *European Security Strategy: A Secure Europe in a Better World*. Brussels: European Union. Accessed on 23 September 2014, http://www.consilium.europa.eu/uedocs/cmsUpload/78367.pdf.

European Commission. 2015. *The European Agenda on Security* Strasbourg: European Commission. Accessed on 6 January 2017, https://ec.europa.eu/home-affairs/sites/homeaffairs/files/e-library/documents/basic-documents/docs/eu_agenda_on_security_en.pdf.

European Council. 2016a. "The European Union in a changing global environment: A more connected, contested and complex world." Brussels: European Union. Accessed on 15 May 2017, http://www.europarl.europa.eu/sides/getDoc.do?pubRef=-//EP//NONSGML+REPORT+A8-2016-0069+0+DOC+PDF+V0//EN.

European Council. 2016b. "Implementation Plan on Security and Defence." In *EU global strategy on foreign and security policy*. Brussels: European Union. Accessed on 20 November 2016, http://www.consilium.europa.eu/en/press/press-releases/2016/11/14-conclusions-eu-global-strategy-security-defence/.

Foley, Catherine, and Andrew Holland. 2012. *Climate Security Report*. Washington, DC. Accessed on 23 November 2016, http://www.americansecurityproject.org/climate-security-report/.

German Government. 2016. *White Paper: On German Security Policy and The Future of the Bundeswehr*. Ministry of Defence. Berlin: Federal Ministry of Defence. Accessed on 22 November 2016, http://www.new-york-un.diplo.de/contentblob/4847754/Daten/6718448/160713weibuchEN.pdf.

Guggenheim, Davis. 2006. An Inconvenient Truth. Paramount Classics.

Halden, Peter. 2007. *The Geopolitics of Climate Change*. Stockholm: Swedish Defence Research Agency.

Hansen, J., D. Johnson, A. Lacis, S. Lebedeff, P. Lee, D. Rind, and G. Russell. 1981. "Climate Impact of Increasing Atmospheric Carbon Dioxode." *Science*, vol. 213, no. 4511:958–966.

Hardin, Garrett. 1968. "The Tragedy of the Commons." *Science*, vol. 162, no. 13:1243–1248.

Hartmann, Betsy. 2010. "Rethinking Climate Refugees and Climate Conflict: Rhetoric, Reality and the Politics of Policy Discourse." *Journal of International Development*, vol. 22, no. 2:233–246. doi:10.1002/jid.1676.

Hayes, Jarrod, and Janelle Knox-Hayes. 2014. "Security in Climate Change Discourse: Analysing the Divergence between US and EU Approaches to Policy." *Global Environmental Politics*, vol. 14, no. 2:82–101. doi:10.1162/GLEP_a_00230.

HM Government. 2010. *Foreign and Commonwealth Office Annual Report 2008–09: Fifth Report of Session 2009–10*. London: Foreign and Commonwealth Office. Accessed on 14 April 2013, http://www.publications.parliament.uk/pa/cm200910/cmselect/cmfaff/145/145.pdf.

HM Government. 2015. *National Security Strategic and Strategic Defence and Security Review 2015*. London: HM Government. Accessed on 22 November 2016, https://www.gov.uk/government/uploads/system/uploads/attachment_data/file/478936/52309_Cm_9161_NSS_SD_Review_PRINT_only.pdf.

Homer-Dixon, Thomas F. 1991. "On the Threshold: Environmental Changes as Causes of Acute Conflict." *International Security*, vol. 16, no. 2:76–116. doi:10.2307/2539061.

Homer-Dixon, Thomas F. 1994. "Environmental scarcities and violent conflict: Evidence from cases." *International Security*, vol. 19, no. 1:5–40. doi:10.2307/2539147.

Homer-Dixon, Thomas F. 1996. "Strategies for Studying Causation in Complex Ecological-Political Systems." *The Journal of Environment and Development*, vol. 5, no. 2:132–148. doi:10.1177/107049659600500202.

Homer-Dixon, Thomas F. 1999. *Environment, Scarcity and Violence*. Princeton: Princeton University Press.

Homer-Dixon, Thomas F. 2007. "Terror in the Weather Forecast." *New York Times*, Arthur Ochs Sulzberger Jr., 24 April 2007.

Huebert, Rob, Heather Exner-Pirot, Adam Lajeunesse, and Jay Gulledge. 2012. *Climate Change & International Security: The Arctic as a Bellwether*. Arlington, Virginia. Accessed on 23 November 2016, http://www.c2es.org/publications/climate-change-international-arctic-security.

Kaplan, Robert D. 1994. "The Coming Anarchy." *World Policy Journal*, vol. 17, no. 2:95–96.
Klare, Michael T. 2007. "Global Warming Battlefields: How Climate Change Threatens Security." *Current History*, vol. 106, no. 703:355–361.
Klötzli, Stefan. 1994. "The Water and Soil Crisis in Central Asia: A Source for Future Conflicts?" In *Environment and Conflicts Project ENCOP Occasional Papers*, edited by Günter Bächler and Kurt R. Spillmann, vol. 11. no. 1. Zurich: Center for Security Studies.
Lang, Christoph I. 1995. "Environmental Degradation in Kenya as a Cause of Political Conflict, Social Stress, and Ethnic Tensions." In *CSS Environment and Conflicts Project (ENCOP)*, edited by Günter Bächler and Kurt R. Spillmann, vol. 12 no. 1. Zurich: Centre for Security Studies.
Levy, Marc A. 1995. "Is the Environment a National Security Issue?" *International Security*, vol. 20, no. 2:35–62.
Libiszewski, Stephan. 1995. "Water Disputes in the Jordan Basin Region and their Role in the Resolution of the Arab-Israeli Conflict." In *CSS Environment and Conflicts Project (ENCOP)*, edited by Günter Bächler and Kurt R. Spillmann, vol. 13, no. 1. Zurich: Center for Security Studies. Accessed on 3 May 2017, http://www.mideastweb.org/Mew_water95.pdf.
Mathews, Jessican Tuchman. 1989. "Redefining Security." *Foreign Affairs*, vol. 68, no. 2:162–177.
McDonald, Matt. 2012. "The Failed Securitization of Climate Change in Australia." *Australian Journal of Political Science*, vol. 47, no. 4:579–592. doi:10.1080/10361146.2012.731487.
Meadows, Donella H., Dennis L Meadows, Jorgen Randers, and William W. Behrens. 1972. *The Limits to Growth*. New York: Universe Books.
Ministry of Defence. 2013. *White Paper on Defence and National Security*. Ministry of Defence. France: French Government. Accessed on 22 November 2016, http://www.rpfrance-otan.org/White-Paper-on-defence-and.
Ministry of Defense. 2008. *National Strategy of Defense*. Ministry of Defense. Brazil: Ministry of Defense. Accessed on 18 November 2016, http://www.defesa.gov.br/projetosweb/estrategia/arquivos/estrategia_defesa_nacional_ingles.pdf.
Ministry of Ecology. 2009. *French Climate Plan*. Sustainable Development and Energy Ministry of Ecology. Paris: Government of France. Accessed on 7 January 2017, http://ec.europa.eu/environment/archives/networks/greenspider/doc/climate_change_campaigns/ccc_france.pdf.
Ministry of Ecology. 2011. *National Plan: Climate Change Adaptation*. Sustainable Development and Energy Ministry of Ecology. Paris: Government of France. Accessed on 7 January 2017, http://www.developpement-durable.gouv.fr/IMG/pdf/ONERC_PNACC_Eng_part_1.pdf.
Ministry of Interior. 2014. *National Internal Security Policy 2014–2018*. Ministry of Interior. Government of Pakistan. http://nacta.gov.pk/policies/NISP.pdf.
Mullen, Michael G. 2011. "From the Chairman." *Joint Force Quarterly*, no. 60:2–3.
Myers, Norman. 1994. *Ultimate Security: the environmental basis of political stability*. New York: W. W. Norton.
An Act to establish a comprehensive and coordinated national climate policy and program, and for other purposes. National Climate Program Act (95th Congress, USC 2901, Public Law 95-367), 17 September 1978. Accessed on 21 September 2015, http://www.gpo.gov/fdsys/pkg/STATUTE-92/pdf/STATUTE-92-Pg601.pdf.
National Defense Authorization Act for Fiscal Year 2008. National Defense Authorization Act for Fiscal Year 2008 (110th US Congress, Public Law 110-181), 28 January 2008. Accessed on 21 September 2013, http://www.gpo.gov/fdsys/pkg/PLAW-110publ181/html/PLAW-110publ181.htm.
NIC. 2008. "National Intelligence Assessment on the National Security Implications of Global Climate Change to 2030", Thomas Fingar. Washington D.C.: National Intelligence Council. Accessed on 17 February 2013, http://online.wsj.com/public/resources/documents/FingarSFR.GCC06252008.pdf.
NRC. 2013. *Climate and Social Stress: Implications for Security Analysis*. Committee on Assessing the Impacts of Climate Change on Social and Political Stresses: Washington, DC:

The National Academies Press. Accessed on 23 November 2016, https://www.nap.edu/catalog/14682/climate-and-social-stress-implications-for-security-analysis.

Obama, Barack H. 2009a. *Executive Order 13514: Federal Leadership in Environmental, Energy, and Economic Performance*. Washington, DC: White House. Accessed on 20 May 2017, https://www.gpo.gov/fdsys/pkg/FR-2009-10-08/pdf/E9-24518.pdf.

Obama, Barack H. 2013. *Executive Order 13653: Preparing the United States for the Impacts of Climate Change*. Washington, DC: White House. Accessed on 6 January 2017, https://www.gpo.gov/fdsys/pkg/FR-2013-11-06/pdf/2013-26785.pdf.

Obama, Barack H. 2015. *Executive Order 13693: Planning for Federal Sustainability in the Next Decade*. Washington, DC: White House. Accessed on 6 January 2017, https://www.whitehouse.gov/the-press-office/2015/03/19/executive-order-planning-federal-sustainability-next-decade.

Obama, Barack H. 2016. *Presidential Memorandum: Climate Change and National Security*. Washington, DC: White House. Accessed on 23 December 2016, https://www.whitehouse.gov/the-press-office/2016/09/21/presidential-memorandum-climate-change-and-national-security.

Oels, Angela. 2012. "From 'Securitization' of Climate Change to 'Climatization' of the Security Field: Comparing Three Theoretical Perspectives." In *Climate Change, Human Security and Violent Conflict: Challenges for Societal Stability*, edited by Jürgen Scheffran, Michael Brzoska, Hans Günter Brauch, Peter Michael Link and Janpeter Schilling, vol. 8, pp. 185–206. Berlin: Springer.

Paul Ehrlich and Anne Ehrlich. 1988. "The Environmental Dimensions of National Security." In *Global Problems and Common Security: Annals of Pugwash 1988*, edited by J. Rotblat and V. I. Goldanskii, pp. 180–190. Heidelberg: Springer.

Penny, Christopher K. 2007. *Climate Change and the Security Council: A Preliminary Framework for Implementing Remedial Measures through Chapter VII of the UN Charter*. CISDL Legal Working Paper Series on Climate Change Law and Policy: Montreal: Centre for International Sustainable Development Law. Accessed on 12 June 2013, http://cisdl.org/public/docs/legal/Penny%20-%20Climate%20Change%20and%20the%20Security%20Council.pdf.

PLA. 2015. *China's Military Strategy*. Beijing: The State Council Information Office of the People's Republic of China. Accessed on 25 July 2015, http://news.xinhuanet.com/english/china/2015-05/26/c_134271001.htm.

Revelle, Roger R., Robert B. Bacastow, James Baker, Charles D. Keeling, Hans H. Landsberg, Ralph A. Llewellyn, Murray Mitchell, Walter H. Munk, Harry Perry, George D. Robinson, Joseph Smagorinsky, Verner E. Suomi, and Warren M. Washington. 1977. *Studies in Geophysics: Energy and Climate*. Washington, DC: National Academy of Sciences. Accessed on 20 May 2017, https://www.nap.edu/read/12024/chapter/1.

Russian Federation. 2009. *Russia's National Security Strategy to 2020*. Moscow. Accessed on 22 November 2016, http://rustrans.wikidot.com/russia-s-national-security-strategy-to-2020.

Russian Federation. 2015. *Russian National Security Strategy*. Moscow. Accessed on 22 November 2016, http://www.ieee.es/Galerias/fichero/OtrasPublicaciones/Internacional/2016/Russian-National-Security-Strategy-31Dec2015.pdf.

Schubert, Renate, Hans Joachim Schellnhuber, Nina Buchmann, Astrid Epiney, Rainer Grießhammer, Margareta E. Kulessa, Dirk Messner, Stefan Rahmstorf, and Jürgen Schmid. 2008. *Climate Change as a Security Risk*. Berlin: Earthscan. Accessed on 20 May 2017, http://ccsl.iccip.net/wbgu_jg2007_engl.pdf.

Schwartz, Peter, and Doug Randall. 2003. *An Abrupt Climate Change Scenario and Its Implications for United States National Security*. Washington, DC: Department of Defense. Accessed on 23 October 2010, http://stephenschneider.stanford.edu/Publications/PDF_Papers/SchwartzRandall2004.pdf.

Scott, Steven, and Sarah Vogler. 2014. "G20 Brisbane: Tony Abbott clashes with Barack Obama over Climate Change." *The Courier Mail*, News Corp, 17 November 2014. Accessed on 10 March 2015, http://www.couriermail.com.au/news/queensland/g20-brisbane-tony-abbott-clashes-with-barack-obama-over-climate-change/story-fnmd7bxx-1227125019686.

Shabecoff, Philip. 1988. "Global Warming Has Begun, Expert Tells Senate." *The New York Times*, The New York Times, 24 June 1988. Accessed on 20 May 2017, http://www.nytimes.com/1988/06/24/us/global-warming-has-begun-expert-tells-senate.html?pagewanted=all.

Sindico, Francesco. 2007. "Climate Change: A Security (Council) Issue?." *Carbon and Climate Law Review*, vol. 1, no.:26–31.

Slaughter, Anne-Marie, Troy Sternberg, Sarah Johnstone, Jeffrey Mazo, Francesco Femia, Caitlin Werrell, Michael Werz, Max Hoffman, David Michel, and Mona Yacoubian. 2013. *The Arab Spring and Climate Change*. A Climate and Security Correlations Series: Washington, DC: The Center for American Progress.

Stang, Gerald. 2014. "Climate change and EU Security: When and how they intersect." Online: European Union Institute for Security Studies. Accessed on 20 November 2016, http://www.iss.europa.eu/uploads/media/Brief_32_climate_change.pdf.

Tanzler, Dennis, Alexander Carius, and Sebastian Oberthur. 2002. *Climate Change and Conflict: Can Climate Impacts Increase Conflict Potentials? What is the Relevance of this Issue for the International Process on Climate Change?* Nature Conservation and Nuclear Safety Federal Ministry for the Environment. Berlin: Federal Ministry for the Environment, Nature Conservation and Nuclear Safety. Accessed on 10 February 2013, http://www.afes-press.de/pdf/Brauch_ClimateChange_BMU.pdf.

The New Daily. 2014. "Runaway climate debate keeps Abbott hopping." In *The New Daily*, 16 November 2014. Published online: The New Daily Pty. Ltd. Accessed on 12 January 2015, http://thenewdaily.com.au/news/2014/11/16/climate-debate-tony-abbott-g20/.

Titley, David. 2015. "The Australian Climate Security Summit." Canberra. Personal communicaton. 29 October 2015.

Torney, Diarmuid. 2015. *European Climate Leadership in Question: Policies toward China and India*. Online: MIT Press.

Townsend, Mark, and Paul Harris. 2004. "Now the Pentagon tell Bush: climate change will destroy us." *The Observer*, Guardian News Accessed on 20 May 2017, https://www.theguardian.com/environment/2004/feb/22/usnews.theobserver.

Trombetta, Maria J. 2008. "Environmental Security and Climate Change: Analysing the discourse." *Cambridge Review of International Affairs*, vol. 21, no. 4:585–602. doi:10.1080/09557570802452920.

Trombetta, Maria J. 2012. "Climate Change and the Environmental Conflict Discourse." In *Climate Change, Human Security and Voilent Conflict: Challenges for Societal Stability*, edited by Jürgen Scheffran, Michael Brzoska, Hans Günter Brauch, Peter Michael Link and Janpeter Schilling, vol. 8, pp. 151–164. Berlin: Springer.

UK. 2007. "Letter dated 5 April 2007 from the Permanent Representative of the United Kingdom of Great Britain and Northern Ireland to the United Nations addressed to the President of the Security Council." New York: United Nations. Accessed on 20 May 2017, http://www.securitycouncilreport.org/atf/cf/%7B65BFCF9B-6D27-4E9C-8CD3-CF6E4FF96FF9%7D/CC%20S2007%20186.pdf.

UK MoD. 2010a. *MOD Climate Change Delivery Plan*. Ministry of Defence. London: Ministry of Defence. Accessed on 10 April 2014, https://www.gov.uk/government/publications/mods-climate-change-strategy. (Document archived, see website. In possession of author).

UK MoD. 2010b. *MoD Climate Change Strategy 2010*. Ministry of Defence. London: Ministry of Defence UK. Accessed on 10 April 2014, https://www.gov.uk/government/publications/mods-climate-change-strategy. (Document archived, see website. In possession of author).

UK MoD. 2013. *Annex D to Sustainability Report—MOD Annual Report and Accounts 2011–2012*. Ministry of Defence. London: UK Ministry of Defence. Accessed on 12 November 2016, https://www.gov.uk/government/uploads/system/uploads/attachment_data/file/35010/20121205_mod_annualreport_Annex_D.pdf.

UK MoD. 2015. *Sustainable MOD: Annual Report 2014/15: Sustainability in the Ministry of Defence*. UK Ministry of Defence. London: UK Government. Accessed on 12 November 2016, https://www.gov.uk/government/uploads/system/uploads/attachment_data/file/447951/20150723-Sustainable_MOD_Annual_Report-internet-ver.pdf.

Ullman, Richard H. 1983. “Redefining Security.” *International Security*, vol. 8, no. 1:129–153. doi:10.2307/2538489.
UN. 1988. “Our Common Future: Report of the World Commission on Environment and Development.” In *Secretary General A/42/427—Development and International Co-operation: Environment*, 4 August 1987. New York: United Nations. Accessed on 10 May 2013, http://www.un-documents.net/a42-427.htm.
UN. 2002. “Background Note on the “Arria-Formula” Meetings of the Security Council Members.” United Nations Secretariat. New York: United Nations. Accessed on 19 January 2017, http://www.un.org/en/sc/about/methods/bgarriaformula.shtml.
UN. 2016. “Leading on climate change.” Online: United Nations. Accessed on 17 January 2017, http://www.un.org/sustainabledevelopment/ban-ki-moon-climate-change/.
UNGA. 2009. “Resolution 63/281: Climate change and its possible security implications.” In *United Nations General Assembly 63rd Session* Agenda item 107, A/RES/63/291, 11 June 2009. New York: United Nations. Accessed on 20 May 2017, http://www.securitycouncilreport.org/atf/cf/%7B65BFCF9B-6D27-4E9C-8CD3-CF6E4FF96FF9%7D/res%2063%20281.pdf.
UNSC. 2007. “Letter dated 5 April 2007 from the Permanent Representative of the United Kingdom of Great Britain and Northern Ireland to the United Nations addressed to the President of the Security Council (S/2007/186).” In *United Nations Security Council 5663rd Meeting*, S/PV.5663, 17 April 2007. New York: United Nations.
UNSC. 2011. “Maintenance of international peace and security: Impact of climate change.” In *United Nations Security Council 6587th meeting*, S/PV.6587, 20 July 2011. New York: United Nations. Accessed on 20 May 2017, http://www.circleofblue.org/wp-content/uploads/2012/09/UN-Security-Council_climate-change-July-2011.pdf.
UNSG. 2008. “We have moved climate change up to top of the agenda.” In *Secretary-General Statements and Messages, Press Release*, vol. SGM/SM/11408. New York: United Nations. Accessed on 18 April 2012, http://www.un.org/News/Press/docs/2008/sgsm11408.doc.htm.
UNSG. 2009. “Report of the Secretary-General: Climate change and its possible security implications (A/64/350).” In *United Nations General Assembly 64th session*, Item 114, A/64/350, 11 September 2009. New York: United Nations. Accessed on 20 May 2017, http://www.un.org/ga/search/view_doc.asp?symbol=A/64/350.
UNSG. 2011. “Possible Security Implications of Climate Change.” New York: United Nations. Accessed on 1 March 2013, http://www.un.org/News/Press/docs/2011/sc10332.doc.htm.
UNSG. 2012. “Secretary-General’s video message to Planet Under Pressure Conference.” In *Statement by the Secretary-General*, 29 March 2012. New York: United Nations. Accessed on 16 March 2013, http://www.un.org/sg/statements/index.asp?nid=5960.
US DoD. 2010. *Quadrennial Defense Review*. Washington, DC: Department of Defense. Accessed on 20 May 2017, https://www.defense.gov/Portals/1/features/defenseReviews/QDR/QDR_as_of_29JAN10_1600.pdf.
US DoD. 2012a. “Department of Defense FY-2012 Climate Change Adaptation Roadmap.” In *Department of Defense Strategic Sustainability Performance Plan FY-2012*. Washington, DC: Department of Defense. Accessed on 20 May 2017, http://www.denix.osd.mil/sustainability/dod-sspp/unassigned/dod-sspp-fy-2012/.
US DoD. 2012b. *Department of Defense Strategic Sustainability Performance Plan FY-2012*. Washington, DC: Department of Defense. Accessed on 9 November 2016, http://www.denix.osd.mil/sustainability/dod-sspp/unassigned/dod-sspp-fy-2012/.
US DoD. 2013a. *Department of Defense Strategic Sustainability Performance Plan FY-2013*. Washington, DC: Department of Defense. Accessed on 5 March 2014, http://www.denix.osd.mil/sustainability/loader.cfm?csModule=security/getfile&pageid=35931.
US DoD. 2014a. “2014 Climate Change Adaptation Roadmap.” In *Department of Defense Strategic Sustainability Performance Plan FY-2014*. Washington, DC: Department of Defense.

Accessed on 20 May 2017, http://www.denix.osd.mil/sustainability/dod-sspp/unassigned/dod-sspp-fy-2014/.

US DoD. 2014b. *Department of Defense Strategic Sustainability Performance Plan FY 2014*. Defense. Alexandria: Office of the Deputy Secretary of Defense for Installations and Environment.

US DoD. 2014c. *Quadrennial Defense Review 2014*. Washington, DC: Department of Defense. Accessed on 20 May 2017, http://archive.defense.gov/pubs/2014_Quadrennial_Defense_Review.pdf.

US DoD. 2015. *National Security Implications of Climate-Related Risks and a Changing Climate*. US DoD. Washington, DC: US Government. Accessed on 11 February 2016, http://archive.defense.gov/pubs/150724-congressional-report-on-national-implications-of-climate-change.pdf?source=govdelivery.

US DoD. 2016a. *Department of Defense Strategic Sustainability Performance Plan FY 2016*. Washington, DC: Department of Defense. Accessed on 9 November 2016, http://www.denix.osd.mil/sustainability/dod-sspp/unassigned/department-of-defense-strategic-sustainability-performance-plan-fy-2016/ and http://www.denix.osd.mil/sustainability/dod-sspp/.

US DoD. 2016b. *DoD Directive 4715.21: Climate Change Adaptation and Resilience*. US DoD. Washington, DC: Office of the Under Secretary of Defence for Acquisition, Technology, and Logistics. Accessed on 11 February 2016, http://www.dtic.mil/whs/directives/corres/pdf/471521p.pdf.

US GAO. 2014. *Climate Change Adaptation: DOD Can Improve Infrastructure Planning and Processes to Better Account for Potential Impacts*. US Government Accountability Office. Washington, DC: US Government. Accessed on 7 January 2017, http://www.gao.gov/assets/670/663734.pdf.

US GAO. 2015. "Limiting the Federal Government's Fiscal Exposure by Better Managing Climate Change Risks." Washington DC: US Government Accountability Office. Accessed on 23 July 2015, http://www.gao.gov/highrisk/limiting_federal_government_fiscal_exposure/why_did_study#t=1.

US House of Representatives. 2007a. "Global Climate Change Security Oversight Act." In H.R. 1961 (110th), 19 April 2007. Washington, DC: US Congress. Accessed on 23 December 2016, https://www.govtrack.us/congress/bills/110/hr1961.

US House of Representatives. 2007b. "The National Security Implications of Climate Change." In *Hearing Before the Subcommittee on Investigations and Oversight Committee on Science and Technology*, 110th Congress, First session, Serial No. 110-58, 27 September 2007. Washington, DC: US Government Printing Office. Accessed on 16 March 2013, http://www.gpo.gov/fdsys/pkg/CHRG-110hhrg34720/pdf/CHRG-110hhrg34720.pdf.

US House of Representatives. 2008a. "National Intelligence Assessment on the National Security Implications of Global Climate Change." In *Joint Hearing Before the Select Committee on Energy Independence and Global Warming and Subcommittee on Intelligence, Community Management, Permanent Select Committee on Intelligence House of Representatives*, 110th Congress, Serial No. 110-41, 25 June 2008. Washington, DC: US Government Printing Office. Accessed on 1 September 2013, https://www.fas.org/irp/congress/2008_hr/climate.pdf.

US House of Representatives. 2008b. "National Security Implications of Global Climate Change." In *Joint Hearing before the Select Committee on Energy Independence and Global Warming and Subcommittee on Intelligence, Community Management, Permanent Select Committee on Intelligence*, 110th Congress, Second session, Serial No. 110-41, 25 June 2008. Washington, DC: US Government Printing Office. Accessed on 6 March 2014, http://www.gpo.gov/fdsys/pkg/CHRG-110hhrg62523/html/CHRG-110hhrg62523.htm.

US Navy. 2009a. *Task Force Climate Change Charter*. Washington, DC: US Navy. Accessed on 6 March 2013, http://greenfleet.dodlive.mil/files/2010/09/Task-Force-Climate-Change-Charter.pdf.

US Navy. 2009b. *US Navy Arctic Roadmap*. Washington, DC: US Navy. Accessed on 3 August 2013, http://www.navy.mil/navydata/documents/USN_artic_roadmap.pdf.

US Navy. 2010b. *Navy Climate Change Roadmap*. Washington, DC: US Navy. Accessed on 6 March 2012, http://www.navy.mil/navydata/documents/CCR.pdf.

US Senate. 2006. "The Lugar-Biden Climate Change Resolution." In *U.S. Senate Foreign Relations Committee*, S. Res. 312, 23 May 2006. Washington, DC: US Senate Accessed on 14 November 2016, http://www.c2es.org/docUploads/S%20Res%20312%20as%20passed%20by%20SFRC%20May%2023%2006.pdf, http://www.c2es.org/federal/congress/109/summary-lugar-biden-climate-change-senate-resolution-312.

US Senate. 2007. "Climate Change: National Security Threats." In *Hearing Before the Committee on Foreign Relations*, 110th Congress, Hearing No. 110-402, 9 May 2007. Washington, DC: US Government Printing Office. Accessed on 14 November 2016, https://www.gpo.gov/fdsys/pkg/CHRG-110shrg42725/pdf/CHRG-110shrg42725.pdf.

US Senate. 2009. "Climate Change and Global Security: Challenges, Threats and Diplomatic Opportunities." In *Hearing Before the Committee on Foreign Relations*, 111th Congress, Senate Hearing No. 111-207, 21 July 2009. Washington, DC: US Government Printing Office. Accessed on 14 November 2016, https://www.gpo.gov/fdsys/pkg/CHRG-111shrg54243/html/CHRG-111shrg54243.htm.

USACE. 2014. Adaption Policy Statement. In *USACE Climate Preparedness and Resilience Policy Statement*, edited by Jo-Ellen Darcy: US Army Corps of Engineers.

Vis, Peter, and Jos Delbeke. 2015. *EU Climate Policy Explained*. Abingdon: Routledge.

Warren, Dane. 2015. *Climate change and International Peace and Security: Possible Roles for the U.N. Security Council in Addressing Climate Change*. New York: Columbia Law School. Accessed on 18 January 2017, https://web.law.columbia.edu/sites/default/files/microsites/climate-change/warren_-_cc_and_international_peace_and_security_-_roles_for_the_un_security_council.pdf.

Werrell, Caitlin, and Francesco Femia. 2014. "Must Watch: PACOM Commander on the Climate Change Threat to the Asia-Pacific." Online: The Center for Climate and Security. Accessed on 10 October 2015, http://climateandsecurity.org/2014/03/07/must-watch-pacom-commander-on-the-climate-change-threat-to-the-asia-pacific/.

Werz, Michael, and Laura Conley. 2012. *Climate Change, Migration, and Conflict in Northwest Africa: Rising Dangers and Policy Options Across the Arc of Tension*. Climate Migration Series: Washington, DC. Accessed on 20 November 2016, https://cdn.americanprogress.org/wp-content/uploads/issues/2012/04/pdf/climate_migration_nwafrica.pdf.

White House. 2010. *The National Security Strategy of the United States of America*. White House. Washington, DC: Government of the United States of America. Accessed on 20 May 2017, http://nssarchive.us/NSSR/2010.pdf.

White House. 2015. *National Security Strategy*. Washington, DC: Government of the United States of America. Accessed on 20 May 2017, http://nssarchive.us/wp-content/uploads/2015/02/2015.pdf.

WMO. 1988. "The Changing Atmosphere: Implications for Global Security." In *Toronto Conference on the Changing Atmosphere*, Series No. 710, pp. 292–304. Toronto: Secretariat of the World Meteorological Organization.

Youngs, Richard. 2014. *Climate Change and EU Security Policy: An Unmet Challenge*. Washington, DC: Carnegie Endowment for International Peace.

Youngs, Richard. 2015. *Climate Change and European Security*. London: Taylor and Francis.

Part III
Case Studies in Climate Securitisation

Chapter 4
Case Study Aim and Method

Abstract This chapter outlines the securitisation case study aim, method and its key limitations. Climate security case studies were undertaken with the aim of examining the process of climate securitisation within the Australian and US political-military sector between 2003 and 2013. This was a key period that spanned centre-right and centre-left governments in both countries and coincided with securitisation moves in the international arena aimed at influencing a binding global commitment to reduce emissions. Three lines of inquiry were developed for the case studies that addressed how climate change was framed, the urgency of the response, and what types of response measures were put in place to address climate change by the political-military establishments of the US and Australia. Securitisation frameworks were developed by the author based on securitisation theories from the Copenhagen and Paris Schools.

Keywords Securitisation · Climate case study · Climate security method · Securitisation method

4.1 Aim

The case studies presented in this book were undertaken with the aim of examining the process of climate securitisation in the militaries of Australia and the United States between 2003 and 2013. In order to provide greater context and insight, the scope was extended to include the political sector of each country; or, more specifically, those political actors directly involved in the formal national security decision-making process.[1] To achieve this aim, the case studies established three principal lines of inquiry:

---

[1]The term 'political-military' sector is used throughout to describe both the military sector and the political sector. In reality, only parts of these institutions are examined. The boundaries are defined in the case studies.

M.D. Thomas, *The Securitization of Climate Change: Australian and United States' Military Responses (2003–2013)*, The Anthropocene: Politik—Economics—Society—Science 10, DOI 10.1007/978-3-319-49658-0_4

- *Contextual Framing*. In what context did the Australian and US political-military sectors frame climate change? This question examined whether or not climate change was framed as a security- or non-security-related issue. It also sought to categorise the different referent objects it was securing or otherwise the 'type' of security category (e.g., national security, human security, environmental security).
- *Temporal Framing*. To what level of urgency did the Australian and US political-military sectors frame climate change? This question examined whether climate change was framed as an urgent issue to be dealt with immediately (thereby justifying emergency measures) or as strategic issue prevalent across decades (thereby not justifying an emergency response).
- *Response Measures/Actions*. What response measures (actions) did each military undertake in specific response to address climate change? Did any of these measures break free of 'normal' rules and promote emergency measures? Similarly, to what extent did the measures become part of the normal practices and processes of the military sector? Which departments, sections or programmes within the military-bureaucratic structures responded to climate change? How enduring was this response? How deep was the organisational penetration of climate change within the Australian and US military-bureaucracies?

4.2 Analytical Framework and Method

4.2.1 Analytical Framework

The case studies used two conceptual frameworks to address the research aims. The main framework, Copenhagen School's 'securitisation theory', was originally outlined by Ole Wæver in 1995 but received its fullest treatment in *Security: A New Framework for Analysis* (1998) as a means to broaden the scope of security studies and to distinguish the process of securitisation from that of politicisation. In their seminal work, Buzan et al. (1998) assert that although threats and vulnerabilities may arise in many different sectors (political, economic, environment, societal and military), most are successfully dealt with as either non-politicised issues (no government involvement) or as politicised issues (some form of government involvement). For a threat to become 'securitised' and therefore 'above politics' it needs to be framed as an '*existential threat* to a referent object by a securitizing actor' who then 'generates endorsement of *emergency measures*' that cause actors to move '*beyond rules* that would otherwise bind' (Buzan et al. 1998: 5 and 23; emphasis added). These three critical criteria (presented as an existential threat, justifying emergency measures and breaking free of normal rules) formed the core elements identified in the three lines of inquiry listed above and were used as fundamental indicators during the case studies.

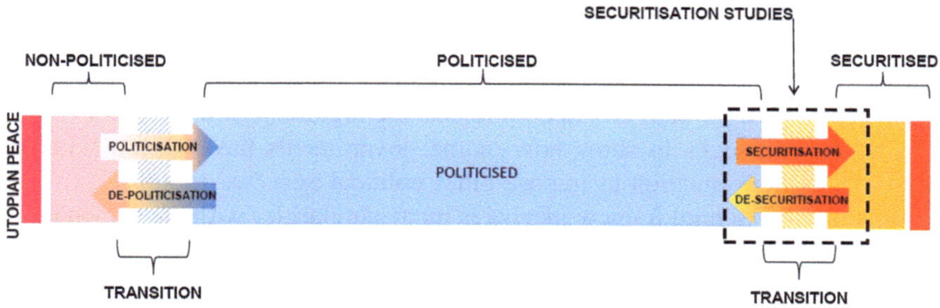

Fig. 4.1 Copenhagen School modalities (non-politicised, politicised and securitised). *Source* The author.

'Actions' or 'events' (such as certain speech-acts, policy measures or even military build-up and security 'incidents') that contribute to emergency measures beyond normal rules might generally be considered as securitising moves. Such actions might 'move' an issue beyond what political scientist Oels (2012: 201) has called the 'threshold of exceptionality' and into a securitised modality; 'de-securitising' moves bring the issue back to 'normal' politics. Copenhagen School theorists place a premium on the politicised modality, 'security [is a] negative … de-securitization is the optimal' (Buzan et al. 1998: 29). Figure 4.1 illustrates (in a modified form) the Copenhagen framework where issues can oscillate between a non-politicised, politicised or securitised modality. The most dominant mode is the politicised modality and it is graphically presented as the largest of the three. The case studies primarily focus on what took place inside the area of the dashed box, and where securitisation studies are typically focused (and specifically the *process* of securitisation that is denoted by the rightwards arrow within the dashed box). More broadly, the case studies also draw on the Copenhagen School concept of 'sectors' by focusing on the political-military sector. This was useful in confining the scope of inquiry to 'manageable proportions' (Buzan et al. 1998: 8).

The second conceptual perspective used was the so-called 'Paris School' that examines the governance of security issues below the threshold of exceptionality. Grounded in the earlier works of French sociologists, Bourdieu (1969) and Foucault (1982), and more recently by Huysmans (2004, 2006) and particularly Bigo (2006, 2008), the Paris School focuses on revealing the everyday (bureaucratic and routine) practices of the 'professionals of (in)security' (Oels 2012: 197). In general terms (and as interpreted and used throughout this research), the Paris School examines the governance and bureaucratisation—the translation into new policies and initiatives—of (security) issues, especially by the agencies of the police, intelligence, military and associated security professionals. Integral to the Paris School framework is a critical perspective that seeks to understand how security monoliths position themselves to advantage via various institutional and governance mechanisms and transnational networks to control security agendas

(Bigo et al. 2008). In this framework, issues become securitised to suit the agenda of institutional elites, enhancing their power and relegating democratic process. By doing so, illiberal practices become the standard practices of liberal democracies. As was noted by Bigo, the Paris School rejects the argument of a 'permanent state of exception' and seeks to show how 'some governments have played with the opportunities of the situation to impose other political agendas' (2008: 5).

While the Paris School framework shares many similarities with the Copenhagen framework, it differs in its interpretation of the exceptional. For instance, while the Copenhagen School tends to view securitisation as a distinct modality (Fig. 4.1), this is not the case with the Paris School, where the 'existential threat and the politics of terror cannot be distinguished so easily' (Bigo/Tsoukala 2008: 5). Furthermore, rather than focus on the speech-acts of the main actors ('[they] are not decisive'), the Paris School views security as a 'social field' which focuses on security institutions and lower level practitioners whom Bigo regards as the real architects of (in)security (Bigo/Tsoukala 2008: 4). Finally, whereas the Copenhagen School adopts a 'threat-based' construction of security (in terms of how issues are framed and the actual 'grammar' used by the securitising actor), the Paris School is rather focused on governance practices as exemplified by risk-based approaches; what Olaf Corry has labelled 'risk security' (Corry 2011). In this view, society has become '(over-) controlled by cumulative security practices using 'risk' as a major tool and strategy of legitimation' (Corry 2011: 243). For Corry, the Paris School approach is defined by this perspective where 'risks (understood as conditions of possibility for harm) to a referent object [lead] to programmes for permanent changes aimed at reducing vulnerability and boosting governance-capacity of the valued referent object itself' (Corry 2011: 248). Thus, risk is a governance technique and one that 'promotes long-term precautionary' measures (ibid: 235).[2]

Without wishing to enter into the theoretical nuances, the case studies used the Paris School more as a conceptual basis for examining a range of 'everyday' bureaucratic processes, practices and programmes by the political-military establishment in their response to climate change; an examination of the *un*exceptional. In this sense, the case studies examined governance techniques that included institutional processes, practices and programmes below the threshold of exceptionality. Moreover, while the Copenhagen theory was mainly used to develop the securitisation frameworks (see below), the Paris School acted as a guiding principle during the analytical and discussion phase (presented at Chap. 7).

[2]This book does not intend to enter into particular scholarly debates surrounding the competing definitions of 'risk' or 'threat-based' approaches to security. Nor does it specifically examine climate change framed in the 'grammar' of risk; where Corry and others (Lucke et al. 2013) distinguish 'risks' from 'threats' and one which requires 'a different grammar' or 'a different kind of speech act' (Corry 2011: 236). Rather, the research prioritises Copenhagen 'threat-based' approaches to security.

4.2.2 Method

The case studies examined a total of 1,569 documents from the Australian political-military sector and 2,093 from the US political-military sector, across the period 2003–2013. Tables 5.1 and 6.1 in the respective case study chapters detail *where* (in terms of location within the respective political-military bureaucratic structure) the documents were sourced from. Once the documents were sourced and downloaded, each was individually formatted into Portable Document Format (.pdf) in Adobe Acrobat and then uploaded into the content analysis software tool, *NVivo*. During uploading, all documents were individually classified according to their *year* (of publication), *author*, document *type* (e.g., speech, policy, media or journal) and political *affiliation* (if applicable).

Once the data was prepared, quantitative content analysis was then applied through an examination of the artefacts (directed by 'word search' functions) and 'coding' of appropriate passages of text into one of two 'securitisation frameworks' that were developed by the author. The first framework (at Table 4.1), was used to analyse the *contextual* framing of climate change. This framework involved three steps. Step one was to decide whether climate change had been framed in a securitised or non-securitised manner (Level 1 'Broad', at Table 4.1). Step two was to identify the specific securitised or non-securitised category/referent object (Level 2 'Refined', at Table 4.1). For example, was climate change framed as an energy security issue or as a national security issue? Step three was to identify whether climate change had been framed as a *threat*, *challenge*, in *neutral* or benign terms, or as an *opportunity* (Level 3 'Specific', at Table 4.1). An overall example of how this method was applied can be made by examining the following passage from a 2008 speech to parliament by the former Australian Prime Minister, Kevin Rudd. During this speech, Rudd (2008g) stated: 'climate change represents a most fundamental national security challenge to the long-term future' (2008g). This passage was coded as a *challenge* (Level 3) within the realm of *national security* (Level 2) in the context of a *securitised* frame (Level 1).

The second securitisation framework (Table 4.2) examined *temporal* framing. The intent here was to determine whether political elite and military forces framed climate change as an urgent issue requiring emergency measures or as a longer-term issue justifying an incremental response (or no response at all). An example of temporal framing can be given by examining a passage of text from a response given by the Australian Deputy Secretary Strategy Michael Pezzullo in 2007 to the Standing Committee on Foreign Affairs, Defence and Trade. In response to a question, Pezzullo commented that 'the strategic implication of climate change in a security sense … [is] likely to be felt more over a 40–50 year period' (Michael Pezzullo quoted in Commonwealth of Australia 2007a: 102). This passage was coded as framing climate change as a long-term (security) issue.

During the coding process, coding rules were established that limited the number of codes per document and for bodies of text that might have been repetitious across or between levels in the bureaucracy (intended to remove skewing or bias).

Table 4.1 Securitisation framework for analysing contextual framing

Level 1: Broad		Level 2: Refined	Level 3: Specific
Securitised frames	1	General security reference	a. Opportunity
	2	Global and regional security	b. Neutral
	3	National security	c. Challenge
	4	Human security	d. Threat
	5	Energy security	
	6	Environmental security	
	7	Biosecurity	
	8	Convergence/Multiple security impacts	
	9	Resource security(excluding energy)	
	10	Non-traditional, new, emerging security issue	
Non-securitised frames	1	Economic	
	2	Agriculture	
	3	Forestry	
	4	Energy and resources	
	5	Environment and sustainability	
	6	Government and regulatory	
	7	International and global	
	8	Health	
	9	Transport	
	10	Other/Non-aligned	

Source The author

Table 4.2 Securitisation framework for analysing temporal framing

Temporal frame	Qualifying description	Timeframe (years)
Urgent	An urgent issue requiring immediate action	0–2
Short term	Response required in decadal timeframe	3–10
Medium term	Response required but not in short term, response required/effects felt decade plus	10–20
Long-term	Strategic response required or taking effect over decades	20+

Source The author

Qualitative content analysis (or, more commonly, textual analysis) of important speeches, policies and other such documents was also undertaken to supplement the codification process. This enabled the quantitative findings established from the securitisation frameworks described above to be placed in a much broader context, and helped to build a more complete outlook on how the political elite and military institutions approached climate change as a security issue.

Overall, the use of securitisation frameworks in combination with *NVivo* enabled securitisation to be *quantified* and represented in graphical form, making interpretation readily accessible and easy to grasp. Since each individual document was also codified, distinct patterns emerged that made it possible to analyse *when* securitisation had occurred, *who* was responsible for it (including an assessment of which parts of the bureaucracy) and from *what* type of artefact it had originated.

4.3 Limitations

The case studies presented in this book are not without fault and include a number of limitations. For the purpose of alerting the reader and to guide those who might follow a similar pathway in the future, these include:

- *Theoretical boundaries*. While the case studies broadly adhered to the main tenets of the Copenhagen and Paris School approaches, they did not address every aspect. Prominent aspects not dealt with during the analysis included Copenhagen School concepts of *audience* and of *desecuritisation*.
- *Unit of Analysis*. The broad unit of analysis was the political and military sector of the US and Australia. The size and scale of each of these respective establishments, however, made it necessary to 'ring-fence' the bureaucratic structures and include some parts but exclude others. For example, the US military was mainly focused on the US Navy. Each case study defines the boundary in more detail.
- *Artefacts*. An 'artefact' was considered a document, policy, email, text, speech, media release, article or some other form of communication that an actor or institution has used to convey information to an audience. Artefacts examined during the course of the case studies constituted the most fundamental unit-of-analysis considered. Limitations on the selection of artefacts included: (1) Artefacts had to be publicly available, open-source and published by either the government or military institution; (2) Artefacts had to be unclassified or contain no classification label; and (3) Artefacts had to be published between January 2003 and June 2013.

4.4 Chapter Conclusion

The case studies in this book were undertaken with the aim of examining the process of climate securitisation within the Australian and US political-military sector. Three lines of inquiry were developed that addressed how climate change was framed, the urgency of the response, and what types of response measures were put in place to address climate change. Securitisation frameworks were developed by the author based on interpretations of securitisation theories from the Copenhagen and Paris Schools.

References

Bigo, Didier. 2006. "Security, Exception, Ban and Surveillance." In *Theorizing Surveillance: The panopticon and beyond*, edited by D. Lyon, pp. 46–68. Devon: Willan Publishing.

Bigo, Didier. 2008. "Globalized (in)security: The field and the ban-opticon." In *Terror, Insecurity and Liberty: Illiberal practices of liberal regimes after 9/11*, edited by Didier Bigo, Elspeth Guild and R. B. J. Walker, pp. 10–48. Abingdon: Routledge.

Bigo, Didier, and Anastassia Tsoukala. 2008. "Understanding (in)security." In *Terror, Insecurity and Liberty: Illiberal practices of liberal regimes after 9/11*, edited by Didier Bigo, Elspeth Guild and R. B. J. Walker, pp. 1–9. Abingdon: Routledge.

Bigo, Didier, Anastassia Tsoukala, Laurent Bonelli, Emmanuel-Pierra Guittet, and Christian Olsson. 2008. *Terror, Insecurity and Liberty: Illiberal practices of liberal regimes after 9/11*, edited by Didier Bigo, Elspeth Guild and R. B. J. Walker, *Routledge Studies in Liberty and Security*. Abingdon: Routledge.

Bourdieu, Pierre. 1969. "Intellectual field and creative project." *Social Science Information*, vol. 8, no. 2:89–119. doi:10.1177/053901846900800205.

Buzan, Barry, Ole Wæver, and Jaap de-Wilde. 1998. *Security: A New Framework for Analysis*. Boulder: Lynne Rienner Publishers.

Commonwealth of Australia. 2007a. "Budget Estimates." In *Standing Committee on Foreign Affairs, Defence and Trade Hansard*, 31 May 2007. Canberra: Commonwealth of Australia. Accessed on 13 February 2011, http://www.aph.gov.au/~/media/Estimates/Live/commttee/S10269.ashx.

Corry, Olaf. 2011. "Securitisation and 'Riskification': Second-order Security and the Politics of Climate Change." *Millennium: Journal of International Studies*, vol. 40, no. 2:235–258. doi:10.1177/0305829811419444.

Foucault, Michel. 1982. "Afterword: The Subject and Power." In *Michel Foucault: Beyond structuralism and hermeneutics*, edited by Hubert L. Dreyfus and Paul Rabinow, pp. 208–225. Chicago: University of Chicago Press.

Huysmans, Jef. 2004. "A Foucaultian view on spill-over: freedom and security in the EU." *Journal of International Relations and Development*, vol. 7, no.:294–318. doi:10.1057/palgrave.jird.1800018.

Huysmans, Jef. 2006. *The Politics of Insecurity: Fear, Migration and Asylum in the EU*. London: Routledge.

Lucke, Franziskus von, Thomas Diez, and Zehra Wellmann. 2013. "What's at Stake in Securitising Climate Change? Towards a Differentiated Approach." European Conference on International Relations, Warsaw, September 2013.

Oels, Angela. 2012. "From 'Securitization' of Climate Change to 'Climatization' of the Security Field: Comparing Three Theoretical Perspectives." In *Climate Change, Human Security and Violent Conflict: Challenges for Societal Stability*, edited by Jürgen Scheffran, Michael Brzoska, Hans Günter Brauch, Peter Michael Link and Janpeter Schilling, vol. 8, pp. 185–206. Berlin: Springer.

Rudd, Kevin M. 2008g. "The First National Security Statement to the Parliament Address by the Prime Minister of Australia The Hon. Kevin Rudd MP." In *PM Transcripts: Transcripts of the Prime Ministers of Australia*, 4 December 2008. Canberra: Department of Prime Minister and Cabinet. Accessed on 5 March 2013, http://pmtranscripts.pmc.gov.au/release/transcript-16289.

Chapter 5
Climate Security Case Study: Australia

Abstract This chapter analyses the response by the Australian political-military establishment to climate change through the lens of securitisation theory. Mixed content analysis techniques were applied to systematically examine more than 1,500 speech-acts, policies, doctrinal articles and other publications between 2003 and 2013. The key findings reveal that climate change was not a security issue in Australia until 2007. Prior to that, climate change was a marginal issue and was not conceived as a security issue by the centre-right government under John Howard. This changed in 2007 with the election of a centre-left government who elevated climate change as a foremost policy issue. Subsequently, climate change was framed as a security issue and it registered in major strategic Defence documents. However, the twin failures of Copenhagen and the Carbon Pollution Reduction Scheme (CPRS), saw climate change become highly politicised and then marginalised. This stalled wider securitisation efforts and curtailed ADF climate responses. The chapter concludes with an update on climate security initiatives within the ADF since 2013.

Keywords Climate security · Case study · Climate change · Australian military · Australian Defence Force

5.1 Introduction

This case study examines, through the prism of securitisation theory, how the Australian Defence Force (ADF) and selected areas of the Australian political establishment responded to climate change between 2003 and 2013. In total, 1,569 documents were examined that consisted mostly of speech-acts but also included media publications, journal articles, military doctrine and departmental policy. A final Sect. 5.4 has been included in this chapter to provide an update on climate security discourse in the Australian political-military sector over the period 2013–2017 (this section uses a general literature review only and does not apply any of the methods described in the previous chapter).

M.D. Thomas, *The Securitization of Climate Change: Australian and United States' Military Responses (2003–2013)*, The Anthropocene: Politik—Economics—Society—Science 10, DOI 10.1007/978-3-319-49658-0_5

The research boundary for this case study was grouped into three separate levels or programmes: Strategic, Operational and Tactical. Table 5.1 shows how the case study was structured and enumerates the artefacts/documents by type that were examined for each level. This approach identified any evidence of securitisation and highlighted divergence within and between the different levels of the political and military establishments and it also made a consistent framework for the US case study. The Strategic level for this case study included examination of documents from the Prime Minister, Minister for Defence (MINDEF), Chief of Defence Force (CDF) and Secretary for Defence (SECDEF). The Operational level included seven areas within the Australian Department of Defence including Defence Support Group, Chief Information Officer Group, Defence People Group, Defence Science and Technology Organisation, Vice Chief of Defence Force (VCDF), Capability

Table 5.1 Australian case study: sources, domains and numbers of documents examined

	Programme	Document type					Total
		SPR	DTN	S&A	MDA	JNP	
Strategic	Prime Minister	3	0	153	20	0	176
	National Security Advisor	1	0	4	0	0	5
	Minister of Defence	5	0	148	28	5	186
	CDF and Secretary	29	0	196	15	39	279
							(646)
Operational	Defence Support	5	0	8	1	0	14
	Chief Information Officer	2	1	1	0	0	4
	Defence People Group	0	1	0	0	24	25
	Defence Science and Technology	0	0	15	2	71	88
	Vice Chief of Defence Force	0	7	5	1	57	70
	Capability Development	5	1	0	1	0	7
	Defence Materiel Organisation	1	0	14	2	75	92
							(300)
Tactical	Navy	0	3	22	1	213	239
	Army	4	1	51	4	143	203
	Air Force	0	9	14	1	124	148
	Intelligence	1	0	1	0	0	2
	Joint Operations Command	0	12	17	0	2	31
							(623)
	Total	56	35	649	76	753	1,569

Strategic, *Operational* and *Tactical* represent the three different domains of the national security establishment. Each domain consisted of various *Programmes* from where each document originated. (Document Type: *SPR* Strategic Policy, *DTN* Doctrine, *S&A* Speeches and Announcements, *MDA* Media, *JNP* Military Journals, Newspapers and Other publications.) *Source* The author

Development and Defence Materiel Organisation.[1] The final level considered by the case study was the service arms of Army, Navy and Air Force as well as examination of documents from Intelligence and Joint Operations Command.

This case study is presented in two main sections. Section 5.2 presents results and discussion of the Australian Strategic Programmes while Sect. 5.3 presents results and discussion of the Operational and Tactical level Programmes. The format for each section is a combination of results derived from the coding process combined with an initial discussion and interpretation of the findings (a more complete synthesis is undertaken in Chap. 7). The results and discussion are also supplemented throughout by a description of key events, quotes and actions (e.g., policies) that attempt to provide a richer, more accurate and contextualised account of climate securitisation.

5.2 Climate Security in Australian Political and Military Strategic Programmes

5.2.1 Introduction

To better understand the climate securitisation actions of the military, it was important to examine how Australia's political leaders responded to climate change. For the purpose of ensuring the research remained within an achievable scope, a decision was taken which narrowed examination of speeches in the political realm to the offices of the Prime Minister and Minister for Defence. The military-strategic areas considered by this case study in the Strategic Programme were Chief of Defence Force (CDF), Departmental Secretary (Defence) and (a very limited number) from the National Security Advisor. Total artefacts considered for the Australian Strategic level was 646 (see Table 5.1).

The strategic level—commonly referred throughout as *Strategic Programmes*—represented the *ends* through which national strategy was implemented via *ways* (operational level) and *means* (tactical level). Collectively, each of the areas within Strategic Programmes were considered instrumental in the formulation, development, direction and articulation of Australian strategic and defence policy across the period 2003–2013. *Ipso facto*, these programmes formed the basis of Australian strategic policy and their rhetoric and actions in relation to climate change were considered crucial in setting Australia's climate security agenda.

A word search for *climate change* and like terms within all 646 artefacts in the Strategic Programmes returned 165 separate sources containing 945 direct references. Of these 945 references 76 were coded in a securitised frame while 285 were

[1]Capability Development and the Defence Materiel Organisation are now a combined division called Capability Acquisition and Sustainment Group (CASG).

Table 5.2 Climate change framing in Australian Strategic Programmes, 2003–2013

Frame	Strategic programme				
	PM	NSA	MINDEF	CDF	Total
Securitised	37	1	27	11	76
Non securitised	246	6	14	19	285
Total	283	7	41	30	361

Source The author

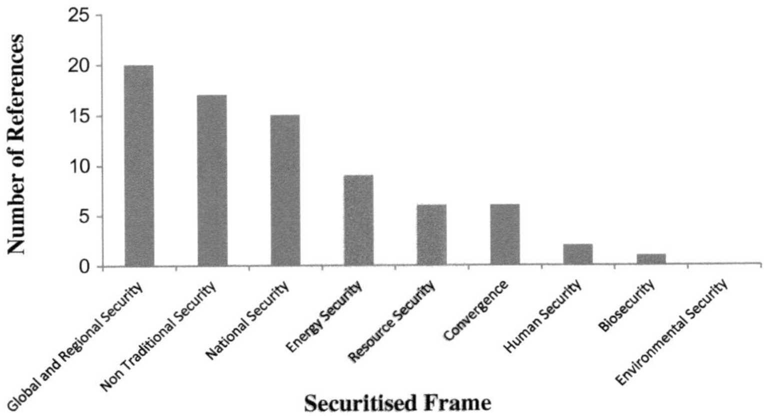

Fig. 5.1 Contextual framing of climate security, strategic level 2003–2013 (AU). *Source* The author

coded in a non-securitised frame (Table 5.2).[2] The following passages provide an overview of the strategic level (results are aggregated to discern any large scale trends).

Contextual Framing. The most common securitised categories are shown at Fig. 5.1. This illustrates that when climate change was framed as a security issue within the Strategic Programmes, it was predominately done so as: (1) a global and regional security issue; (2) as a non-traditional security issue; and (3) as a national security issue. Human security and environmental security were not major frames/referent objects. In terms of non-securitised context, climate change was mainly framed as: (1) a global issue; (2) an economic issue; and (3) an energy/resource issue (illustration not shown).

Table 5.3 shows that when climate change was framed as a security issue it was predominately done so with forceful language (*challenge* and *threat* were coded 43 times against a *neutral* context which was coded 33 times). Significantly, climate

[2]Non-securitised references consist of 'Other Context' and 'Defence Context'. A total of 584 references were not coded as they were considered not relevant or were repetitious.

Table 5.3 Language used to frame climate security issues, strategic levels (AU)

Frame	Strategic programme				
	PM	NSA	MINDEF	CDF	Total
Opportunity	0	0	0	0	0
Neutral	16	0	12	5	33
Challenge	15	1	13	4	33
Threat	6	0	2	2	10
Total	37	1	27	11	76

Source The author

change was never found by this research to be framed as an *opportunity* within a securitised context. This might suggest that there was a lack of consideration of the 'upside' risk, or of the strategic opportunities climate change may yield. In a non-securitised context, climate change was slightly framed as a heightened threat (*challenge* and *threat* were coded 145 times against *neutral* and *opportunity*, coded 140 times). Climate change was framed as an *opportunity* on four occasions in a non-securitised context.

These results, collectively describing the outcomes of how the strategic levels framed climate change, are useful in gauging broader trends but are limited in discerning the finer points of the securitisation discourse. This was evident particularly in relation to the differences that emerged through time and across different actors. In order to gain a better understanding, the following sections describe the results for the individual Strategic Programmes.

5.2.2 Prime Minister(s)

The Prime Minister Programme analysed 176 artefacts, and found 87 separate sources containing 657 references to *climate change*. Table 5.2 shows that for the Prime Ministerial Programme, 37 references were coded in a securitised context and 246 were coded in a non-securitised context. These results indicated that climate change was collectively framed as a non-securitised issue by Prime Ministers John Howard, Kevin Rudd and Julia Gillard.

However, further analysis of the Prime Ministerial Programme revealed large differences in approaches to climate change and climate security between the centre-right and centre-left of Australian politics—specifically, between centre-right prime ministership of John Howard (1996–2007) of the Liberal Party of Australia and against the centre-left prime ministerships of Kevin Rudd (2007–2010) and Julia Gillard (2010–2013) of the Australian Labor Party. The following sections examine this in further detail. The empirical results are supplemented by detailed quotes that serve to build a securitisation narrative and provide greater understanding, context and robustness.

5.2.2.1 Prime Minister John Howard

I retain some degree of scepticism about … climate change.

Howard (2006d)

In order to understand how Prime Minister John Howard approached climate change as a security issue, it is critical to understand the broader context within which he framed climate change. To do this, the following section starts by examining the non-securitised aspects of climate change. Indeed, this research found that Prime Minister John Howard had primarily framed climate change as a (non-securitised) economic and energy-related issue (Table 5.4). Howard's early rhetoric was decisively more neutral than Rudd and Gillard (Table 5.5) and his self-declared 'balanced approach' to climate change was characterised by long-term strategies that would not undermine Australia's economic competitiveness. Though he accepted he was often cast as a climate sceptic, Howard's (2006a) central ambivalence related to 'the intensity and pace of damage being done by climate change'. For Howard, the *rate* of climate change meant 'long-term climate change' (Howard 2007a). On this basis, Howard (2006a) defaulted to his conservative instincts in which it was 'important to keep the challenge of climate change in perspective' and respond with 'prudent risk management' (Howard 2007h).

Table 5.4 Framing of climate issues (non-securitised), Prime Ministers 2003–2013

Non-securitised frame	Howard	Rudd/Gillard
General economic reference	21	30
Agriculture, farming, water, fish	0	9
Forestry	4	5
Energy and resources	20	15
Environmental, sustainability, development	0	11
Government and regulatory	10	27
International, global, bi-lateral	7	68
General reference	4	15
Total	66	180

Source The author

Table 5.5 Language used to frame climate issues (non-securitised), Prime Ministers 2003–2013

Frame	Howard	Rudd/Gillard
Opportunity	2	2
Neutral	40	71
Challenge	22	92
Threat	2	15
Total	66	180

Source The author

These prevailing sentiments were captured in a 2006 speech Howard gave to the Queensland Division of the Liberal Party:

> The point I want to make to you this morning is that [we] must bring a sense of balance to this debate going on in the community about global warming and climate change. We need a sense of balance in relation to this debate. The climate is changing, we do need to reduce greenhouse gas emissions … but we need to avoid the danger of going overboard in the other direction. *We must reject panicky, knee jerk reactions that run the risk of hurting our prosperity* (Howard 2006c; emphasis added).

Although Howard spoke of a 'balanced', 'measured', 'commonsense', 'sensible' and 'practical' approach, it is clear that he would not pursue emission reductions at the risk of jeopardising Australian industry (or the Australian economy more broadly). As such, the 'balance' appeared inarguably in favour of maintaining, exploiting and even expanding, Australia's use of fossil fuels. Status quo interpretations of Australia's national interest was therefore forefront in Howard's calculations:

> [W]e must remember the great contribution that gas and coal and other fossil fuels are making to the prosperity of this country. One of the reasons Australia is doing so well at the moment is that Providence has given us these vast natural resources … they [fossil fuels] are fuelling the huge economic growth, the historic expansion of China … Japan and Korea (Howard 2006c).

More directly:

> Clearly a country such as Australia, with its extraordinary supplies of fossil fuels, being the largest coal exporter in [the] world and having large reservoirs of LNG and other energy, clearly Australia is a country that is not going to lightly walk away from the *sensible exploitation of fossil fuels* (Howard 2006f; emphasis added).

Again, in 2007, Howard argued that 'I am not going to embrace an approach to climate change that damages our great resource industries' (Howard 2007f). A 2007 media release by John Howard summed up his approach:

> The Government's priority is to tackle climate change without damaging Australian jobs and living standards (Howard 2007g).

At the heart of this lay a desire to leverage Australia's comparative advantages with a focus on immediate economic growth and full employment:

> [I]t would be a colossal act of self-injury for this country to be panicked into changes in response to climate change that unfairly disadvantaged our great resource industries, robbed those industries of investment and trade opportunities and resulted in the unemployment of many thousands of Australians (Howard 2006c).

Thus, John Howard styled himself as a 'climate change realist' (Howard 2007h) which, for him, meant avoiding 'knee-jerk reactions', the temptation to be 'mesmerised by individual reports' (Howard 2006c) and rather 'looking at the evidence as it emerges and responding with policies that preserve Australia's competitiveness and play to her strengths' (Howard 2007a).

Central to this outlook was a suspicion of UN and European efforts to reduce emissions via the Kyoto Protocol. In this respect, Howard chided the Kyoto Protocol as 'ideological posturing' (Howard 2007e) that set 'arbitrary targets' (Howard 2006e) which 'gave the impression that all you've got to do is put a signature on a bit of paper, and hey presto, the world stops getting warm' (Howard 2006b). Though he often cited the exclusion of major emitting nations the US, China and India as being a major flaw ['we will not ratify something that does not encompass the world's largest emitters' (Howard 2004)], he ultimately framed non-ratification of Kyoto on economic rationalist grounds:

> [T]he whole debate surrounding the Kyoto protocol has been driven out of Europe rather than out of countries whose economic circumstances are similar to our own and we have to be very careful to keep those things in mind ... we must make sure that the adjustment that inevitably will have to be undertaken does not occur in a way that puts us at a competitive disadvantage to the rest of the world.
>
> We must be certain that any additional costs borne by our resource industries are also borne by competitor resource industries in other countries, otherwise we are going to lose the competitive advantage that we now have and that is the reason why we never ratified the existing Kyoto protocol (Howard 2006c).

The totality of Howard's approach was then brusquely summed up:

> We've made it very clear that we won't ratify the Kyoto agreement – we took that decision some years ago because we feared that ratifying that agreement in the form in which it then and still largely existed could have damaged the comparative advantage of this country enjoyed as a result of fossil fuels ... nothing has changed to alter that fact (Howard 2006a).

These views were previously outlined in the Howard Government's 2004 Energy White Paper, *Securing Australia's Energy Future*. This policy paper placed emphasis on ensuring development of Australia's energy resources, reforming domestic energy markets, improving energy security and energy efficiency while investing (on a proportionally weaker basis) in renewable and clean energy. Although the policy noted climate change as a 'major issue' with the potential to 'raise global temperatures, resulting in deleterious effects to people and the natural world', it also emphasised that Australian greenhouse gas emissions were only 1.6% of world emissions and therefore 'too small ... to make a difference on its own' (Commonwealth of Australia 2004: 24).[3]

---

[3]*Energy Future* proceeded to fund the balance of 'new initiatives' on promoting the use of fossil fuels. The most significant of these was a $1.5 billion (AUD) fuel excise reform measure that exempted all fuels used for off-road for all business purposes as excise-free between 1 July 2006 and 1 July 2015. Other measures promoting Howard's 'sensible exploitation of fossil fuels' approach included incentives for petroleum exploration in frontier offshore areas and a $500 million low emission technology fund that was ultimately directed to two carbon sequestration projects (Chevron Pty Ltd., $60 million; CS Energy Ltd., $50 million), a clean coal demonstration project (HRL Ltd., $100 million) and a large-scale solar concentrator project (Solar Systems, $75 million) (Department of Industry 2011).

Securing Australia's Energy Future was also indicative of the broader Howard government perspective that climate change was a long-term issue that 'requir[ed] action over a long period, well beyond the normal commercial and political horizons' (Commonwealth of Australia 2004: 140). On Kyoto, *Energy Future* reiterated Howard's view that ratification was 'not in the national interest' since it 'did not include all the largest emitters in the world, nor does it include a pathway for addressing developing countries whose emissions will soon overtake those of industrialised countries' (Commonwealth of Australia 2004: 24).

Emphatically, Howard (2013) rejected a call to adopt a mandatory target of 20% of electricity to be sourced from renewables by 2020. Though some measures were taken to support renewable energy initiatives, Howard viewed them as making a 'valuable contribution' but one that 'would never take the place of fossil fuels' (Howard 2006d). Looking toward technological solutions, Howard preferred addressing climate change through development of clean coal technology and via establishment of an Australian nuclear power industry. Put simply, Howard backed technologies that complemented Australia's natural resource advantages, particularly of coal and uranium. Thus, Howard (2006a) argued to 'accelerate the development of clean coal technologies' while articulating the case for nuclear: 'If Australia is serious about addressing climate change it must consider adding nuclear power to its energy mix' (Howard 2007g). He further argued:

> Now we're all against nuclear weapons but to be against nuclear power in 2007 is about as out of date and old in your ideas as one can imagine because if we are ever to fix this problem of climate change, the challenge of it, we need to have all the options open to us (Howard 2007d).

Linking the requirement to maximise Australia's comparative advantage, Howard also noted:

> I believe that Australian attitudes towards nuclear power are changing. Nuclear power is potentially the cleanest and the greenest of them all and we would be foolish from a national interest point of view with our vast reserves of uranium to say that we are not going to consider nuclear power (Howard 2006c).

Surprisingly, despite raising the prospect of an Australian nuclear industry to address climate change, no Howard Government speeches or policies were identified by this research that addressed the potential security or geo-strategic aspects involved in such a move.[4]

The overall picture therefore—until at least 2006—of Howard's approach to climate change was characterised by a self-proclaimed balanced approach but that in substance and rhetoric favoured immediately the fossil fuel industries, advancement of clean coal and nuclear technology options and a rejection of international agreements that may have threatened Australia's comparative

[4]This point may not be 'surprising' if it were the intent of Prime Minister Howard and the nuclear lobby to deliberately disassociate nuclear energy from its security implications (see Han, C. C. 2014. 'Demarketing fear: Bring the nuclear issue back to rational discourse.' *Energy Policy*, 64:183–192).

advantage. Pervasive throughout this was the belief that any action on climate change must not jeopardise Australia's economy and, above all else, any response must be in the national interest.

This approach changed somewhat from 2006 but particularly across 2007 when Howard announced a number of policies under the broad remit of responding to climate change. The expanded measures to address climate change indicated a general trend that portrayed Howard as viewing climate change as a much more prominent and, on occasion, urgent issue.[5] Thus, although Howard (2006d) 'retained some degree of scepticism … about climate change', a prime ministerial press release in 2007 now heralded that his government had spent $3.4 billion on the 'climate change challenge' since 1996 (Howard 2007c). He even championed a domestic emissions trading scheme (albeit one that 'protected the export sector' and would 'not have to sacrifice economic prosperity to tackle the problem') as well as various forms of government assistance including a $336 million green voucher investment to help Australian schools install solar hot-water systems, rainwater tanks and 'improve energy and water efficiency' (Howard 2007c). In the international arena, Howard now raised climate change as a 'priority' on the agenda of the 2007 Asia-Pacific Economic Cooperation forum (Australian Bureau of Statistics 2008: 167).

All this marked a significant, even contradictory, change from his previously held positions. For example, while he now set out to promote an emissions trading scheme, he had once derided: '[I]f Australia is forced to impose a discriminatory impost on carbon emissions … the global environment would be no better—indeed it would be worse—and Australia will have lost out' (Howard 2004). By July 2007 Howard (2007h) even began referring to 'the *threat* of climate change' (emphasis added). His political reframing, therefore, was also accompanied by a marked escalation of rhetoric. In a September 2007 press release, Howard (2007b) again noted the 'threat of climate change' and sought a 'new international consensus' on global emissions via the 'Sydney Declaration'.[6] Not to be overstated however, these were the *only* occasions identified by this research when Howard framed climate change as a threat. Notably, however, this research found that Howard never once framed climate change explicitly as a *security* threat (Table 5.7).

Many of these initiatives, however, simply reflected his original views as they also clung tightly to Howard's unshakeable principle of not ceding economic

[5]These included a domestic 'cap and trade' emissions scheme, the development of an Australian nuclear energy industry as well as numerous climate change partnerships and initiatives (e.g., the Asia-Pacific Partnership on Clean Development and Climate, Global Initiatives on Forests and Climate, Clean Coal Partnership with China and a $336 million initiative in which every school in Australia would be eligible for 'green vouchers' of up to $50,000 to help install solar hot water systems and rainwater tanks). Howard's revised outlook on climate change culminated in the 2007 APEC Summit 'Sydney Declaration' in which he arranged a UNFCCC 'umbrella group' to help establish a new international framework 'beyond Kyoto'.

[6]Although in his *speech* on the APEC declaration, Howard labelled climate change a 'great challenge'. Thus, in spoken form, this research found Howard only framed climate change as a 'threat' on one occasion (during his weekly radio address in July 2007).

advantage. Very likely they also represented a political response to negate the challenge presented by the ascendancy of Labor's Kevin Rudd who, in Howard's view, was 'happy to risk damage to the economy in the pursuit of Green voting preferences' (Howard 2007h). Reflecting on the issue in late 2013, Howard added his government was hit by a 'perfect storm' of events that included long-term drought, release of the *Stern Review*, Al Gore's movie *An Inconvenient Truth*, and a prevailing sense amongst Australians that 'we could afford it' (Howard 2013). There was also a healthy degree of political realism. For Howard, 'the atmosphere —political that is—was certainly conducive' to increasingly bold responses symptomatic of his political opposition (Howard 2013).

Howard's shift towards a more expanded climate policy response in 2007, however, did not alter his view that climate change remained a long-term issue requiring long-term responses. In total, this research identified seven such references that supported this outlook (Table 5.8). The corollary to this was that Howard *never* framed climate change as an urgent or immediate issue. As will be shown, this stood in clear contrast to his successor. Even by late 2007, after Howard had committed vast policy attention to the issue, he only considered climate change *and* energy security as 'priority' issues. Rhetorically, this was as close as Howard came to upgrading the policy urgency of his climate change response that had otherwise 'wallowed at the edges of mainstream public debate' during his term in office (Macintosh et al. 2010: 200).

The conviction that climate change represented a long-term problem requiring long-term responses probably contributed to Howard never mentioning climate change as a major security issue. This was reflected in the fact that climate change was *not referenced at all* (either as a security issue, risk, challenge or threat) in any of the major national strategic defence policy documents published during Howard's tenure as Prime Minister. Of the three major strategic guidance documents examined during this period (*A Defence Update 2003*, *A Defence Update 2005* and *A Defence Update 2007*) all placed and retained prominence on the War on Terrorism accompanying military commitments to Afghanistan and Iraq and proliferation of Weapons of Mass Destruction (WMD). These themes dominated Howard's security agenda throughout his prime ministership and were central to speeches he made relating to Australia's national security.

Although it should not be overstated, the one notable exception to this was the conflation of climate change as an energy security concern. Building on ideas established in the 2004 Energy White Paper, Howard declared climate change and energy security as flipsides of the same coin in which:

> [N]o country that is seeking to expand and lift its living standards is going to forsake the availability of cheap resources and cheap supplies of energy. And equally no country is going to imperil its energy security as part of the process of reducing the negative effects of climate change.

Table 5.6 Framing of climate as a security issue, Prime Ministers 2003–2013

Securitised frame	Howard	Rudd/Gillard
Global and regional security	0	3
National security	0	9
Human security	0	1
Energy security	9	0
Environmental security	0	0
Biosecurity	0	1
Convergence	0	2
Resource security	0	2
Non-traditional, new security	0	10
Total	9	28

Source The author

Continuing on in the same speech, he asserted:

> I do want to say something about the related issues of climate change and energy security. And I deliberately link the two of them because you can't think of the reaction of relevant countries to climate change without understanding the importance to them of energy security … put bluntly, *there is no way that a country is going to embrace climate change measures or responses to the growth of greenhouse gas emissions, which in anyway imperil the energy security of that country* (Howard 2006d; emphasis added).

These sentiments linked strongly to his position on protecting Australia's comparative economic advantage in the resource sector and general unwillingness to impose costs on the Australian populace or promote measures that aroused energy security concerns.

Beyond this, however, climate change arguably failed to register as a security concern. Compared with his successor, Howard undoubtedly possessed a narrow interpretation of how climate change would affect global, regional, domestic and human security. Table 5.6, for instance, shows that of the nine coded securitised references found for Howard during this research, all were allocated under concepts of energy security. In contrast, of the twenty-eight securitised references identified under Rudd-Gillard, climate change was coded across *seven* distinct securitised categories that included *global security*, *national security*, *human security*, *biosecurity*, *convergence security* (defined by a multiplicity of cross-cutting issues), *resource security* and as a *new-security* concern.

5.2.2.2 Prime Ministers Kevin Rudd and Julia Gillard

The consequences for Australia of failing to act … on climate change are severe.

Rudd (2009)

No nation can accept such threats [from climate change]. We must act. And we will act.

Gillard (2011)

Having examined the speeches of John Howard, the following analysis shifts to examining the prime ministerships of Kevin Rudd and Julia Gillard. This research examined documents (predominately speeches) from the prime ministership of Kevin Rudd from 3 December 2007 and into the prime ministership of Julia Gillard, until April 2013. Although the following passages refer to this period as the Rudd-Gillard era, where required, distinctions are drawn between each Labor Prime Minister.

Tables 5.4 and 5.6 show that Prime Minister Kevin Rudd (and to a lesser extent Julia Gillard) predominately framed climate change as a non-securitised issue (180 non-securitised references against 28 securitised). In a non-securitised context, this research found Rudd-Gillard framed climate change primarily as a global issue, an economic issue and a domestic regulatory issue.

In one sense, this represented a similarity with Howard insomuch as climate change was most often addressed in a non-securitised setting. In this instance, similar themes such as the economy, regulation and the prospect of an international agreement dominated the political narrative. Both politicians also spoke of opportunities relating to climate change. Howard spoke of 'maximising opportunities from our response to climate change' that included development of low emissions technologies (somewhat paradoxically, this was particularly applied to the coal sector) and as a 'regional carbon trading hub' (Howard 2007h). Rudd-Gillard also regularly stressed the opportunities, particularly in relation to developing a clean energy economy: 'I am doing this because I see a great opportunity we can seize. I see a great clean energy future for our great country' (Gillard 2011b).

Beyond this, however, the differences were stark. Howard opposed Kyoto; Rudd and Gillard supported it. While Howard was arguably out-manoeuvred (politically) into supporting an emissions trading scheme, Rudd initially embraced it and a version was legislated by Julia Gillard.[7] Howard did not view climate change as an urgent issue; Rudd and Gillard spoke frequently as though it was an immediate issue requiring swift, resolute action. To bolster their cause, Rudd and Gillard constantly described climate change as a social, moral and economic 'threat' in which Australia was 'on the front line to suffer' (Rudd 2008f). Howard—except on the two occasions identified above—never used such strong language. Whereas Howard was wary of 'individual reports', Rudd and Gillard regularly evoked 'the overwhelming global scientific evidence' (Rudd 2009c) amassed by the world's various institutions and academies. Rudd and Gillard also cast themselves as

[7]Prime Minister Howard was arguably out-manoeuvred into supporting an emissions trading scheme in 2007 in response to the rise of Kevin Rudd and his platform of acting on climate change. Howard's language until this moment had indicated that he would not introduce legislation that threatened Australia's resource industries. In his memoirs, Howard noted how he had perceived the public considered the Labor Party as 'more committed to taking decisive action against climate change than was the Coalition'. Thus, Howard 'concluded that the Government would need to shift its position on climate change' and, following a report by a task-force, he announced at a meeting of the federal council of the Liberal Party (3 June 2007) the framework for an Emissions Trading Scheme. This was six months prior to the election date. See Howard (2010: 551 and 552).

champions of renewable energy, expanding on Howard policies and implementing a raft of new initiatives (e.g., an expanded renewable energy target, a clean energy initiative and energy efficiency measures).[8] Whereas Howard was overt about offering protection to Australian industry against international action that might threaten its resource advantage, Rudd countered:

> If we allow our actions to be dictated by what we falsely conclude to be in our narrow self-interest, then we harm not just others but ourselves as well because climate change inaction harms us as well (Rudd 2009c).

For Rudd, 'the economic cost of inaction on climate change is far greater than the cost of action' and 'failure to engage with the global community on climate change would exclude us from the chance to shape the global response in ways consistent with our national interest' (Rudd 2008a).[9] This aligned with Rudd's broader strategic idea of Australia as an activist Middle Power in which he believed 'those that share the benefits of these systems must also share the responsibilities of supporting and enhancing them' (Rudd 2008g). Lastly, Rudd and Gillard regularly framed climate change as a threat to security. Apart from associating it under the broad umbrella of energy security, Howard never framed climate change in these terms.

Before examining the security dimension in more detail, it is worth examining the rhetoric deployed by Rudd-Gillard on climate change more broadly. This is important, since it demonstrated how Rudd-Gillard framed climate change in such a manner that then made it politically plausible to present it as a securitised issue to the Australian public. This is significant, since it underlines the *politically constructed nature* of climate change as a security threat. This conclusion tends to debase Paris School ideas which may portray the military as deliberately legitimising (climate) threats for their own advantage (that is, it was the *government* itself who used securitisation to pursue political advantage). As will be shown later in this chapter, in the Australian context, climate change was framed as a security threat first by the politicians and second by the 'generals'.

On Kyoto, whereas Howard (2006a) viewed it as a 'bit of paper' fashioned to 'accommodate the environmental goals and position of European countries', Kevin Rudd viewed it as a necessary commitment to promoting a rules-based international order. Noting these differences, Rudd signed the Kyoto Protocol 'within minutes of

---

[8]Examples included expanded Renewable Energy Target (20% by 2020), the Clean Energy Initiative (including the Solar Flagships programme to generate 1000 MW of solar electricity), and the Energy Efficient Homes package to install ceiling insulation in up to 2.9 million homes. See archived PM website at NLA at: http://pandora.nla.gov.au/pan/79983/20100322-1756/www.pm.gov.au/Policy_Priorities/Future/Priorities.html#Climate (20 May 2017).

[9]This is not to suggest that Rudd did not place the economy at a premium or that he was uninterested in maintaining Australia's natural competitive advantages. Speaking to the US Chamber of Commerce, Rudd declared he was 'committed to responsible economic management and is unashamedly pro-market, pro-business and pro-globalisation'. In arguing for a shift to a low carbon economy, Rudd was also adamant this must done in such a way that 'maintains our competitiveness'.

taking office' (Rudd 2008c) and made that the 'first official act of the Government' (Rudd 2008f).

The urgency to act on climate change defined Labor's political narrative and was constantly referenced in the speeches of Rudd and Gillard. This was born out in the research results in Table 5.8 where Rudd and Gillard framed climate change as an urgent issue on *twenty-seven* separate occasions. To create a sense of urgency, Rudd and Gillard portrayed Australia as being particularly vulnerable to climate change, frequently enlisting droughts, cyclones, bush-fires and extreme weather events as part of the causal chain of events. This aspect contrasted with Howard, who rather framed these events as enduring features of the Australian landscape akin to Dorothea Mackellar's 1908 poem *My Country*, 'Of droughts and flooding rains' (Mackellar 1908).

Another feature of the Rudd and Gillard rhetoric was to portray climate change as a vast global ('civilizational') threat of unprecedented proportions that demanded an historic response. The following series of quotes made to different audiences capture this narrative:

- "There is no greater challenge now facing our world and our nations than dangerous climate change. Australia is on the front line to suffer the impacts of climate change", Prime Minister's website (Rudd 2008f).
- "Climate change is the greatest moral, economic and social challenge of our time", Address to Kyoto University (Rudd 2008d).
- "Globally, the change that we must achieve over coming decades in response to the climate threat is as great in scale as the Industrial Revolution", Address to US Chamber of Commerce (Rudd 2009e). And,
- "We can no longer afford to wait for action on climate change; the time for action is now", Doorstop at the United Nations New York (Rudd 2009g).

On the destructive vastness of climate change, Rudd envisioned Australia's fate:

> As one of the hottest and driest continents on earth, Australia's environment and economy will be amongst the hardest and fastest hit by climate change if we do not act now.
>
> The scientific evidence from the CSIRO and other expert bodies have outlined the implications for Australia … temperatures rising by around 5 degrees … 40 per cent more drought months … [90 per cent] fall in irrigated agricultural production … storm surges and rising sea levels – putting at risk over 700,000 homes and businesses … the risk of land being inundated or eroded by rising sea-levels range from $50 billion to $150 billion … our GDP dropping by two and a half per cent (Rudd 2009c).

For Rudd, such a scenario would destroy 'jobs, our houses, our farms, our reefs, our economy and our future' (ibid). Such catastrophist views were never expressed by Howard.

As the political time frame tightened on an international agreement at the Copenhagen Summit in 2009, Rudd spoke of a 'Grand Bargain' between developed and developing nations that reconciled the 'three great challenges' of binding emissions targets, financing arrangements and diffusion of low-emission technologies (Rudd 2009a). Largely, his domestic success depended on safe passage of the proposed Carbon Pollution Reduction Scheme (CPRS) legislation which, in and of itself, was partly dependent on a successful international outcome at Copenhagen.[10] Thus, as both deadlines loomed, Rudd ratcheted up the rhetoric:

> The latest scientific research on climate change confirms our worst fears. Climate change is happening faster than we previously thought, creating a more *serious threat* to our economy, our environment and to future generations (Rudd 2009d; emphasis added).

Becoming more desperate, and recognising the 'danger [of] collapsing political momentum towards national and global action on climate change', Rudd (2009c) turned on 'sceptics' and 'political conservatives' who challenged the science and derailed action on his proposed climate policies. In a withering attack some twenty days before the centre-right-dominated Australian Senate was to vote on CPRS passage, Rudd raged into full invective during a six-thousand-word speech at the Lowy Institute:

> This is a profoundly important time for our nation, for our world and for our planet. In Australia, we *must* pass our Carbon Pollution Reduction Scheme … The truth is this is hard, because climate change sceptics, the climate change deniers, the opponents of climate change action are active in every country. They are a minority. They are powerful. And invariably driven by vested interests. Powerful enough to so far block domestic legislation in Australia … to threaten a deal on global climate change … holding the world to ransom, provoking fear campaigns in every country (Rudd 2009c).

In the same speech, Rudd argued that the failure to act would have 'severe consequences', he upped the rhetoric accordingly and in doing so exposed the political imperative:

> It is time to be totally blunt about the agenda of the climate change sceptics in all their colours … It is to destroy the CPRS at home, and it is to destroy agreed global action on climate change abroad, and our children's fate – and our grandchildren's fate – will lie entirely with them … The stakes are that high …
>
> The clock is ticking for the planet, but the climate sceptics simply do not care. The vested interests at work are simply too great …
>
> For people who claim to hold the conservative torch, their scepticism is in fact radical in its riskiness and recklessness. By deliberately undermining and eroding the capacity to achieve both domestic and international action on climate change the sceptics are attempting to force the world to take the most reckless bet in our long history (Rudd 2009c).

Similar sentiments are recorded in the rhetoric of Julia Gillard. Following the failure by Rudd to pass the CPRS, and preparing to pass a new Clean Energy initiative, Gillard argued:

> No nation can accept such threats [from climate change]. We must act. And we will act. The time for words ended yesterday. The time for deeds begins today (Gillard 2011a).

---

[10]The CPRS was first voted down by the Senate on 13 August 2009. Rudd sought a number of amendments to the CPRS Bill to garner Coalition Party support prior to seeking reintroducing it again in 2010. For a comprehensive overview of the CPRS policy development, see Macintosh et al. (2010).

Table 5.7 Language used to frame climate security issues, Prime Ministers 2003–2013

Frame	Howard	Rudd/Gillard
Opportunity	0	0
Neutral	6	10
Challenge	3	12
Threat	0	6
Total	9	28

Source The author

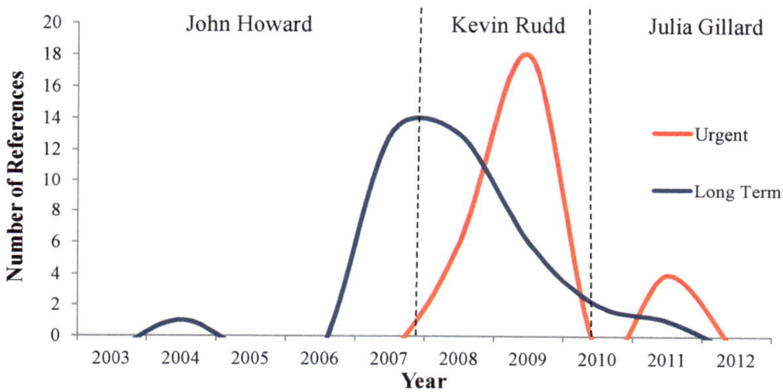

Fig. 5.2 Changing temporal frames of three Prime Ministers to climate change. During the height of Rudd's climate change agenda (2009), he increasingly framed climate change as an urgent issue (see '*red line*' for urgent frame), though not in a security sense. *Blue line* is climate change framed as a long-term issue. Data lines have been smoothed

The intensification of language used by Rudd and Gillard to frame climate change is shown by research results at Tables 5.5 and 5.7.[11] This shows that when Rudd-Gillard framed climate change as a non-securitised issue, they invariably framed it as a *challenge* or *threat* on more than a hundred separate occasions across five years. This contrasted sharply with John Howard, who framed climate change as a non-securitised *challenge* on just twenty-two occasions and *threat* just twice over an equivalent time period. Importantly, whereas Howard *never* framed climate change as a security threat, Rudd and Gillard did so with some regularity. In addition to this, as the Copenhagen conference drew nearer, Rudd increasingly framed climate change as an urgent issue, rather than as a long-term issue (Fig. 5.2).

[11]The majority of this analysis was coded by references attributed to Kevin Rudd rather than Julia Gillard.

Moreover, the rhetoric deployed by Rudd and Gillard revealed much about the political spitefulness that had come to characterise the debate in Australia during this period. As noted by the Australian Broadcasting Corporation's lead political programme, *7.30*:

> [Climate change] has long since dissolved into a toxic political debate. It was a key factor in Malcolm Turnbull losing the Liberal leadership, in Mr Rudd being [elected then] dumped as Prime Minister and replaced by Julia Gillard in 2010 and then in Ms Gillard's removal in June (Sales 2013).

This 'toxic' political aspect cannot be underestimated, particularly in its effect on the Australian military and broader national security establishment. Moreover, the Australian military has long positioned itself as an *apolitical* institution that avoids political debate.[12] Thus, climate change as a security issue *and* as a political issue presented a conundrum for the Australian military since any articulation on the subject was highly likely to provoke a heavy response from a divided body politic. To maintain its non-partisan (apolitical) position there is some sense that the ADF avoided discussion on the matter and did nothing more than was immediately required by the government of the day. (These points are further taken up in the discussion in Chap. 7.)

Having outlined the non-securitised dimension of climate change of the Rudd-Gillard era, attention now shifts to examining its securitisation during this period. In short, the significant political capital invested by Rudd and Gillard to promote their climate change agenda was accompanied by equal determination to frame climate change as a strategic security threat. With the exception of climate change having some connection to energy security, this research found that John Howard never invoked climate change as a security issue (refer to Table 5.7).

This research coded 28 securitisation references in the speeches of Rudd-Gillard, with a preference to frame it as a new and emerging security issue or as one concerning national security. Most references to climate security were made by Kevin Rudd in 2008 and 2009. As previously noted, Rudd and Gillard cast a wide net, offering up seven distinct security categories in which climate change would impact: *national security*, *global security*, *biosecurity*, *human security*, *resource security*, *non-traditional* security and *convergence security* (multiple/overlaid security concerns resulting from climate change in the one coded reference). Not surprisingly, Rudd and Gillard tended to use evocative language by framing climate change as a security *challenge* or security *threat* (which was used on 18 occasions against 10 neutral references, see Table 5.7).

Rudd's broad approach to framing climate change as a security issue was evident as early as April 2008. In a speech to the London School of Economics, Rudd stated he had met with the Prime Minister of Great Britain (Gordon Brown) and discussed the security implications of climate change. Expanding on this, Rudd argued:

---

[12]See the discussion below by the Deputy Secretary Strategy Michael Pezzullo during the Senate Estimates hearing in 2008, who noted Defence 'steer[s] well clear' of political debate on climate change.

> Climate change will aggravate existing strains and create new tensions within and between states as the supply of natural resources, including food and water, becomes increasingly unpredictable. *Climate change is not just an environmental, economic and moral challenge. It is also a security challenge* (Rudd 2008b; emphasis added).

Seeking to cast his political opposition as fixated on narrow security agendas (e.g., marked by Howard's fixation on terrorism), Rudd regularly spoke on broadening the national security discourse to include climate change. In the same speech cited above, Rudd made the case:

> A third major international change is the evolution of the concept of 'national security'. Clearly the traditional concept of the term endures … But the concept has to be broadened to include … non-state actors … food security, water security and energy security as well as security from threat of health pandemics.
>
> Climate change will lead to changed rainfall patterns and, therefore, to changes in agricultural production that will have an impact on food security. Severe weather events will occur more frequently in some regions, making it more important that nations are ready to respond to natural disasters (Rudd 2008e: 11).

Expanding on the seriousness of the climate threat, Rudd framed it to the Pacific Island Forum as a matter of national survival:

> On climate change, we – the leaders of the Pacific Islands Forum – have put out a call to action to other global leaders urgently seeking their support to deal with the threat of climate change.
>
> For so many of my colleagues here on the platform before you, this is not just a matter of importance, it is not just a matter of urgency. *For many of them, it is a matter of national survival.*
>
> The very viability of certain small island states is at stake on the question of how we deal or choose not to deal with climate change (Rudd 2009f; emphasis added).

As someone who placed national security as the 'foremost responsibility of government', Rudd moved quickly to formally assimilate climate change into Australia's national security agenda. During a major speech to Parliament in late 2008 (*The First National Security Statement to the Australian Parliament*), Rudd argued:

> [A] new and emerging challenge [that] unless properly dealt with by effective policy action, will have long-term security impacts – locally, regionally or globally. *Over the long-term, climate change represents a most fundamental national security challenge* (Rudd 2008g; emphasis added).

Distancing himself from his political opposition, Rudd argued that climate change had been given less attention than traditional security threats and identified that significant climate change would bring about 'unregulated population movements, declining food production, reductions in arable land, violent weather patterns and resulting catastrophic events' (2008g).

Signalling the domestic national security community, including the ADF, Rudd further argued that such consequences would require formal incorporation into Australia's national security policy and analysis process. This moment represented a marked divergence from Howard era thinking in which climate change was not considered a threat, challenge or issue to Australia's national security. These utterances by Rudd effectively marked the formal incorporation of climate change as a governance issue into Australia's national security agenda and its security institutions. It can be regarded, in some respects, as a signal to the military-bureaucracy that it was henceforth permissible to develop climate policies in accordance with government direction. Evidence of this shift was the inclusion of climate change within all subsequent major national strategic policy documents from 2008 (including *2009 Defence White Paper*, *2013 National Security Strategy* and the *2013 Defence White Paper*).

Having laid out the basic climate securitisation agenda, Rudd subsequently spoke on the emerging involvement of the Australian military to respond to climate change security threats. In Rudd's view, since '[o]ur defence force lies at the heart of our national security framework' it made sense to graft the military into the climate securitisation agenda (Rudd 2009b). In a 2009 speech to the Australia-Israeli Chamber of Commerce, Rudd argued:

> Increasing security risks associated with climate change, resource security and energy security are likely to exacerbate existing problems in developing states.
>
> Many countries in our immediate region will be especially vulnerable to rising sea levels, droughts and reduced access to food and energy. Natural disasters, compounded in some cases by the impact of climate change, also pose an increasing demand on our defence and wider national security assets (Rudd 2009b).

Thus, rather than simply being incorporated into 'policy and analysis process', Rudd now envisaged that responding to climate change might, at some stage, require the actual deployment of the ADF.[13] Once again, this represented a profound shift in how climate change had been cast by the former Prime Minister, John Howard.

Notwithstanding this assessment, further nuance becomes evident on closer examination of the coding outcomes. First, Table 5.8 reveals a paradox: that Rudd-Gillard framed climate change as both an urgent *and* as a long-term issue. Closer scrutiny shows that climate change was mainly framed as an urgent issue in 2008 and 2009 and again in 2011 (Fig. 5.2). In 2008 and 2009 Rudd framed climate change as an urgent and immediate issue on twenty-four separate occasions (eighteen of these occurring in 2009). Julia Gillard framed it likewise on four separate occasions in 2011. No other instances of such urgency were recorded in all other years examined. The key insight from this is that climate change was deliberately framed as an urgent issue for the sole purpose of garnering support for the passage of legislation, Rudd's CPRS during 2009 and Gillard's Clean Energy legislation in 2011.

In contrast to this (and apart from the Pacific Island Forum example given earlier), when Rudd and Gillard framed climate change as a securitised issue it was

[13]Further analysis of this aspect in the Defence White Papers is addressed in the following section under MINDEF.

Table 5.8 Temporal framing of climate change John Howard and Rudd-Gillard

Prime Minister	Urgent	Short term	Medium term	Long-term
Howard	0	1	1	7
Rudd/Gillard	27	0	0	14

John Howard speeches from 2003–2007, Kevin Rudd and Julia Gillard from 2007–2013. *Source* The author

predominately presented as a *long-term* issue. Rudd (2008g) made this point particularly clear when he stated '[o]ver the long-term, climate change represents a most fundamental national security challenge, over the long run'. The subsequent 2009 Defence White Paper affirmed this when it declared 'the large-scale strategic consequences of climate change are, however, not likely to be felt before 2030' (Defence 2009b: 31).

This finding has some significance. It indicated that although Rudd-Gillard invoked climate change as a securitised issue, there was no intent to adopt emergency measures on *security grounds* to deal with the consequences of climate change. Indeed, although Rudd and Gillard spoke forcibly on the urgency and imperative of acting on climate change, it is clear that they did not employ anything approaching emergency measures. This applied equally to non-securitised aspects. For Rudd, an *emergency measure* (in a non-securitised sense) may have resembled a decision to pursue a double-dissolution election in early 2010 on the issue of passing the CPRS legislation. That he did not indicated an unwillingness to adopt emergency measures of any sort.

A further point of discussion concerned the motive behind securitisation: if climate change was considered to be a long-term *security* issue by Rudd that would be manifest in twenty-year's time, why bother framing it as a security issue at all? Indeed, was it possible that Rudd could have pursued a strong political climate change agenda (through social and economic policies) without also pursuing a climate security agenda?

Four possibilities for this are (briefly) explored. Firstly, attempts by the Rudd-Gillard government to frame climate change as a security matter—and therefore concerning the military—may have been done as a political strategy to bolster public support behind Labor's broader political climate change agenda. The military repeatedly polled as one of the most trusted organisations by the Australian public. In 2011, for example, a survey by the Australian Survey of Social Attitudes (Australian National University) identified 91% of respondents having 'a great deal' or 'quite a lot' of confidence in the military (more than 11% points ahead of its nearest rival) (Evans 2012). Drawing the military into the climate change agenda might have been an act to leverage this credibility. Thus, securitising climate change might be viewed as a political strategy that simply intended to further the legitimacy of Labor's approach.

A second reason may be the recognition that militaries are one of the few departments of state that have the actual capacity to conduct long-range strategic planning and where the long-term capability planning cycles actually coincide with the likely onset of climate change (see, for example, discussion below by Deputy Secretary, Michael Pezzullo). On these grounds, Rudd and Gillard might have deliberately acted on an otherwise reluctant military by imposing the issue upon the ADF's strategic planning. Moreover, this reason related to a genuine *belief* by Rudd and Gillard that they were acting in the national interest by engaging military planners to deal with the likelihood of climate threats based on the warnings of scientific institutions and the scientific community more broadly.

A third possibility was that climate change fitted neatly into the Rudd Government's broader strategic policy narrative that witnessed a move away from Howard era security (terrorism, WMD, proliferation and expeditionary missions) toward a continental and regionally focused security architecture (e.g., Asia-Pacific Community) of which humanitarian aid and disaster relief had greater prominence. In this regard, climate change was co-opted as one of many new emerging threats that provided justification for the new strategic approach. Indeed, some argued—as Hans Günter Brauch and Úrsula Oswald Spring observed in 2011—that this new rationale represented the thin edge of the climate militarisation wedge to justify larger military budgets. This was also raised by Chinese academic, Dr. Chun Zhang, who argued that climate change (amongst other threats) had been used by the Rudd Government as a means to justify an increased defence spending 'including the ready transmission from the traditional to the non-traditional security field'. Zhang added:

> For the military, however, trouble means opportunity. That's why the 2009 Australian Defence White Paper emphasises various threats … climate change. To justify increased defence spending … troubled times ask for a stronger military force … Or, in other words, they ask for rebuilding of Australian middle power leadership (Zhang 2009: 8).

This point, however, was hard to sustain in the face of evidence. Put simply, despite the rhetoric of Rudd to lift Defence spending to 3% real growth and the aspirations of the 2009 Defence White Paper, Australian Defence spending *decreased* as a percentage of gross domestic product across the period 2009–2013. By mid-2013, Defence spending was at its lowest point since before World War Two at 1.59% of GDP (Thomson 2013: vi).

Less likely, but related to this point, was the possibility that this strategy was chosen since it arguably presented a more *affordable* one and was in line with Labor's historical inclination to favour continental/regional national security approaches. In 2011–12, for example, Operation SLIPPER (ADF military contribution in Afghanistan and the wider Middle East) cost $1,221.8 million. By comparison, Operations ASTUTE (East Timor) and ANODE (Solomon Islands) cost roughly six times less at $203.5 million (Watt 2012). It was possible that securitising climate change was a strategy designed to place renewed emphasis on HADR and stability operations that was more affordable, but, equally important, it suited Labor's enduring strategic focus on the Asia-Pacific. Although such a cost-saving strategy would have been appreciated following the Global Financial Crisis, it is probably the least likely reason given—at face value—the constant declaration by Rudd (2009b) that 'there is no higher priority than the national security of our country' and his personal inclination towards more internationalist approaches that diverged somewhat from traditional Labor Party strategic thinking.

The final possibility to securitise climate change by the Labor Party related to geo-politics and the maintenance of a credible military force following the anticipated drawdown in the Middle East. In this context, HADR operations as a consequence of climate change may have represented an attractive strategy since they may have been used as a legitimate means to maintain military readiness, exercise capabilities, improve military-military relations in a non-threatening manner, build partnerships, enhance interoperability and enhance military-civilian relationships. In this context, climate change became another 'threat' or 'adversary' while HADR operations became the military way of addressing it or contributing to 'defeating it'. Many ADF leaders had identified the likelihood of a long-peace (or 'peace dividend') following withdrawal from the Middle East. In this sense, a focus on HADR may have been considered a low-cost, politically plausible and military-supported means of using an otherwise under-utilised (military) workforce. This may have represented a good outcome for the Australian military establishment since it would always necessitate *some* funding and provide an aspirational future, thereby preventing the type of atrophy, hollowing-out and demoralisation that occurred within the ADF following the last period of sustained combat in Vietnam.[14]

Having examined the Rudd-Gillard securitisation, a concluding observation can be made about the timing and frequency with which prime ministers over this period discussed climate change. Figure 5.3 shows that references to climate change (and global warming) in prime ministerial speeches and statements gradually increased through 2006, peaked in 2009 and then declined sharply in 2010 before a final, albeit brief, resurgence in 2011.

This empirical finding matches the general observation that climate change had become an important issue for both Howard and Rudd during the 2007 election campaign and remained so until overtaken by the Global Financial Crisis and then the twin failures of Rudd's CPRS legislation and Copenhagen Conference in December 2009.[15] The 2011 resurgence can be explained by Julia Gillard discussing climate change in the light of the Clean Energy legislation.

Further examination of 2009 and 2010 reveals a dramatic change of direction by Kevin Rudd. In the twenty speeches analysed in 2009, Rudd directly referred to *climate change* or *global warming* on more than *220* occasions. In ten speeches he gave in 2010, before being deposed as Prime Minister, Rudd only managed to reference climate change on *fourteen* occasions. The contrast in language was sharp and his new outlook was highlighted by a major speech he gave at the National

[14]See for instance, the speech given by Lieutenant General David Morrison to the Australian Strategic Policy Institute on 11 April 2012 discussing the post-Vietnam malaise and requirement to look to developing the Army for future conflict. [Author note: speech no longer available online, in possession of author].

[15]Lowy Institute polling in 2010 and 2011 shows that climate change ranked third-last amongst Australia's twelve foreign policy priorities (down 29% since 2007, when it was considered the most important domestic policy priority) (Wesley 2012).

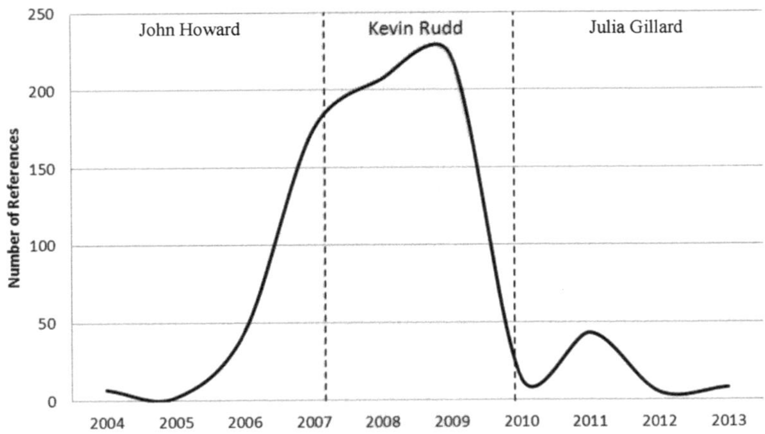

Fig. 5.3 Total references to *climate change* for Australian Prime Ministers 2003–2013. *Source* The author

Security College at the Australian National University. In this speech, Rudd glossed over climate change, referencing it on just two occasions and downgrading its urgency to an 'emerging' one (Rudd 2010a).

A corresponding drop-off in securitised references to climate change was also recorded by this research. Figure 5.4 shows the number of securitised references coded by year for *all* Strategic Programmes. Again, this result from the political elite aligns with findings elsewhere that climate change had peaked in (Australia's) public interest around 2008 (Wesley 2012).

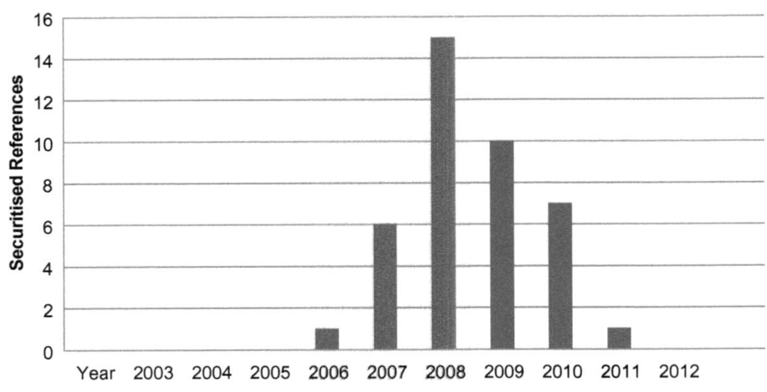

Fig. 5.4 Frequency of *climate change* framed as a security issue, 2003–2013. Data taken from the speeches and policies for *all* Australian Strategic Programmes 2003–2013. *Source* The author

5.2.3 *Minister for Defence*

Terrorism ... is the defining issue of my generation and that of my children.

Nelson (2007b)

Defence takes its obligations to reduce its greenhouse gas emissions seriously.

Fitzgibbon (2008a)

The Minister for Defence (MINDEF) programme analysed 186 artefacts of which twenty-eight contained eighty-seven direct references to *climate change*. Table 5.2 shows that of these eighty-seven references, twenty-seven were coded in a securitised context and fourteen were coded into a non-securitised context. In this respect, the MINDEF framed climate change more often as a security issue than as a non-securitised issue. A list of the MINDEFs examined by this research is at Table 5.9.

As will be shown, the (overwhelming) majority of securitisation coded references in the MINDEF programme are attributable to the Labor MINDEFs from (late) 2007 onwards. That the MINDEF programme made proportionally more securitised references than the prime ministerial programme can be accounted for by the nature of the Defence portfolio which focuses on security matters rather than broader socio-economic and political matters.

In terms of specific language used during securitised framing, MINDEFs tended to adopt heightened language (*challenge* and *threat* were coded fifteen times against *neutral* which was coded twelve times). Similar to the Prime Minister Programme, climate change was not framed as an *opportunity* within a securitised context. The first reference to climate change in a securitised context identified by this research was by Liberal Minister for Defence, Brendan Nelson, on 14 June 2007 in a speech to the Centre for Economic Development of Australia. During this speech Nelson noted that Australia was 'gearing up' for the potential security impacts of climate change over the long-term (2007a). He specifically identified population shifts as a consequence of changes to the climate.

Despite this, the overwhelming focus for MINDEF under the Howard Government during period 2003–2007 remained the threat of terrorism, Weapons of Mass Destruction (WMD) and other traditional security issues. In a speech in August 2007, Nelson maintained this point when he declared terrorism the 'defining

Table 5.9 Names and tenure of Defence Ministers examined by this research

Minister for Defence	Took office	Left office	Prime Minister
Robert Hill	26 November 2001	20 January 2006	John W. Howard
Brendan Nelson	20 January 2006	3 December 2007	John W. Howard
Joel Fitzgibbon	3 December 2007	9 June 2009	Kevin M. Rudd
John Faulkner	9 June 2009	24 June 2010	Kevin M. Rudd
Stephen Smith	24 June 2010	27 June 2013	Julia E. Gillard

Source Wikipedia (2017f)

security issue for my generation and that of my children' (2007b). Citing that he did not share an uncritical 'apocalyptic view' of climate change, Nelson focused on hard, traditional security threats:

> Many things have happened [since 2005] … the Middle-East and Iraq … a resurgent Taliban, the continuing struggle against Terrorism … the North Koreans … an ascendant Iran … a coup in Thailand … in Fiji (Nelson 2007b).

Empirical evidence of the importance placed on addressing terrorism was reflected in a text search of all examined Liberal Party MINDEF speeches and documents across 2003 until the end of 2007. During this text search, 480 direct references to *terrorism* (and associated words, *terror* and *terrorist*) from forty-two separate artefacts were found.[16] This contrasted sharply with references to *climate change* or *global warming*, which yielded just four direct references in the sixty-six speeches and policy documents examined between 2003 and the end of 2007.

Defence Update 2007 was the last major Australian national security policy document (i.e., a Defence White Paper, Defence Update or National Security Strategy) that did not address the security implications of climate change. Rather—and in keeping with the strategic outlook by the Howard Government—*Defence Update 2007* identified the security impacts of globalisation, terrorism, fragile states and WMD proliferation as the primary national security issues. That climate change was not addressed in any major national security documents between 2003 and 2007 reflected Howard's outlook that climate change was *not* a security issue.

As was previously argued, the election of the Rudd Government brought a new focus to climate change. Accordingly, the subject was also taken up by a series of Labor Defence Ministers. This point was reflected in the research results which showed that of the twenty-eight securitised references to climate change attributed to the position of MINDEF, twenty-seven were made under Labor MINDEFs from 2008 through 2013 (Joel Fitzgibbon (2007–2009), John Faulkner (2009–2010) and Stephen Smith (2010–2013)). In pure quantitative terms, *climate change* or *global warming* was directly referenced eighty-seven times in twenty-five speeches from Labor MINDEFs from 2008 through 2013 (sixty-five occasions when major policy documents are excluded). Following the prime-ministerial lead, Labor MINDEFs were found to frame climate change as a long-term security threat (Table 5.10).

However, in contrast to Rudd's confidence that climate change was a serious threat to security and stability, the initial speeches delivered by the first Labor Defence Minister, Joel Fitzgibbon, indicated a degree of uncertainty. In a speech to the Brookings Institute, Fitzgibbon (2008c) posed:

> What risks and threats will we face in the emerging strategic environment? Will changes in the planet's climate and environment create new sources of tension and conflict?

---

[16]Excluding the major policy documents (e.g., Defence Updates), the number of references was 380.

Table 5.10 Research results of temporal framing by MINDEFs 2002–2013

	Urgent	Short term	Medium term	Long-term
MINDEF	1	0	0	6

Source The author

To the Shangri-la Conference in Singapore, Fitzgibbon contemplated:

[W]e can't be sure exactly how climate change will affect security but we know that it will have some impact on our world (Fitzgibbon 2008d).

Recycling previous speeches, Fitzgibbon at the National Press Club again asked: 'Will changes in the planet's climate and environment create new sources of tension and conflict?' (2008b). As a Minister in an electorate (Hunter) encompassing major 'aluminium smelting, coal mining [and] major [coal] power stations', the lack of enthusiasm and critical outlook was understandable (Fitzgibbon 2014). Consequently, this research found that Fitzgibbon spoke little (on the public record at least) and with scant detail on the security implications (or otherwise) of climate change.

Nonetheless, it was Fitzgibbon (in March 2008) who announced that the Department of Defence would join *Earth Hour* to improve its energy efficiency as part of Energy Efficiency in Government Operations (EEGO). Fitzgibbon (2008a) further added, 'Defence takes its obligations to reduce its greenhouse gas emissions seriously … In addition, a number of Defence bases will be holding Earth Hour events such as tree plantings'. Beyond this, it was the Department—not Fitzgibbon—that developed an internal departmental level *Combat Climate Change* initiative.[17] Placed in perspective (particularly when compared against what occurred in the US military), Fitzgibbon's one-page media release that Defence would contribute to *Earth Hour* was more symbolic than meaningful.

To aid Fitzgibbon and to gauge national opinion on security matters in preparation for the 2009 Defence White Paper, the Rudd Government commissioned a community consultation panel that received 450 written submissions and undertook thirty public meetings across Australia with over 600 people having attended. The final report, *Looking Over the Horizon: Australians Consider Defence*, provided a more concrete answer to Fitzgibbon's questions:

[Climate change] *will have* strategic implications including causing greater regional instability, pressure upon natural resources, and severe weather events in the region (Defence 2008b: 1; emphasis added).

Nonetheless, strategic uncertainty regarding the impact of climate change remained a dominant theme in the 2009 Defence White Paper (*Defending Australia in the Asia-Pacific Century: Force 2030*). Although *Force 2030* was the first of its type in Australian strategic policy history to address the security dimension of climate change, it reflected Labor's mixed outlook. Although parts appeared clear,

[17]Discussed in more detail in ADF Programmes (below).

portraying climate change as a 'new security risk … likely to exacerbate already significant population, infrastructure and governance problems in developing countries, straining their capacity to adapt' (Defence 2009b: 31), other sections were more cautious:

> Uncertainty about the effects of climate change and the period of time over which potential impacts may develop makes it difficult to assess its strategic consequences (Defence 2009b: 39).

One of the more interesting passages in *Force 2030*, however, concerned the manner in which the Government sought to employ the ADF if diplomatic and policy measures failed to curb dangerous climate change. Describing the deleterious consequences of climate change across the region, *Force 2030* argued that Australia's main effort remained 'coordinated international climate change mitigation and economic assistance strategies' (Defence 2009b: 40). But then, in a somewhat ambiguous passage, *Force 2030* declared the ADF a live option:

> Should these and other strategies fail to mitigate the strains resulting from climate change, or resource security issues, and they exacerbate existing precursors for conflict, the Government would possibly have to use the ADF as an instrument to deal with any threats inimical to our interests (Defence 2009b: 40).

Given the sensitivities to nations who were perhaps suspicious of the West's action on climate change—particularly concerning the idea of military options to restrict industrial emissions—this passage has remained decidedly overlooked. It is highly dubious however, that the Government ever intended the ADF to be considered an 'instrument' to 'deal with' the *true* source of climate change: namely industrial emissions of other countries. Perhaps most obviously, and as noted by the British Foreign Secretary Margaret Beckett (2007), 'we are all our own enemies'.

More realistic was the idea that the ADF would need to respond to the security implications of a changing climate in the event that the 'planet's changing climate … gives rise to very old forms of confrontation', thereby precipitating 'stabilisation operations' (Defence 2009b: 40). The other alternative provided was 'more frequent and severe natural disasters and extreme weather', thereby occasioning an increase in ADF deployments for humanitarian and disaster relief (HADR) (Defence 2009b: 40).

Thus, despite the uncertainty and expectation that the strategic consequences would not be felt for decades, *Force 2030* crystallised two major security implications of climate change. Firstly, that climate change may exacerbate the severity and frequency of natural disasters, thereby increasing the requirement for humanitarian and disaster relief (HADR) operations. And secondly, that climate change may exacerbate existing precursors to conflict whereby vulnerable states (with limited capacity to respond) may dissolve into violent conflict, thus requiring the ADF to conduct a stabilising intervention.

These facets were reiterated with greater confidence in the 2013 Defence White Paper (*Defending Australia and its National Interests*), stating a direct causal link between the effects of climate change and the increasing demand of both HADR and stabilisation operations:

> The risks associated with resource insecurity may be exacerbated by changes in the global climate system. The inundation of low-lying regions, more frequent and severe natural disasters and shifts in rainfall patterns would lead to loss of agricultural production in some areas and potentially large-scale human migration.
>
> The combination of the effects of *climate change and resource pressures will increase the risk of insecurity and conflict*, particularly internal instability in fragile states, many of which have increasingly large populations in areas that will be affected by climate change. *These factors, taken together, point to an increasing demand for humanitarian assistance, disaster relief and stabilisation operations over coming decades* (Defence 2013b: 18–19; emphasis added).

This was particularly identified to be the case across the South-Pacific. Thus, in *Defending Australia*, the formal government outlook shifted to assess that climate change *will* increase the risk of insecurity and conflict thereby affecting an *increase* in HADR and stability operations. This position marked a much clearer and more certain assessment than earlier iterations by Rudd, successive Defence Ministers and *Force 2030*. Arguably, when this particular assessment is combined with the *2012 Force Posture Review*—which posited that the second most important force structure determinant is 'contributing to stability and security in the South Pacific and East Timor' (Defence 2009b)—it is feasible to consider climate change an indirect determinant of ADF force structure and capability planning.

Some evidence to partly support this claim was contained in the rhetoric of Labor government ministers who increasingly referred to existing and new capability in the context of its ability to support HADR and stability operations. It was also partly found in the actions of a number of Defence acquisitions that were specifically designed to bolster ADF HADR capability. This came to particular prominence following Cyclone Yasi (February 2011), during which the Royal Australian Navy (RAN) was exposed for an inability to provide sufficient amphibious capability. As a consequence, Labor quickly procured two amphibious vessels specifically 'in support of humanitarian and disaster relief operations domestically and in the region' (Smith 2012).

The increased attention on HADR and regional stability operations demarcated the Rudd Government from the Howard era dominated by terrorism, WMD and expeditionary military deployments in support of US operations in Afghanistan and Iraq. The difference is evident when comparing the priorities of Defence Updates 2003 and 2005 against the 2009 and 2013 White Papers. While *Defence Update 2005* (2005a: 12) prioritised terrorism, WMD proliferation and declared Australian security interests are 'not defined by geography alone', the 2013 White Paper provided a contemporary account of Professor Paul Dibb's 1986 so-called concentric circles approach by citing 'the defence of Australia [then] the security, stability and cohesion of our immediate neighbourhood [then] the stability of the Indo-Pacific [and lastly] international order' (Defence 2013b: 24–27).

Furthermore, although both Rudd and Howard placed emphasis on maintenance of a strong US alliance, Rudd possessed a desire for greater activism in regional affairs. This was evident in Rudd's National Security Strategy speech to Parliament, in which he advanced a 'creative middle power diplomacy' that prioritised the Asia-Pacific ('regional engagement is crucial') through 'strengthening bilateral relationships', 'effective engagement in regional institutions', 'shap[ing] the future of regional architecture' and a policy of '[regional] security policy cooperation' (Rudd 2008g; emphasis in original). In this context, Rudd proposed an Asia-Pacific Community by 2020 as a means of 'strengthening political, economic and security cooperation in the region' (ibid).

Thus the strategic differentiation between Howard and Rudd may ultimately be mapped to the differing outlooks that have dominated Australia's strategic culture since Federation. Although it is not the intent of this chapter to go into detail on the two competing camps, both are well summarised by Michael Evan's study paper *Tyranny of Dissonance: Australia's Strategic Culture and Way of War 1901–2005*. The main point here was that HADR and stability operations were arguably manifestations of a continental and regionally focused strategic doctrine favoured by the Rudd-Gillard governments. Climate change—portrayed as a threat likely to increase homeland and regional humanitarian and stability missions—fitted neatly into this broader strategic narrative. As has been shown, it also fitted neatly with Rudd's broader political narrative.

The groundwork for presenting Labor's differentiated outlook is witnessed in the rhetoric of Labor Ministers describing a *new* approach for military forces expected to execute the government's revised outlook. The following passages illustrate this through the example of HADR. In a cornerstone speech early in May 2008, Joel Fitzgibbon opined:

> [Defence policy] is more complex and challenging than ever before … the world has changed so much … But the role of military forces is also changing – today, defence forces find themselves participating in a wide range of non-traditional operations, such as disaster relief and stabilisation and reconstruction (Fitzgibbon 2008d).

Fitzgibbon repeated this theme in numerous speeches to regional neighbours. To the US and Japan: 'greater practical cooperation between our government and Defence forces in areas such as humanitarian and disaster relief, peace keeping and maritime security' (Fitzgibbon 2008c). To Indonesia: 'further cooperate in this area through medical training for disaster situations, disaster response management courses, enhanced logistics cooperation and planning toward a bilateral disaster relief exercise' (Fitzgibbon 2008e).

Other Ministers read from the same script. Minister for Defence Science and Personnel (Warren Snowden) highlighted the expanding role of defence forces in which 'Governments are asking their military forces to undertake a new range of missions—such as humanitarian assistance and post-conflict stabilisation' (Snowden 2008). Parliamentary Secretary to the Minister of Defence, Dr. Mike Kelly, spoke of Labor's 'new thinking':

> To address the challenges of this contemporary and future international security environment, we must expand our domestic capability beyond individual agency planning and

> operations to integrate our military, police, political, humanitarian, economic and development goals (Kelly 2008).

In the same speech, Kelly cited the establishment of the Asia-Pacific Civil-Military Centre of Excellence (notably in his own electorate) as a Centre that would focus—amongst other things—on HADR operations. Expanding on this, Kelly even saw the Centre as having direct planning involvement and participation in executing future military operations, 'there will be "dotted lines" from the Centre to Headquarters Joint Operations Command, where they will assist in the development of the ADF operational plans' (Kelly 2008).

Similar rhetoric was also reflected in *Looking Over the Horizon* and, as already noted, the 2009 and 2013 Defence White Papers. *Force 2030*, for instance, identified the security of the South Pacific and East Timor as the second most important national security priority. In doing so, emphasis was placed on the importance of HADR missions:

> After ensuring the defence of Australia from direct attack, the second priority task for the ADF is to contribute to stability and security in the South Pacific and East Timor. This involves conducting military operations, in coalition with others as required, including in relation to protecting our nationals, providing disaster relief and humanitarian assistance, and on occasion by way of stabilisation interventions …
>
> Australia will continue to have particular responsibilities to assist our neighbours in dealing with humanitarian and disaster relief needs, and to support their stability and security (Defence 2009b: 13).

Thus, by political design, HADR increasingly gained a foothold in Australia's national security consciousness through rhetoric (highlighted above) but also through action such as training exercises, doctrinal development, bureaucratic policies and strategic partnerships. The following examples emphasised the strategic *actions*:

- The first desktop regional disaster-relief exercise co-hosted by Indonesia and Australia in May 2008;
- The first field activity for ASEAN Regional Forum (ARF) that witnessed an ADF contingent join other South East Asian, Pacific Island, the US and other countries in demonstrating disaster relief capabilities;
- Establishment in 2008 of the Asia-Pacific Centre for Civil Military Co-operation with the aim to improve Australia's effectiveness in civil-military collaboration for conflict and HADR management (administered by the Vice Chief of Defence Force);
- Strategic Partnership Agreement (2009) between the ADF and AUSAID, providing a closer framework for cooperation between the two agencies in recognition of shared strategic interests; and
- Increased international strategic agreements and activities with Asia-Pacific nations in which HADR was emphasised. Examples included: Exercise COOPERATION SPIRIT, a HADR activity hosted by the ADF in participation with the Chinese People Liberation Army and the New Zealand

Defence Force (2012); US–Australian increased cooperation on HADR as instanced by the Regional Leaders Seminar in Cairns (2011) that featured HADR; Defence Cooperation Agreement with Vietnam emphasising HADR (2010).

This design, however, was also reinforced by an apparent upsurge in operational HADR deployments due to natural disasters. Thus, circumstance played its part. Although Labor sought to link these events to the strategic and political narrative of the Rudd-Gillard government, many HADR events were not specifically climate-related (e.g., volcanoes, earthquakes and tsunami); nevertheless '[i]n an era of climate change, we can only expect the incidence of climate-related natural disasters to increase' (Kelly 2012). Examples of HADR operations, including those caused by non-climatic events, were:

- Increasing regional ADF contingents deployed in support of HADR activities including (since 2008) Operations PNG ASSIST, PADANG ASSIST, SAMOA ASSIST, HAITI ASSIST, PAKISTAN ASSIST II, CHRISTCHURCH ASSIST and PACIFIC ASSIST; and
- Increasing domestic deployments by the ADF in support of (climate-change-related) disasters. Between 2005 and 2013 the ADF responded to 275 domestic disaster relief missions, some of which (outside conflict and peace deployments) were the largest in its history (ANAO 2014).

HADR therefore became the tangible (practical) dimension of ADF climate response. But despite this assessment, a general observation for the MINDEF portfolio was a distinct *lack* of climate change initiatives instigated by Australian Ministers of Defence between 2003 and 2013. This research, therefore, identified an important gap. The lack of any major programmes or policy initiated by the Minister for Defence—such as large scale Defence mitigation or adaptation programmes—stood in sharp contrast to those instigated by Australia's two main historical allies, the US and UK militaries.

5.2.4 ADF Executive (Chief of Defence Force and Secretary of Defence)

[Has Defence done any analysis on the security implications of climate change?] No.

Ric Smith, Secretary of Defence (2006)

[On climate change] Defence is the agency of last resort.

Peter Jennings, Deputy Secretary Strategy (2011)

The CDF and SECDEF programme (also identified in this chapter as the ADF Executive) analysed 279 artefacts of which thirty-five contained 154 references to *climate change*. Table 5.2 shows that of these 154 references, eleven were coded

Table 5.11 Names and tenure of CDF examined by this research

Chief of Defence Force	Took office	Left office
Peter Cosgrove	4 July 2002	3 July 2005
Angus Houston	4 July 2005	3 July 2011
David Hurley	4 July 2011	30 June 2014

Source Wikipedia (2017c)

Table 5.12 Names and tenure of Defence Secretaries examined by this research

Defence Secretary	Took office	Left office
Ric Smith	11 November 2002	3 December 2006
Nick Warner	3 December 2006	13 August 2009
Ian Watt	13 August 2009	5 September 2011
Duncan Lewis	5 September 2011	18 October 2012
Dennis Richardson	18 October 2012	Incumbent

Source Wikipedia (2017d)

within a securitised context and nineteen were coded within non-securitised context. In this respect, the CDF Programme framed climate change predominately as a non-securitised issue than as a securitised one[18] (Tables 5.11 and 5.12).

In terms of specific language used during securitised framing, the CDF Programme adopted forceful language (*challenge* and *threat* were coded six times against *neutral* which was coded five times). Where climate change was framed in a non-securitised manner, the outlook was significantly more neutral (there were 16 coded references for *neutral* against three for *challenge* and *threat*). Significantly—but perhaps not surprising—climate change was never framed as an *opportunity* within a securitised or non-securitised context.

Several points regarding the CDF and SECDEF Programme warrant further discussion. First is the general lack of discussion on climate change by the ADF executive until at least 2007. Evidence of a seeming lack of interest in the subject until this time existed during a 2006 Senate Estimates hearing involving Green's Senator Kerry Nettle and then Secretary of Defence, Ric Smith (Commonwealth of Australia 2006: 70):

> Senator Nettle: This is a general question to Defence as to whether you have done any analysis examining the Defence and security implications of climate change?
>
> Mr Smith: Not that I am aware of. I am aware of studies that have been done by consultants that we have seen, but I do not recall that we have done any in-house.

In a subsequent exchange during the same hearing, Smith was adamant that Defence had *not* examined the subject in any detail (Commonwealth of Australia 2006: 88–89):

[18]This research also included some statements from executives other than the CDF and Secretary, namely the Deputy Secretary for Strategy. In this sense, the research describes the ADF Executive to include the CDF, Secretary and Deputy Secretary of Strategy.

Senator Nettle: I wanted to go back to your answer before, Mr Smith, when you talked about consultant's reports and check if we were talking about the same consultant's reports in relation to defence and the security implications of climate change.

Mr Smith: I have seen one from the Lowy Institute.

Senator Nettle: Has there been any defence analysis of that one?

Mr Smith: Let me ask Mr Pezzullo [Deputy Secretary Strategy]. He advises that he is not [sic].

Senator Nettle: Has Defence commissioned any?

Mr Smith: No.

Senator Nettle: Has there been any Defence analysis of the Pentagon one, the UN one or the CSIRO one?

Mr Smith: Not that I know of, no.

Senator Nettle: There is an ONA report as well. Has Defence been involved in that?

Mr Smith: We are aware of the ONA report. We are part of government with them.[19]

Excluding the above discussion, between 2003 until 2006 (from the sources examined) there was only *one* direct reference to climate change emanating from a *single* source within the CDF/SECDEF Programme (the 2003 *Defence Annual Review*). Put another way, in the years 2004, 2005 and 2006 no sources examined by this research (excepting the above discussion) within the ADF Executive referred to climate change.[20]

This all changed in 2007. In total, six sources were found that contained thirty-one direct references to *climate change* (unfiltered). Similar numbers were sustained across 2008 through to 2012.[21] Figure 5.5 shows the number of direct references to *climate change* for the PM Programme compared with the ADF Executive. It shows that the ADF Executive began to refer to climate change about one year *after* the PM Programme begun to increasingly refer to the issue. This 'strategic lag' is interesting, since it suggested that the highest echelons of the ADF (the CDF and Secretary) had seemingly avoided to publicly call-out climate change

[19]This research sought the Office of National Assessment (ONA) reports but was informed by ONA through personal correspondence that 'assessments are classified and not subject to release until the closed period of 20 years expires. Additionally, ONA is an exempt agency under the Freedom of Information Act 1982 and is therefore not subject to FOI requests'. Despite the lack of access, an exchange during Senate Estimates in May 2007 revealed that ONA were assessing the security impacts of climate change—of the five reports produced by ONA between January and May 2007, one was a major national assessment on climate change.

[20]The search covered sixty-five separate sources examined between 2004 and 2006.

[21]This research examined no documents beyond April 2013 and therefore only contained five source artefacts for that year. This contributed to the lack of references in the 2013 CDF Programme.

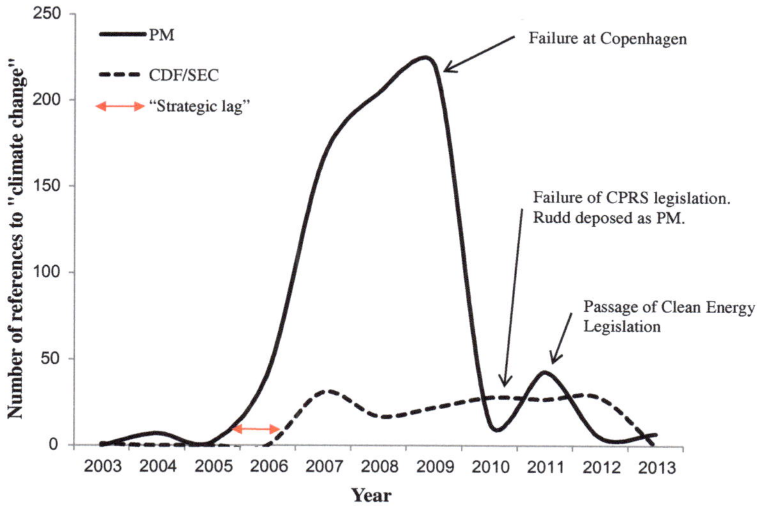

Fig. 5.5 Temporal variation in references to *climate change* by Australian Prime Ministers and military executive, 2003–2013. *Solid line* denotes references by Prime Ministers (PM). *Dashed line* denotes references by the military executive (CDF/SEC). *Source* The author

as a strategic issue of significance for the ADF (or of national security) until well after the Prime Minister.

This was largely reflected in the cautious tone of the CDF, Air Marshall Angus Houston (2005–2011) who stated that climate change was an 'entirely new' challenge (Houston 2007: 57). In 2008, Houston subsequently argued that although climate change posed 'security consequences' they were 'out in the long-term', adding that 'a lot more analysis needs to be done before we can come up with firm deductions and conclusions' (Houston quoted in Commonwealth of Australia 2008a: 47). In 2010, Houston again reiterated the long-term outlook, stating '[c] limate change is not expected to have large-scale consequences for the strategic environment before 2030 … it is an incremental process' (Houston 2010).

An exchange during the Senate Estimates on 31 May 2007 is particularly instructive about how climate change was framed within the ADF executive. This early assessment set the strategic tone for Defence's climate change response for the years following 2007. In the view of this author, it is perhaps the single most important public articulation by a senior Departmental Defence official on the ADF's position concerning climate change (at least until a 2016 speech by the Chief of Army, Lieutenant General Angus Campbell).[22]

[22]For the purpose of this study, the Deputy Secretary Strategy was included in the CDF Programme since he works closely with the CDF and Secretary to develop ADF strategy.

The exchange occurred between Green's Senator Kerry Nettle and then Deputy Secretary of Strategy, Mr. Michael Pezzullo, responsible for developing strategic policy within the ADF (emphasis is used to highlight key aspects). Due to the importance of this statement (and rarity of this kind of well-recorded discussion) the majority of the exchange is included (Commonwealth of Australia 2007a: 101–103; emphasis added):

> Senator Nettle: [C]ould somebody explain what challenges specifically Defence believes that climate change presents for Defence and therefore what adjustments need to be made to the way Defence operates in order to deal with the challenge of climate change?
>
> Mr Pezzullo: … To answer your question, I need to talk a little bit about our planning process, but I will start by saying that it is not really a question of our beliefs. We will need to conduct some evaluations and assessments in the years to come. I would like to start by referring you to the Director-General of ONA's evidence the other night … that ONA has produced this year five reports – one of which was a major national assessment on climate change ….
>
> He made the point that the strategic implications of climate change in a security sense – I am not talking with respect to the responsibilities of other portfolios; I am here as a senior Defence official – are likely to be felt more over a 40 to 50-year period. That is relevant to us because that is the kind of planning horizon that, although a bit long, is the kind we need to look at in terms of the acquisition of major pieces of equipment, which in some cases have lives of 20, 30 and 40 years. He did indicate that in ONA's judgment the impacts of climate change may not be so material in a security sense in the next two to three years; *it is very much a long-run set of challenges.*
>
> I have given direction to my planning staff to scope in our planning guidance process a series of what the literature would call *non-traditional security threats, of which climate change is one.* There are also issues around water resources, resource depletion – for example, in relation to fisheries – demographic changes, the movements of people across traditional state boundaries, and other issues such as pandemic threats, HIV-AIDS and the rest …
>
> Clearly, the climate change issue is more relevant to the latter part of that planning horizon, as Mr Varghese indicated the other night. Looking at climate change, we are just starting to scope this in our minds, along with our intelligence colleagues. In due course we will be engaging our capability development colleagues, because that is where the rubber will hit the road. This is going to be a very complex issue to assess, because what you are trying to assess are changes in what is itself a very complex system – namely, global weather – how that is going to interact with the global strategic system, and whether that is going to create strategic contingencies over the next 30, 40 or 50 years where those contingencies would necessitate a change or a set of changes that government would have to consider in the way that they structured the Australian Defence Force. That goes to how we structure the ADF, how we equip it and how we operate it.
>
> If there is evidence to our mind starting to emerge that there will be more conflict over things like water resources, for instance, as and when climate changes – *and I make no judgment about the rate at which the climate might change; there is a very dramatic and active debate going on politically around all of that, so we steer well clear of that* – and we will start to make judgments about the kind of force that we will need to develop beyond the force that we are developing now. I think the implicit judgement or the intuitive judgement that we can make now is that *the climate change factors that the scientists believe are at*

> *play will probably not affect the force that is being developed over the current 10–year period, which is our acquisition period of 2007-17.* Beyond that, *we do need to give consideration to those kinds of non-traditional security dynamics in relation to the force that we will develop beyond that period.*[23]

This statement outlines the main ADF position on climate change from 2007–2013. Overwhelmingly, the impression was of a Defence organisation that viewed climate change as a long-term, non-traditional security threat that would likely have no impact on Defence operations and force structure in the medium term.

The comment by Pezzullo that Defence avoids commenting on the *rate* of climate change because of the political sensitivities ('we steer well clear of that') also highlighted the challenges of an apolitical organisation attempting to deal with the consequences of a politicised security threat. Arguably, analysis on the *rate* of climate change goes to the heart of what Defence should have provided comment on. For, if Defence (collectively including the military services and intelligence community) were to judge that climate change were increasing in its rate-of-change, this could have profound consequences for how national security, force structure and capability planning were (and *are*) conceived. It might also be particularly relevant for the broader Australian public, who arguably place more trust in assessments by established national institutions than those provided from global institutions such as the UN or of a more partisan nature such as NGOs.

The other inescapable aspect of Pezzullo's comments is the reactionary role envisioned by Defence in relation to climate change. For Pezzullo 'the rubber hits the road' only when Defence was forced to consider capability or force structure implications of climate change in '30, 40 or 50 years'. This is instructive, since it provided no sense of a requirement for proactive Defence planning for the more short-term aspects, including: the possibility of tipping points, mitigation of its own departmental emissions, likelihood of a surge in climate-induced natural disasters (both nationally and regionally) or the impact that legislation (such as carbon pricing) might have on Defence procurement, industry and operations.

This posture was somewhat confirmed in 2011 when then Deputy Secretary of Strategy Peter Jennings noted that the primary effort rested with other government agencies:

[23]Several important points are revealed by this assessment. First, Defence consider climate change a long-term event where the security implications of climate change will be felt over a forty- to fifty-year period. Second, Defence consider climate change a non-traditional security threat that will be compounded by a range of other resource security issues. Third—and most critical—while Defence do not expect climate change to impact on near term capability development (2017) it will be crucial to assess how it might impact future force structure.

> The lead agency for climate change policy is the Department of Climate Change, and Defence notes other Government agencies like AusAID have the lead in helping societies in our region adapt to and mitigate the effects of climate change. *Defence is the agency of last resort* in this respect (quoted in Alexander 2011: 30; emphasis added).

The final observation on this statement by Pezzullo is that it remained significant since it was *not* made by either the CDF or the Secretary. Indeed, no CDF or Secretary was identified by this research as making such a detailed and lengthy assessment on ADF's outlook concerning climate change.

While Air Marshall Houston spoke of climate change in a general sense, other CDFs were more circumspect. Indeed, of the speeches, articles, interviews and Hansard records examined by this research, Air Marshall Houston was the only CDF between 2003 and 2013 to make any comment on climate change. For CDF General Cosgrove (2002–2005) this was perhaps understandable given the low emphasis by the political establishment. But for CDF General David Hurley (2011–2014), this was a surprising result (particularly considering that the research analysed seventy-two documents covering his speeches, interviews, written articles and Senate Estimates between 2011 and 2013).

Despite the lack of rhetoric by the various CDFs and Secretaries regarding climate change, incremental change was evident. While the perception of climate change as a long-term security issue remained largely consistent over the period examined (Table 5.13), other military perceptions of climate change shifted (particularly across the Rudd-Gillard years).

Firstly, whereas climate change was initially viewed as a non-traditional security issue, it came to be framed by the CDF Programme also as a global, regional and national security issue. Air Marshall Houston particularly identified the South Pacific as a key region to be affected. In one of his more extensive quotes on the subject, Houston (2010) highlighted 'weak governance, crime and social discord' across the South Pacific that 'will only be exacerbated by climate change'. Houston added:

> With already weak economies and governance, these island nations have little capacity to adapt. The changing rainfall patterns and ocean dynamics, extreme weather, and rising sea levels that climate change will bring threaten [sic] the agriculture and fisheries on which the South Pacific is dependent. From there, it is a small step to political instability and social disorder (Houston 2010).

Nonetheless, the range of security categories (Table 5.14) remained markedly less than those identified by Prime Ministers Rudd and Gillard (Table 5.6).

Table 5.13 Temporal framing by CDF and SECDEF Programme, 2003–2013

	Urgent	Short-term	Medium-term	Long-term
CDF/SECDEF	0	0	0	10

Source The author

Table 5.14 Distribution of securitised categories for CDF and SECDEF Programme, 2003–2013

Securitised frame	CDF/SECDEF
Global and regional security	5
National security	4
Energy security	0
Environmental security	0
Biosecurity	0
Convergence	0
Resource security	0
Non-traditional/New security	2
Total	11

Source The author

Another facet of the shifting views of climate change held by the CDF and SECDEF Programme was that it would necessitate an increase in the frequency and intensity of HADR operations (as a consequence of an increase in climate change related natural disasters). This point was again taken up by Deputy Secretary Strategy, Peter Jennings, in 2011:

> Defence is aware that climate change may bring more extreme weather events that could cause humanitarian crises at home and in our immediate region (quoted in Alexander 2011: 30).

Jennings further identified rising sea levels and changes in rainfall patterns as placing additional strains on fragile societies that could lead to conflict. Downplaying the requirement for new capabilities, Jennings noted that the ADF was already well positioned to support such operations on account of new amphibious ships, a strategic air-lift ship and helicopter transport aircraft as well as the 'expansion of the land force by two infantry battalions for conducting stabilisation operations' (Jennings quoted in Alexander 2011: 30).

That said, Jennings presented a more detailed account of how Defence had progressively bureaucratised the issue of climate change. In contrast to the *single* branch examining climate change for Defence in 2007, Jennings (by 2011) had identified three distinct areas within Defence responsible for addressing climate change. In this context, Jennings cited Strategic Policy Division as the 'lead in addressing the repercussions for Defence', Infrastructure Division as responsible for managing ADF greenhouse gas emissions and adapting Defence infrastructure, and the Commander of Joint Logistics as leading on ADF energy and fuel supply (Jennings quoted in Alexander 2011: 31). Arguably, there were likely to be other areas (such as finance and governance sections) that also dealt with administrative and financial implications resulting from ADF carbon emissions exposed to the Clean Energy legislation of 2011. Another area was Defence procurement agencies, particularly the Defence Materiel Organisation (DMO) (now the Capability Acquisition and Sustainment Group). In 2012, for example, small changes were made to the Defence Procurement Guidelines that identified 'environmental sustainability' (including energy efficiency) as a basis for consideration (Defence 2014: 1.2-2).

Importantly, Jennings also noted that the strategic impacts of climate change would now be considered as part of annual Defence planning. But he also pointed to very heart of the (Australian) climate security conundrum:

> Defence will continue to monitor developments to ensure our assessments of the future impact of climate change on security *are consistent with Government policy* and the prevailing scientific position (quoted in Alexander 2011: 31; emphasis added).

Once again, this comment highlighted the challenge of apolitical institutions assessing politicised security threats. It exposed the potentially conflicted scenario in which government policy may be incongruent with the scientific position. Put another way, if a new government believed that climate change was *not* a threat (as with the centre-right government of Prime Minister Tony Abbott), but the scientific community assessed that it was one, where might this leave Defence assessment? Perhaps more to the point, where might this leave Defence's public position on climate change? Despite such contradictions and challenges, the broader point from these passages is that Defence had made some attempts to render climate change governable as a security issue. It had set about implementing what Rudd had previously described in his 2008 *National Security Statement* by incorporating climate change in both policy and process.

The final evidence of Defence making incremental change was the establishment of a number of climate-change-related initiatives. While it was not clear if these were initiated by the ADF Executive, since no references were made by the CDF or Secretary in examined artefacts, it was likely they had some knowledge and involvement—more so since it was tendered during a Senate Estimates hearing. In response to a series of questions during Senate Estimates, the ADF described (as at 2012) the major actions it had taken in relation to climate change. These included (Defence 2012: 260 and 261):

- A $2 million study titled *Adaptation and Planning Strategies to Mitigate the Impact of Climate Change Induced Sea Level Rise, Flooding and Erosion at Selected Sites* to identify possible threats to Defence bases and infrastructure from climate change and the impacts of these threats on Defence capability.
- Defence membership of an Inter-Departmental Committee on Climate Change, involvement with the Department of Climate Change and Energy Efficiency local councils and 'other interested parties' to 'instigate a coordinated approach to the effects of climate change on Defence bases'. And,
- Engagement with the clean technology industry to 'create cost effective energy solutions' in Defence projects.[24]

[24]An example of this is the Perth Wave Energy Project providing renewable energy for HMAS Stirling. See, for example at: http://carnegiewave.com/projects/perth-project-2/ (20 May 2017).

On the one hand, these measures represented a significant change from 2006 when the Secretary appeared oblivious to climate change activities within his Department. On the other, when compared against the US military initiatives, they were substantially small-scale, low-key and minimalist responses. For instance, of the 279 artefacts examined under ADF Executive, just one contained a reference to the *Adaptation and Planning Strategies* (a single line-item uncovered in an annex to a Budget Estimates submission).

The incremental change by the ADF executive regarding climate change reflected views brought about by a change in government (in 2007). Central to this was that the ADF now had to adapt itself to the new strategic outlook imposed by the Rudd Government. As previously discussed, this entailed a more nuanced approach—not as singularly focused on terrorism and WMD—but informed by a broader definition of national security that included non-traditional security threats. The new requirement was thus framed by then Chief of Defence Force, Angus Houston:

> Traditionally, national security policy making was focused on protecting territory from external attack by other states and promoting our security interests. More recently, national security has become a much broader concept (Houston 2010).

In the same speech, Houston surmised the consequence of this new outlook:

> [W]e are no longer dealing with state-on-state conflict. Our focus must also be on other traditional threats that are now seen to impact on the security interests of the nation – such as natural disasters or piracy. Added to this mix are 'non-traditional' threats (Houston 2010).

The fundamental dilemma facing Houston was balancing 'remote but also more consequential' high-end war-fighting against the more likely requirement of 'stabilisation, counter-terrorism and disaster-relief operations'. Ultimately, Houston's strategic solution—his reconciliation—was to craft 'a balanced force' over 'the next two decades' with 'acquisitions *currently underway* that will better equip the ADF for security and stabilisation operations' but also replacement of 'existing platforms with more potent options' for high-end war-fighting (Houston 2010; emphasis added).

The point to be made here is that Houston did not envisage wholesale changes to ADF force structure or the rapid acquisition of capability to meet new Government priorities, let alone HADR or, even more tenuously, climate change. Rather, Houston largely stuck with a formula he had stated as early as 2007 when he described the need for a 'balanced, networked and deployable [joint] force' capable of both 'low intensity' and 'high-intensity' operations in an 'increasingly complex world' (Houston 2007: 60 and 61).

Overall, the findings of the CDF Programme, are important for what it has *not* found. No major publicly available speeches directly relating to climate change were made by the CDF or Secretary of Defence in the period examined. Similarly, no major policy initiatives relating to climate change mitigation or adaptation were declared by the ADF executive. On the contrary, the CDF Programme revealed a gradual incorporation of climate change considerations following the election of the Rudd Government in late 2007. Though its responsibility to assess the military implications devolved to discreet areas within the Defence bureaucracy, climate change was consistently regarded as a long-range security issue that would most likely increase requirements for Defence humanitarian and disaster relief. In this sense, climate change was not mainstreamed in the ADF and rather considered itself an 'agency of last resort' that was simply 'consistent' with the Rudd-Gillard policy.

5.3 Climate Security in ADF Operational and Tactical Programmes

The operational and tactical areas of the ADF are considered the *ways* and *means* that implement national strategic *ends*. This research initially attempted to treat *ways* and *means* separately; however, it was apparent that—for the purpose of this research—no genuine distinction was possible. For instance, speeches made by the Service Chiefs were initially considered under Tactical Programmes but then some journal articles published by mid-ranking officers were considered under Operational Programmes. Although this made sense during the research design phase, it is clearer for the reader to have the combined perspective (synthesis) of the tactical and operational writings. As a consequence, the empirical results are mostly represented independently while the written narrative is an integrated account of *ways* and *means*. This enabled a much more fluent account of how the ADF at lower levels—outside the political realm—responded to, and framed climate change.

5.3.1 Overview

Cumulatively, this research analysed 924 artefacts from operational and tactical areas within the ADF. Artefacts examined included speeches, official Defence newspapers and magazines, journals and Parliamentary Hansard recordings by senior operations spokesmen. Of the 924 documents examined, this research identified 464 direct references (unfiltered) to the term *climate change* in fifty-five distinct sources. When the documents were analysed for inconsequential references to *climate change* (e.g., in margins, footnotes or otherwise those documents explicitly about climate change), there were 231 direct references to *climate change* from fifty-three distinct sources. Accounting for this, the research coded a total of sixty passages of text, thirty-six of which were coded in a securitised frame and

Table 5.15 Climate change framing in the ADF (operational and tactical levels)

a. ADF Operational Programmes

Frame	**DSG**	**CIOG**	**DPG**	**DSTO**	**VCDF**	**CDG**	**DMO**	**Total**
Securitised	0	0	4	0	13	0	0	17
Non securitised	4	0	8	1	7	0	0	20
Total	4	0	12	1	20	0	0	37

b. ADF Tactical Programmes

Frame	**Navy**	**Army**	**RAAF**	**INT**	**JOC**	**Total**
Securitised	3	12	2	0	2	19
Non securitised	2	0	0	0	2	4
Total	5	12	2	0	4	23

c. ADF Operational and Tactical, combined

Frame	Total
Securitised	36
Non securitised	24
Total	60

Source The author

twenty-four in a non-securitised frame (Table 5.15c). Details of the sub-programmes and empirical data are footnoted.[25]

Before presenting further empirical findings, it is worth commenting on this finding. The main observation, when compared against the Strategic Programme, is

[25]Seven separate programmes were initially analysed under the functional grouping of Operational Programmes. The seven programmes (Defence Support Group, Vice Chief of Defence Force, Chief Information Officer Group, Defence People Group, Defence Science and Technology Organisation, Defence Materiel Organisation and Capability Development Group) were responsible for ADF workforce capability, estate management, capability development and acquisition, information communication technology, science and technology advice including capability technical risk assessments and joint training and doctrine. Operational Programmes analysed 301 artefacts of which fifteen contained 390 direct references to the term ‘climate change’. Of these 390 references, only seventeen were coded within a securitised context and twenty were coded within a non-securitised context. (This data was heavily skewed by two documents (a 2006 Defence People Group document *Defence Personnel Environmental Scan 2025* and a journal paper for the Centre for Defence and Strategic Studies titled *Climate Change in the Asia-Pacific Region: Security Implications for Australia*). When these documents were removed from the word search, there were thirteen sources containing 157 references. To ensure results were not skewed by the high number of references in these two artefacts, only major conclusions and/or mutually exclusive textual passages were coded). Similarly, five separate programmes were analysed under the functional grouping of ‘Tactical Programmes’. The five programmes (Army, Royal Australian Navy, Royal Australian Air Force, Intelligence and Joint Operations Command) were responsible for delivering tactical effects in the ‘battle-space’. Tactical Programmes analysed 623 artefacts of which 40 sources contained 74 direct references to ‘climate change’. Of these 74 references, 19 were coded within a securitised context and four were coded within a non-securitised context.

the reduced number of direct references to the term *climate change*. Moreover, as a proportion of all documents examined, the Strategic Programme yielded a quotient of 1.46 (945 references from 646 documents), while the Operational and Tactical Programme yielded 0.25 (231 references from 924 documents). Similarly, there was a significant reduction in the number of coded securitised or non-securitised passages. Whereas Strategic Programmes contained seventy-six securitised passages and 285 non-securitised passages, Operational and Tactical Programmes contained just thirty-six securitised and twenty non-securitised. Why was this distinction so pronounced?

Four possibilities are identified. The first was the possibility that the operational and tactical levels were simply not focused on climate change (compared with the political-strategic levels), since the threats posed by climate change were long-term (decadal) and were therefore not considered relevant to the operational or tactical planning cycles which act in days, weeks, months and years. In this regard, climate change did not yet 'belong' at the operational or tactical levels and was rather framed as a strategic issue. The second reason relates to the nature of military culture. Sensing that it was a politically and military-strategically divisive issue, the lower levels of the ADF may have simply taken their cues from the Minister of Defence, the CDF and Secretary as well as their direct Service Chiefs. Moreover, it rarely featured as a high priority in any of their speeches or policies. Thus, for an organisation renowned for its hierarchal structure and obedience to the 'chain of command', the operational and tactical levels were simply 'following orders' by refraining from the debate. Pezzullo's remark that 'we steer well clear' of the politics emphasises this point. Linked to this idea is the notion that the lower levels of the ADF are about the implementation of strategic policy rather than its *generation*. Thus, journals, newspapers and speeches at the operational and tactical levels tended to centre on discussion and analysis of issues pertaining to *means* and *ways* rather that *ends*.

A third reason for the reduced attention paid to climate change at the operational and tactical levels relates to one of the limitations of this research method. Put another way, climate change may have been a prominent issue within the lower levels of Defence, but the inability of this research to access classified documents, or documents from a broader cross-section of Defence, precluded a complete assessment. Lastly, it was possible that while the Operational and Tactical levels were concerned about the security and other implications of climate change, they simply lacked the resources, platforms (forums) or executive support to fully consider and discuss its effects.

In terms of language used to frame climate change as a securitised issue, the operational and tactical levels tended to use heightened adjectives (climate change was framed as a *challenge* or *threat* on twenty occasions), Table 5.16. Interestingly, climate change was never framed as an *opportunity* by operational-tactical military personnel but was framed by Defence civilians as such (on two occasions by DSTO and the Defence Business Industry Unit, discussed below).

Table 5.16 Language used to frame climate security in ADF operational/tactical levels

a. ADF Operational Programmes

Frame	DSG	CIOG	DPG	DSTO	VCDF	CDG	DMO	Total
Opportunity	0	0	0	0	0	0	0	0
Neutral	0	0	2	0	2	0	0	4
Challenge	0	0	1	0	4	0	0	5
Threat	0	0	1	0	7	0	0	8
Total	0	0	4	0	13	0	0	17

b. ADF Tactical Programmes

Frame	Navy	Army	RAAF	INT	JOC	Total
Opportunity	0	0	0	0	0	0
Neutral	2	9	1	0	0	12
Challenge	1	1	0	0	1	3
Threat	0	2	1	0	1	4
Total	3	12	2	0	2	19

c. ADF Operational and Tactical, combined

Frame	Total
Opportunity	0
Neutral	16
Challenge	8
Threat	12
Total	36

Source The author

Few occasions were identified by this research in which operational and tactical personnel framed climate change in terms of timing. Of the four instances, three framed climate change as a long-term set of challenges and one framed it as a medium-term challenge. Although this provided only a limited data set, it was consistent with the temporal framing of the ADF Executive.

In terms of security categories, this research found the operational and tactical areas framed climate change as a *global and regional* security issue (eleven coded references), a *non-traditional and new* security issue (nine coded references) and as a *convergence* security issue (seven coded references). Climate change as a national security concern had four coded references with a further two security categories receiving some attention. In total, climate change was framed as a security matter across seven different security categories. This indicated that the operational and tactical areas of the ADF held a wide (broad) perspective on climate security. This differed somewhat from the ADF Executive but was similar to Rudd-Gillard perspective. The following assessment analyses the Operational and Tactical Programmes in further detail.

5.3.2 Senior ADF Operational and Tactical Commanders

This research analysed ninety-three speeches by the three Service Chiefs (Army, Navy and Air Force) between 2003 and 2013.[26] Of the ninety-three speeches, just *six* contained any reference to climate change. Five of these were by the Chief of Army and one was by the Chief of Navy. All references to climate change by the Service Chiefs occurred from 2007 onwards. Most were delivered as fleeting references under the gambit of it being an emerging security issue and in the context of promoting the security outlook espoused by the 2009 White Paper, *Force 2030*.

Consider, for instance, the fifty-five speeches and dozen or more articles that were examined by this research and published by the Chief of the Australian Army between 2003 and 2013. The first mention of climate change—found in the sources examined by this research—was by Chief of Army Lieutenant General Peter Leahy in 2007. He cited climate change as one of many contributing factors in driving migration and the creation of 'mega-cities' that portended an increase in operations 'amongst the people' and in 'complex urban terrain' (Leahy 2008: 13–14). However, reflecting the focus of the Howard years and the Army's on-going role in Iraq and Afghanistan, Leahy 'directed that Army's intellectual resources be focused on COIN [counter-insurgency]' (Leahy 2008: 14). Accordingly, across the totality of his tenure, Leahy's main focus was to develop a 'Hardened and Networked Army' (HNA) and later the 'Enhanced Land Force' (ELF) that he cast as a 'decade-long programme [to] restructure and re-equip Army for operations in the 21st-century battlespace' (Leahy 2004: 12). Climate change had little to no impact on this restructure and expansion.

In a 'big-picture environmental scan', the next Chief of Army, Lieutenant General Kenneth Gillespie, cast climate change under the 'other [new] area' category, and noted the '*potential* impacts climate change' (Gillespie 2008: emphasis added). In a previous role as the Vice Chief of Defence Force, Gillespie had presciently noted concern about the 'combative effect' of climate change but also the 'compliance effect' brought about by potential legislative changes.[27] Beyond these passages, however, no other discussion by General Gillespie on climate change was found by this research. Indeed, in a notable quote, Gillespie articulated the strategic inflection point between the demise of Howard-era security and the rise of Rudd-era security:

---

[26]While the service chiefs are technically not in the operational chain of command, they undoubtedly influence its direction by virtue of setting the broad 'raise, train and sustain' agenda adhered to by each Service on a daily, monthly and annual basis. In this sense, they were considered 'operational'.

[27]Under Labor Clean energy Package legislation, the estimated effect of a carbon price on Defence in 2012–23 was 'in the order' of $80.4 million ('0.32%' of Defence annual budget). See (2012a: Question 111).

> It is an era of ambiguity, in a region in which a number of security issues from piracy, terrorism, water and energy security, to traditional inter-state rivalry are vying for attention (Gillespie 2008).

Ultimately, General Gillespie's outlook—similar to General Leahy—was focused on enhancing Army's capability. Moving on from HNA and ELF, Gillespie implemented the 'Adaptive Army' initiative which he championed, no less, as 'the most significant restructuring of the Australian Army since the implementation of the Hassett reforms in 1973' (Gillespie 2009: 9).

Lieutenant General David Morrison, Chief of Australian Army from 24 July 2011 to 15 May 2015, cited climate change in the context of the military contributing to the preservation of international order and being able to 'effectively manage other risks and threats' (Morrison 2012). Rejecting the idea of a peace dividend following operations in the Middle-East and the near Pacific, Morrison warned of succumbing to the 'insular' outlook pervasive in strategic doctrine between 1976 and 1999. Thus, the focus for Morrison became Plan BEERSHEEBA, a 'vital structural reform' of Army to improve training, force generation and operational rotation. Echoing previous Army chiefs, Morrison (2012) heralded Plan BEERSHEEBA as 'one of the most important developments since 1976 and the beginning of the modern strategic era of Australian strategic policy'.

In summary, no Australian Service Chief in the speeches and articles examined by this research had outlined the security implications of climate change to any significant degree. Not a single Australian Service Chief (examined by this research) decreed any major responses to climate change, either in terms of adaptation or mitigation. This was largely consistent with the most senior Defence executive (CDF and Secretary) but contrasted somewhat with junior and mid-ranking military officers, who periodically contributed ideas and opinions via professional journal articles. Notable also was the absence of any contribution on climate change by the Royal Australian Air Force. Neither the RAAF Service Chief discussed the matter nor was there much evidence that the lower levels of the air force had considered the issue or developed any response.

5.3.3 *Mid-ranking ADF Operational and Tactical Officers*

This research analysed 722 official journal articles, newspapers and media releases from the operational and tactical areas of the ADF between 2003 and 2013. These articles offered a unique account of how climate change was perceived and framed at the operational and tactical levels. Journal articles (in particular) were the most productive of these sources from which to gauge military opinion on climate change, since they were published in forums designed to stimulate discussion 'free of the constraints inherent in normal staff processes' (Shanahan 2013: 6). This differed from the tightly scripted and controlled speeches and statements by the

ADF senior leadership. This facet was somewhat reflected in the research results. Thus, whereas climate change was rarely declared a security threat by senior ADF officers, it was more frequently cited as one by mid-ranking officers. The most prominent example of this was an article (*Climate Change in the Asia-Pacific Region: Security Implications for Australia*) written in 2007 by RAAF Group Captain Rob Lawson and published by the Centre for Defence and Strategic Studies. Citing the 'threats' posed by climate change on eleven separate occasions, Lawson assessed climate change as a 'valid environmental security issue requiring forthright action and prioritisation' (Lawson 2007: 6). Portending the contents of the 2009 Defence White Paper, Lawson added that climate change 'will need to be factored into established national capability and operational planning processes with emphasis on border protection, counter-terrorism and disaster relief' (Lawson 2007: 17).

Lawson's ideas took hold when journal articles began to appear in which climate change was framed as a factor justifying improvements to Defence capability. It was particularly noticeable in Army journals relating to HADR. Writing in the *Australian Army Journal*, Lieutenant Colonel Peter Woodward argued for increased role of the Army Reserve to deal with increased prevalence of natural disasters. For Woodward, '[c]limate change is forecast to increase the incidence of extreme weather events over time. Flow on effects may be water shortages, greater health problems including disease contagion, flooding, coastal erosion and storm surge' (Woodward 2011: 107). Woodward subsequently made eleven recommendations to improve the Army Reserve's HADR response capabilities, including enhanced training and improved integration between ADF and civilian emergency service authorities.

Relatedly, Brigadier Chris Field cited the 'vital role' espoused by the 2009 *White Paper* for the ADF in 'supporting domestic security and emergency response efforts' (Field 2011: 121). Field argued for ten ideas on 'fixing', 'improving' and 'sustaining' ADF HADR capability. Field further identified Defence as a 'national institution with a significant public profile and high degree of community trust', in which he considered domestic disaster relief as being particularly important given the degree of direct interaction with Australian citizens (Field 2011: 121).

In addition to these journal articles, Commander Steven Cole (with the title of 'Environment Manager' in Navy Strategic Command) presented alongside US military counterparts at the 2010 *Climate and Energy Symposium* held by The John Hopkins University. In his presentation to the symposium, Cole (2010) forecast a wide number of operational impacts from climate change including base infrastructure impacts (e.g., reduced wharf deck clearances, loss of 'some' coastal facilities, reduced performance of breakwaters) and regional security issues (e.g., climate refugees, increased HADR, loss of fisheries). In what was largely found to be an isolated statement, Cole (2010) cited climate change as justification for improving Navy's resilience and capability growth: 'We will need bigger, more seaworthy and more capable ships to cope with projected operational requirements and limitations imposed by climate change'. This was a statement typically *not* found expressed by the senior leaders of the three services.

The broader point from these essays—beyond the merits of their suggestions—is the observation of an emerging discussion taking place regarding the tactical implications of climate change on the ADF. By and large, they focused on the ADF as a type of 'first responder' in the event of a major natural disaster and contrasted with the ADF strategic level that framed it as the agency of 'last resort'. It also hinted that the operational and tactical levels were more advanced in their thinking on climate change than the senior ADF leadership. Arguably, these might be construed as a form of institutional climate adaptation—where middle-ranking officers begun to scope how the institution must grapple with the coming challenges. On the preceding evidence, this can be viewed as a bottom-up process.

Associated with these ideas was that some military-civilian authors began to directly frame climate change as an opportunity. This was distinct from the executive elements of the ADF, who tended to frame climate change in neutral terms or as a security challenge or threat. A 2010 journal article by Ben White from the Australian Business Defence Industry Unit proposed that a new paradigm was required that set Defence on a path towards becoming more sustainable. Framing it as an opportunity, White called for the establishment of a sustainable national security model consisting of a sustainable defence capability (largely via development of renewable energy options for platforms) supported by a sustainable defence industry. White argued that by establishing this framework, Australia could play a lead role in responding to the effects of climate change and be in a position to develop a world-leading industry capability reaping 'innovation, employment and economic benefits' (White 2010: 87).[28]

Another opportunity was the idea that Defence should seize the moment presented by public and political momentum to contribute to mitigating climate change through improved fuel efficiency standards and practices. The broader picture, however, concerned Defence ensuring its own energy security (particularly in terms of concerns over so called peak-oil and the ability to provide cost effective and reliable supplies in a domestic and deployed environment).

Climate change, however, was only one facet of the energy security debate. Often it was downplayed as a reason to justify new energy measures. Dr. Jennifer Palmer (2009: 6) from Defence Science and Technology Organisation (DSTO) observed that 'shifting public opinion about environmental issues, changing government policy, and the escalating price and shrinking supply of oil' would be a 'key driver' of any future Defence energy requirements. Major White (2008: 12) was more blunt when he argued that 'fossil fuel depletion … is arguably the most pressing problem facing military commanders today'. Despite this, he downplayed environmental (climate change) justification since 'this approach is unlikely to achieve the necessary traction to promote action within the ADF'. In a telling passage, White argued:

[28]Note: Articles by Mr. Ben White and Chun Zhang were not coded since they are not ADF members.

> [C]limate change and greenhouse gas emissions are not perceptively pressing concerns [for the ADF] that are physically tangible, and consequently not issues that have been transcribed into [ADF] policy and capability development (White 2008: 12).

This comment reflected the wide sentiment that, although climate change was noted as an *issue*, *challenge* or even a *threat* with the beginnings of institutional adaptation, it arguably lacked a larger and more comprehensive strategy that had the backing of the ADF leadership. Arguably, the lack of ADF leadership on climate change by the ADF executive promoted disinterest at the mid-levels of the organisation since there was little prospect of research or policy initiatives being supported. This was not the case with debate regarding Defence energy security, where the publication of Defence Energy policy (2007) and constant policy emphasis (see below) kept discussion and analysis alive in the various ADF journals.[29]

The overall sentiment of intellectual stultification regarding climate change was captured by Army Land Warfare Studies Centre senior analyst, Dr. Albert Palazzo. In an aptly titled 2012 essay, *The Future of War Debate in Australia: Why has there not been one? Has the need for one now arrived?* Palazzo contrasted debate in the Australian military with those underway in the US military: '[T]he Australian Army has been one of silence' (Palazzo 2012: 1). Particularly relevant for this study, Palazzo identified cultural, bureaucratic and operational impediments in the Australian Army and military more broadly as key reasons for the absence of a (domestic) debate on the future of war, including the security implications of climate change. Coming from a senior civilian 'insider' working within the Department of Defence Land Warfare Studies Centre, the assessment by Palazzo offered an acute appreciation. Ultimately, the three reasons he gave provide some insight as to why climate change failed to gain a foothold in the operational and tactical levels of the ADF (Palazzo 2012: 8–18):

- *Cultural impediments*. Palazzo described a culture of military anti-intellectualism which favoured 'doers' over 'thinkers' and 'outputs over outcomes'. Consequently, this prevented a conducive environment for the development of challenging ideas and frank debate.
- *Bureaucratic impediments*. Palazzo cited institutional barriers in which the Department of Defence had 'set in place policies that discourage … forums [and] debates'. Rather, it was one where 'the institutional preference is to have full control of ideas and messages, particularly if they are unorthodox ones'.
- *Operational impediments*. Palazzo suggested that the 'force's leadership' do not see the requirement to debate the future of war, including climate change. Palazzo further argued that the Australian Army had become a niche contributor —possessing well developed tactical and operational nous—but 'compromised' when required to 'interpret the changing character of war'.

[29]See, for example, White, Martin. 2008. 'Compelling Requirement to Energy-Proof the Australian Defence Force.' *Journal of the Australian Profession of Arms*, no. 175; Palmer, Jennifer. 2009. 'Addressing Energy as a Military Cost.' *Journal of the Australian Profession of Arms,* no. 178; Gray, Matthew. 2009. 'Peak Oil theory: Implications for Australia's strategic outlook and the ADF.' *Journal of the Australian Profession of Arms,* no. 180.

5.3.4 Climate Change in ADF Doctrine

This research analysed thirty-six articles of non-classified ADF operational and tactical level doctrine. From this analysis, climate change was found to enter formal Australian military doctrine from 2007. This was consistent with analysis of speeches by the ADF executive and senior ADF operational commanders. In total, six separate sources of doctrine were identified as referencing climate change.

The doctrinal publications analysed by this research framed climate change as a global security threat or as an environmental or resource-competition challenge. Apart from one report (*Scan 2025*) that 'did not represent the views of the Department of Defence', no doctrine identified climate change as a risk to Defence infrastructure, workforce or other fundamental inputs to capability.

The first doctrinal reference to climate change identified by this research occurred in 2007, *Joint Operations For the 21st Century* and in the Navy's *Future Maritime Operating Concept—2025: Maritime Force Projection and Control* (Defence 2007b: 1) that noted the 'entirely new' security challenge of climate change:

> Other threats to Australia's territorial integrity, sovereignty and broader national interests could also challenge Australia's security in a way that requires the application of military force. Global factors (such as terrorism, pandemic disease, resource depletion and the *security impacts of climate change*) … may affect Australia's security interests, both directly and indirectly.
>
> *Singly, any one of these threats could disrupt the military balance between states* or present non-state actors with an opportunity to challenge state power. Together they shape the new security environment within which the ADF must operate (Defence 2007b: 4–5; emphasis added).

Several observations can be made on this statement. First, it indicated that although the ADF executive were not outwardly addressing climate change at this time, the issue was seemingly mature enough to be included within a major capstone doctrinal piece. Although this point should not be overstated, this indicated some dissonance within the ADF between those who considered climate change a threat and those that had simply not considered it an issue. Secondly, the inflated rhetoric (climate change may 'singly … disrupt the military balance between states') appeared incongruent in relation to the actual amount of analysis and discussion occurring within the ADF on climate change as a security concern. Thirdly, the idea that climate change might disrupt the military balance reflected a unique insight into the mind-set of Defence strategists who may have conceived of climate change as presenting strategic benefit should a competitor military be degraded as a consequence of climate change. On balance, however, climate change was largely considered one of many future threats that, in the words of General Gillespie, were simply 'vying for attention' (Gillespie 2008). Thus, apart from the quote identified above, *Joint Operations For the 21st Century* did not expand any further on climate change.

The 2007 *Future Maritime Operating Concept* (FMOC) framed climate change in a slightly different context when it labelled it as an environmental issue that may have longer-term and 'potentially serious repercussions for our region', such as 'submerging smaller islands or archipelagic nations thereby reducing the size of the Exclusive Economic Zone (EEZ) and displacing populations' (Royal Australian Navy 2007: 8). Further, FMOC identified a number of capability implications regarding the requirement for increased ability to respond to 'environmentally driven security and humanitarian issues in urbanised, littoral areas, particularly where infrastructure is poor'. Specifically, the doctrine called for capability 'able to deliver security and assistance capabilities from a sea base … to have to provide mobility, logistic support and hospital services' (Royal Australian Navy 2007: 8). This assessment forecasted the 2009 Defence White Paper in the sense that climate change would equate to an increase in ADF HADR commitments.

Climate change first entered Australian Army doctrine in the 2009 *Army's Future Land Operating Concept*. This framed climate change under 'resource competition' and identified it as likely to have 'wide ranging economic and resource impacts on Australian, its region, and globally' (Australian Army 2009: 11). It specifically identified that 'climate change and the rise of sea-levels threatens the long-term viability of some island nations' (Australian Army 2009: 13). Beyond this, the doctrinal focus by Army reflected the capability initiatives and thinking of its various Service chiefs through *Complex Warfighting* (2004) and *Adaptive Campaigning* (2006). Bluntly, climate change was not a key feature in Australian Army's capstone doctrine.

Of interest, no doctrinal reference to climate change was found in those publications examined from the Royal Australian Air Force (RAAF). This was consistent with a general observation that the RAAF generally contributed little if anything to discussion on climate change. Apart from minor references that quoted passages from the 2009 Defence White Paper, no single reference of climate change was found in any Air Force documentation (fifteen speeches by the Chief of Air Force, 124 official newspapers and nine articles of doctrine).[30] As a broad observation, this correlated with academic literature that—beyond mitigating air-craft greenhouse gas emissions—identified few security or other impacts for Air Forces more generally.

The final article examined by this research within the context of military doctrine was a 2006 publication, *Defence Personnel Environment Scan 2025*. This report was noteworthy for several reasons. Firstly, although it did not represent the official position of Defence, it was the first report by Defence (in any form) that addressed the broader strategic impacts of climate change. It also did this in some detail, dedicating an entire chapter to the science and broader impacts of climate change.[31]

---

[30]The only exception to this was the single essay by GPCAPT Robert Lawson, identified earlier.

[31]At the time when *Scan 2025* was written, it was also *ahead* of the Lowy Institute publication *Heating Up the Planet* by Dupont and Pearman, which similarly addressed the security implications of climate change on Defence.

Secondly, *Scan 2025* identified a range of issues beyond immediate security concerns and pointed toward an erosion of the ADF's personnel capability as a consequence of climate change. On this front, *Scan 2025* identified climate change as leading 'to a number of strategic scenarios requiring Defence intervention' (Reich et al. 2006: 240). These included increased requirement to protect the nation from 'physical or warlike threats'; increased requirement to support national and international efforts to deal with 'climate triggered situations' such as natural disasters, environmental refugees and the spread of diseases and pandemics; contributions to enhanced and efficient use of natural resources; and a requirement to develop and train a 'large workforce' to establishing new technologies and associated infrastructure that help address national energy and water needs. In this sense, *Scan 2025* pre-dated much of the US literature that identified climate change as a burden multiplier.

Thirdly, the report was significant for its heightened use of language. This was unusual to any of the official Defence literature examined by this book. For instance, *Scan 2025* framed climate change as a pending catastrophe and identified it as 'one of the most serious environmental issues facing Australia and the world' which 'can no longer be ignored by any organization' (Reich et al. 2006: 223). On the whole, however, *Scan 2025* was an exception. Furthermore, the point that it was not officially endorsed by the ADF suggested a willingness to distance the institution of Defence from the report's conclusions. This was supported by a source from within the Department of Defence, who observed:

> The point should be made that many climate change 'papers' have not always been published or seen the light of day … it was all 'toned down' … ONA input on climate change was ignored.
>
> [Regarding Scan 2025] my [manager] wanted the entire Chapter on climate change removed but [a separate manager] said 'no' and ensured it stayed in. I would say this challenged the political reality at the time; one person saw the strategic implications and science and need to raise the issue; *the other saw a risk to political acceptance* (Personal Communication 2013; emphasis added).

This account, coming from within the Department of Defence, supports the broader observation by this book of the inherent challenges of a non-partisan organisation assessing partisan political security threats. It also resonates with what Albert Palazzo described as 'bureaucratic impediments' in which internal debate was stifled to the point of institutional conformity (though, in this instance, the chapter on climate change was permitted). Nonetheless, this remark also hinted at a paradox: while the political conditions were favourable (under a centre-left Labor government) for Defence to publish on climate change, there appeared a departmental level reluctance to do so.

5.3.5 ADF Operational and Tactical Climate Change Response Measures

Having examined the speeches and journal articles of ADF personnel, this section highlights the major policies enacted by the ADF operational and tactical levels in response to climate change. This was not a straightforward process. Many of the programmes initiated by the ADF were related to climate change but were not conspicuously promoted and undertaken on the basis of either adapting to or mitigating climate change. Examples of this included numerous 'energy efficiency', 'greenhouse gas' or 'sustainability' programme. In addition, minor responses by the ADF are likewise not included (for example, the minimal ADF contribution to the government response to the 2009 House of Representatives Standing Committee on Climate Change, Water, Environment and the Arts, *Managing our Coastal Zone in a Changing Climate: The Time to Act is Now*).

Working on this premise, the research identified *three* main programme and/or initiatives that directly related to (and were conspicuously promoted as) adapting to or mitigating the impact of climate change at the ADF operational and tactical levels. Notably, all occurred *after* the election of the Rudd Government in 2007. These included:

- 2008–Present. *Combat Climate Change* initiative (Defence Support Group).
- 2011–Present. Global Change and Strategic Military Geography Defence Study (Vice Chief of Defence Force).[32]
- 2011–2013. *Climate Change Adaptation and Mitigation Study* (Defence Support Group).

A number of other climate change programmes were identified but were not able to be considered. For instance, the 2009 *Defence Annual Report* (DAR) promoted an internal circulation of a 'Climate Change Discussion Paper' that would inform a 'Defence Climate Change and Sustainable Development Strategy' (Defence 2009a: 186). To the awareness of this research, the paper was never developed and remained unpublished. The 2009 DAR also stated that "Defence continued to implement initiatives in support of the Government's agenda on climate change. This included work to set climate change and sustainability policies and targets for implementation across the Defence estate" (Defence 2009a: 186). By 2010 however, reference to 'climate change' and 'targets' had been dropped in favour of 'the development of resource efficiency strategies and trials or new technologies' (Defence 2010a: 311).

More prominent was the promotion of Defence 'energy management' strategies in which the ADF was active in promoting a range of energy and sustainability projects between 2003 and 2013. Although climate change was not promoted as a

[32]Now: *Global Change and Energy Sustainability Initiative*.

driving factor for these initiatives it was a related factor and could have been used as justification if the ADF had so desired.[33]

However, the very fact that Defence went to significant lengths to promote itself as a good environmental manager *but then did not significantly include climate change in any part of its overall environmental strategy* appeared incongruent. Certainly, by 2010, climate change could not have been 'missed'. And yet, in its peak environmental strategy document, *Defence Environmental Strategic Plan*, climate change was directly referenced just twice. On one of these occasions it was done in the context of political necessity, 'Defence must be prepared to implement the *Government's agenda* on … climate change' (Defence 2010b; emphasis added). Furthermore, the sense that Defence actively refrained from even using the term *climate change* was evident when the CDF and Secretary otherwise called it 'climate variability' in the plan's introduction. These examples introduce the suggestion that Defence was—by 2010—wary of engaging in climate change discourse on account of the polarised political debate.

Having made this observation, attention now turns to examining the three main Defence climate change programmes. The first, *Combat Climate Change*, was publicly launched by the Minister for Defence, Joel Fitzgibbon, in 2008, but appeared (from the press release at least) to emanate from within the Department itself. *Combat Climate Change* was arguably the most visible initiative whose aim was to increase awareness of climate change and ADF greenhouse gas emissions and to inform individuals how they could 'make a difference'. With some flair, the *Combat Climate Change* website proclaimed:

> One of the greatest challenges for the world is to combat and adapt to climate change. Australia's Defence Force aims to do *whatever it can* to halt its contribution to greenhouse gas emissions across its broad range of activities (Defence 2008a; emphasis added).

But this was more rhetoric than reality. Analysis of ADF greenhouse gas emissions since the commencement of this programme showed an *increase* (see Fig. 5.6). In addition to this, in 2010 the then Chief of Defence Force confirmed that Defence did not calculate its overall greenhouse gas footprint (domestic and offshore). Asked about this during a 2009 Senate Estimates hearing, the CDF

[33]A summary of Defence environmental programmes and Defence Energy policies 2003–2013 is available on request from the author. This research identified fourteen programmes relating to energy efficiency and forty-five relating to general environmental issues (e.g., water, sustainability, waste etc). Defence's capstone Environmental Policy was published (approximately) every five years. The policy document examined by this book, *The Defence Environmental Strategic Plan 2010–2014*, identified six policy objectives that included: implementing best practice approaches to environmental management, integrating sustainable environmental management techniques into Defence business practice, establishing clear lines of accountability, raising 'environmental awareness', measuring environmental performance, maintaining transparency and establishing partnerships. *Defence Environmental Strategic Plan 2010–2014*, contained 68 specific 'commitments' with an associated thirty-two detailed performance metrics (KPIs). The *Plan* covered seven areas of Defence, including ADF Training and Operations, Defence Capability, Defence logistics, Defence Estate, ICT Procurement, Science and Technology and People Governance.

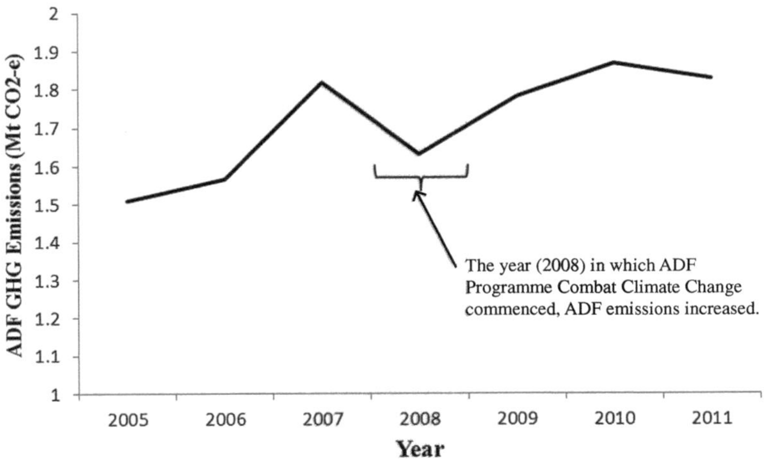

Fig. 5.6 ADF total GHG emissions 2005–2011 (excluding overseas military operations). Note the increase since the inception of Combat Climate Change initiative in 2008. Data compiled from multiple reports and available from *Energy Use in the Australian Government's Operations* publications. *Source* Commonwealth of Australia (2013c, see website for reports at 2005–06: 11, 2006–07: 11, 2007–08: 44, 2008–09: 36, 2009–10: 42, 2010–11: Annex C) (*Energy Use in the Australian Government Operations* ceased publication from 1 July 2013. Energy data including emissions are thereafter reported by individual agencies, at: http://www.industry.gov.au/Energy/EnergyEfficiency/Non-residentialBuildings/GovernmentBuildings/EnergyUseOperations/Pages/default.aspx (13 August 2016).)

described how Defence was contributing to CO_2 reductions through energy efficiency measures such as optimal cruising speeds for planes and ships, but then conceded 'we do not measure it … I guess that is something that we are going to have a close look at in the near future' (Houston quoted in Commonwealth of Australia 2009b: 64).

Furthermore, while many of the goals outlined in the various Defence energy initiatives were admirable, most describe improving energy *intensity* and *efficiency*. Thus, there was no legislative obligation or regulatory requirement for Defence to actually decrease the *quantity* of its emissions footprint. Even more than this, the Government also expressly stated 'there is no policy requirement [for the ADF] to reduce operational fuel consumption' (Commonwealth of Australia 2013a: ii). Given that operational fuels constituted more than 50% of all Defence greenhouse gas emissions, the statement that the ADF was doing 'whatever it can' appeared somewhat hollow.[34] Lastly, and perhaps most tellingly, was the assertion in the *Defence Energy Management Strategy* (2005b) that stated any measures to improve energy efficiency and reduce energy wastage would only be done so 'within the

[34]See Energy Use in the Australian Government's Operations. Current and previous editions available on internet at: http://industry.gov.au/Energy/EnergyEfficiency/Non-residentialBuildings/GovernmentBuildings/EnergyUseOperations/Pages/PreviousEditions.aspx (27 October 2016).

context of *not compromising Defence capability*' (emphasis added). This was not a surprising announcement, but simply a clear articulation of the relative priority between contributions toward the environmental initiatives against the hard practicalities of national security.

These outcomes are perhaps not surprising, considering the finer details of the *Combat Climate Change* initiative. Moreover—and much like the various Defence energy management strategies before them—they arguably represented small-scale energy saving initiatives that did not articulate a broader, systemic approach to reducing energy and greenhouse gas emissions across the wider organisation. Simply put, the focus of the programme had a short-term outlook with an intent to influence individual behavioural change rather than organisational change regarding energy and resource use.

Consider, for instance the three key areas of *Combat Climate Change* regarding how Defence personnel can 'reduce greenhouse gas emissions' by reducing energy use, waste and vehicle emissions through 'turn[ing] off lights … PC monitors … appliances in breakout and common areas … printers, photocopiers and faxes after use' (Defence 2008a). Other suggestions included car-pooling, printing 'double-sided' and even using a more water efficient shower heads at home since '[w]e're light on [sic] for water in Australia so this is a good idea anyway' (Defence 2008a). Other such examples were abundant. Although these initiatives were commendable, they were hardly programmes of substance. In addition, there was no accompanying published policy paper other than the single page media release by Joel Fitzgibbon in 2008. Furthermore, no reference to the programme was found in any of the hundred or more speeches analysed.

A second ADF climate change response identified by this research was a programme administered by the office of the Joint Capability Co-Ordination Division (headed by an Air Vice-Marshall) within the VCDF Group. The programme, *Global Change and Energy Sustainability Initiative* (GCESI), recognised that Defence would not be immune from the 'significant new challenges that will affect national security, operating environments, people and budgets more than ever over the coming years' (Defence 2011). Its key goals included 'enhancing organisational situational awareness, increasing the quality of impact assessments, improving energy resilience, strengthening security strategies and identifying current and future challenges for Defence capability' (Defence 2011). Through three projects (forums, 'flash-jams' and workshops), GCESI also focused on 'climate variability and associated energy issues' with the aim of assessing the impact of these on ADF preparedness (Defence 2011).

GCESI was notable for three reasons. In the first instance it represented one of the first occasions where Defence had outwardly engaged with broader academic, military, scientific and industrial communities on climate change. Secondly, GCESI sought to treat climate change beyond the security realm and rather examined a range of impacts across capability, policy and operating concepts. Thirdly, GCESI existed at a relatively senior level within the Defence organisation, headed by a two-star officer. Overall, the GCESI project also represented the closest approximation to the 'mainstreaming' of climate change within the Department of Defence.[35]

Despite this, a true assessment of the programme's output and effectiveness remained difficult. From the material available via open-source internet, no major policy documents or reform changes were identified as flowing from the GCESI programme. Similarly, no indication of the programme having major influence was found within the broader military literature examined by this research. On these grounds, GCESI appeared to represent something of a large 'scoping study' with climate change embedded within a broader global change framework that examined issues on the longer term strategic horizon.

The third major initiative concerned the *Defence Climate Change Adaptation and Mitigation Study*. This study was considered a more long-term strategic approach and was completed in two phases by Defence industry contractors AECOM Australia Pty. Ltd. between 2011 and 2013. Phase One, *Assessment of the Impact of Climate Change Induced Sea Level Change on Significant Defence Bases*, was a broad study to identify Defence bases most at risk (Commonwealth of Australia 2012a). Phase Two of this assessment was completed in mid-2013 and specifically examined high risk bases (determined during Phase One). The total cost of both assessments was $2.14 million.[36] Although, both reports were classified RESTRICTED and were therefore beyond research scope, the very fact that Defence committed funding to the study indicates a genuine willingness to understand potential infrastructure impacts. On the other hand, it was no advancement on what was already occurring across municipal councils, other State and Federal agencies as well as other major Australian industries (Gurran et al. 2008; Commonwealth of Australia 2009a; Steffen et al. 2009).

Although this analysis focused on describing three main initiatives directly addressing climate change, there were a range of other programmes related to climate change that have not been discussed. Many of these programmes were aimed at improving Defence energy efficiency and were often promoted by Defence as such. Climate change was rarely cited as a key reason for the establishment of these plans, particularly with regards to *energy*.

Lastly, two other initiatives by the Defence Science and Technology Organisation (DSTO) warrant comment. The initiatives were important since they again indicated a broadening of Defence's response to climate change beyond the purely security realm. A 2011 Defence Science and Technology Organisation (DSTO) publication, *National Security Science and Innovation Strategy: Annual Statement of Priorities*, identified the development of an extreme environmental indicator tool to predict, mitigate and prepare for extreme environmental events (such as floods, bushfires and cyclones). DSTO linked the development of this tool to four security outcomes including the improved ability to predict extreme events,

[35]The author spoke at a GCESI seminar on the security implications of climate change for the ADF, 2013.

[36]For the total cost of this contract, see both (2012a) and (2012b).

enhancing ability of first responders and improving community and critical infrastructure resilience. The prediction tool was one of six priorities and arguably represents DSTO's contribution toward implementing the Rudd-Gillard Government priority on addressing climate change. The second was found in a DSTO publication, *Australian Defence Scientist*, that stated 'because of the growing prominence of climate change as an issue for the Australian Government … Defence is placing greater emphasis on fuel efficiency' (DSTO 2008: 4). As such, DSTO identified a number of initiatives to improve Defence energy use that included improved techniques that reduced bio-fouling thereby reducing drag and improving fuel efficiency on ships. Many ADF journal articles focused on improving Defence energy security (through energy efficiency, renewable energy programme and inculcating fully burdened cost of fuel).

In summary, the ADF instigated a number of initiatives that ranged from adaptation (e.g., *Climate Change Adaptation Strategy*) to mitigation (*Combat Climate Change*). However, the vast majority of initiatives that could have been rendered part of the ADF response to climate change were not promoted as such. Rather, the majority were framed as energy efficiency and sustainability measures. The overall impression of the ADF climate change programme, therefore, was one of limited organisational penetration and breadth. Even GCESI, which existed with the endorsement of relatively senior leaders within the ADF, was really an information-scoping initiative rather than a major effort to mainstream climate change via reformation of ADF policies, practices and doctrine. In the absence of conspicuous support by the most senior leaders within the ADF there emerged a number of 'island' initiatives within the bureaucratic structures of Defence; each with similar objectives but isolated for want of a coherent strategic plan from which a common picture could emerge. Seemingly, the ADF had considered the development of such a plan in 2010 but—tempered by the emergence of a federal opposition that stridently opposed the then Labor Government's suite of climate policies—it remained unpublished.

5.4 Climate Security Discourse in Australia Update, 2013–2017

The 2013 election of a right-wing government under the prime ministership of self-professed climate 'weathervane' Tony Abbott resulted in a repudiation and abolition of Australian climate change policies developed during the 2007–2013 Rudd-Gillard-Rudd era (Bolt 2014). Prominent amongst these were the repealing of the Clean Energy Future Act (i.e., Australia's so-called 'carbon-tax'); the appointment of 'climate denier' Dick Warburton to review and then wind back Australia's renewable energy targets; the abolition of the Climate Commission as a source of independent advice; moves to abolish the Clean Energy Finance Corporation and Renewable Energy Agency (White 2014; Jotzo 2013). The ascension of Prime Minister Malcolm Turnbull in September 2015 under a centre-right Liberal-National government initially more prone to climate action was duly tempered by his reduced electoral margin at the July 2016 Australian election. Rather than consolidating his political authority, the election made it increasingly difficult for Turnbull to pass legislation through the Senate, with some commentators noting policy paralysis on account of being held captive to right-wing elements of the Liberal-National coalition who continued to stridently advocate against climate change (Brull 2016; Berman 2016). Thus, while Turnbull ratified the Paris Agreement, he simultaneously supported the planned development of the world's largest coal mine in central Queensland (Adani coalmine), and he backed down on considering any form of emissions intensity scheme or carbon tax in his 2017 review of climate change policies (Butler 2016).

Given this patchy political backdrop, climate security in Australian public policy, media and scholarly discourse developed along a somewhat intermittent trajectory. On the one hand it became an increasing feature for some media outlets and think-tanks to portray climate change as a national security issue and as a clarion call for action. Key publications that advocated for climate change to be factored within Australia's national security considerations and institutions were published by the Australian Strategic Policy Institute (2013), the Centre for Policy Development (2015) and the Climate Council (2015). The latter of these organisations also hosted the first Australian Climate Security Summit that bought together prominent national and international climate and security experts (Barrie/Steffen 2015). Widening interest in the area was also formed with international links, including the Washington based think-tank Center for Climate and Security, which formed the Asia-Pacific Climate Security Working Group with the Australian scholarly, NGO and think-tanks. On the other hand, climate security received scant attention within the Department of Defence and wider security institutions. Apart from the VCDF Global Change programme which continued to inject the issue into departmental strategic thinking and the Royal Australian Navy's associated initiatives to gradually increase the use of bio-fuels across its fleet, the only other developments were a 2013 report examining adaptation and planning strategies to mitigate sea level rise, flooding and erosion at selected military bases and a 2014 study into the exposure of Defence bases to flood, fire and storm events (Defence 2016a). Problematically, all reports remain classified and unavailable for local and state government officials or the business and wider community to examine for holistic integration. In this regard, the ADF continues to assess climate risks from a narrow and limited perspective, islanding itself from wider initiatives that could contribute to greater levels of societal resilience. Beyond these, no new large-scale climate mitigation or adaptation initiatives were developed within the ADF. With the notable exception of a single speech by the Australian Chief of Army in 2016 that identified climate change as a major strategic trend, the issue also continued to be sparingly mentioned in the speeches of ADF senior leadership. Symptomatic of this, lack of uptake occurred when the Chief of the Australian Defence Force, Air Chief Marshall Mark Binskin, stated that the ADF does not 'specifically plan for climate change' and refuted a suggestion to appoint a senior officer as a climate strategist on the basis that 'it might look good, but practically it doesn't give me anything' (Commonwealth of Australia 2015a: 59–60). Discourse on the subject tended to atrophy within the ADF, and suggestions by the broader national security community continued to lament the lack of progress compared with the US military (Bergin 2016). Despite this, the inclusion of climate change as a security challenge and as a risk to Defence infrastructure in the 2016 White Paper cemented it as a bi-partisan security issue. Such incrementalism largely reflected the on-going challenge of addressing the security dimensions of what remained a hyper-politicised issue in the Australian body politic.

5.5 Chapter Conclusion

Climate securitisation in Australia between 2003 and 2013 witnessed two distinct phases. The first phase, occurring under the prime ministership of John Howard, saw climate change wallow at the margins of mainstream public debate. During this period it never registered as a major issue for the ADF, it was absent in all Defence White Papers and Defence Updates, and was never framed by John Howard as a security issue. With few exceptions, it was largely ignored by the Australian scholarly and think-tank community. This changed with the election of a centre-left government that—in part—was installed with a mandate to act on climate change. Under Kevin Rudd and then Julia Gillard, the Labor Government actively pursued a climate security agenda, but within a much wider socio-economic climate change policy setting, framing it as a major global and national security threat and (for the first time) elevating it in the Defence White Paper and the National Security Strategy. Subsequently, the ADF commenced a number of programme and initiatives to address climate change but, following the twin failures of the Copenhagen climate conference and the passage of the CPRS in early 2010, climate change became hyper-politicised and ADF interest and action was subsequently curbed. No speeches, policies or actions by the ADF, or the government for that matter, could be said to have framed climate change sufficiently as a security threat to muster 'emergency measures'. The election of a right-wing government under Prime Minister Tony Abbott then witnessed a deconstruction of climate policies previously developed by the ALP. With a changed political outlook, climate security became even more marginalised, with a general acceptance that the moment of 'securitisation' for the purpose of generating urgent action had passed. The ascension of a centre-right government under Prime Minister Malcolm Turnbull, more prone to action on climate change than his predecessor, was tempered by his narrow 2016 election victory that limited his political freedom to act on the issue. Despite this, climate change was registered in the 2016 Australian Defence White Paper as a security concern and thereby consolidated it as a bi-partisan security issue.

References

Alexander, Natalie. 2011. “Climate Change on radar.” In *Defence Magazine*, edited by Michael Weaver, vol. 3, pp. 30–31. Canberra: Department of Defence. Accessed on 12 April 2013, http://www.defence.gov.au/defencemagazine/editions/2011_03/mag.pdf.

ANAO. 2014. *The Auditor-General, Audit Report No. 24 2013–14 Performance Audit: Emergency Defence Assistance to the Civil Community*. The Auditor-General’s Department. Canberra: Commonwealth of Australia. Accessed on 20 May 2017, https://www.anao.gov.au/sites/g/files/net616/f/AuditReport_2013-2014_24.pdf.

Australian Army. 2009. *Adaptive Campaigning: Army’s Future Land Operating Concept*. Department of Defence. Canberra: Directorate of Future Land Warfare. Accessed on 20 May 2017, https://www.army.gov.au/sites/g/files/net1846/f/acfloc_2012_main.pdf.

Australian Bureau of Statistics. 2008. *Year Book Australia: A comprehensive source of information about Australia*. Vol. 90, *Australian Year Book*. Canberra: Commonwealth of Australia.

Barrie, Chris, and Will Steffen. 2015. “Statement from the co-chairs of the Australian Climate Security Summit.” Sydney: Climate Council. Accessed on 8 January 2017, https://www.climatecouncil.org.au/statement-from-the-co-chairs-of-the-australian-climate-security-summit.

Barrie, Chris, Will Steffen, Alix Pearce, and Michael Thomas. 2015. *Be Prepared: Climate Change, Security and Australia’s Defence Force*. Sydney: Climate Council. Accessed on 20 September 2015, https://www.climatecouncil.org.au/uploads/fa8b3c7d4c6477720434d6d10897af18.pdf.

Beckett, Margaret. 2007. “The Urgent Case for Climate Security.” London: Royal United Services Institute. Published online 10 May 2007. Accessed 15 May 2017, https://rusi.org/publication/urgent-case-climate-security.

Bergin, Anthony. 2016. “Defence must regard climate change as a serious issue.” *The Australian*, News Limited, 2 December 2016. Accessed on 20 May 2017, https://www.aspi.org.au/opinion/defence-must-regard-climate-change-as-a-serious-security-issue.

Berman, Alan. 2016. “A Vote For Turnbull Could Be A Vote For Abbott.” *The Huffington Post*, AOL, 5 July 2016. Accessed on 9 December 2016, http://www.huffingtonpost.com.au/alan-berman/a-vote-for-turnbull-could-be-a-vote-for-abbott/.

Bolt, Andrew. 2014. “Tony Abbott has arrived as a true leader.” *Herald Sun*, News Limited, 3 September 2014. Accessed on 9 December 2016, http://www.heraldsun.com.au/news/opinion/andrew-bolt/tony-abbott-has-arrived-as-a-true-leader/news-story/88238eab8e7da2705b6de64aec04853c.

Brull, Michael. 2016. “Into The Abyss: Right On Cue, Desperate Malcolm Turnbull Turns On Minorities.” *New Matilda*, At Large Media, 31 October 2016. Accessed on 9 December 2016, https://newmatilda.com/2016/10/31/into-the-abyss-right-on-cue-desperate-malcolm-turnbull-turns-on-minorities/.

Butler, Josh. 2016. “It Took Just 35 Hours For A Massive Backflip On Emissions Policy.” *The Guardian*, The Guardian, 7 December 2016. Accessed on 7 January 2017, http://www.huffingtonpost.com.au/2016/12/06/it-took-just-35-hours-for-a-massive-backflip-on-emissions-policy/.

Cole, Steve. 2010. “Climate Change Implications for the Royal Australian Navy.” Climate and Energy Symposium 2010: Proceedings on Climate and Energy, Imperatives for Future Naval Forces, Baltimore, 23 & 24 March 2010. Accessed on 18 October 2012, http://www.jhuapl.edu/ClimateAndEnergy/Presentations/2010/RT5_Cole.pdf.

Commonwealth of Australia. 2004. *Securing Australia’s Energy Future*. Department of the Prime Minister and Cabinet. Canberra: Commonwealth of Australia. Accessed on 15 March 2011, http://www.efa.com.au/Library/cthEnergyWhitePaper.pdf.

Commonwealth of Australia. 2006. “Supplementary Budget Estimates.” In *Standing Committee on Foreign Affairs, Defence and Trade Official Committee Hansard*, 1 November 2006. Canberra: Commonwealth of Australia.

Commonwealth of Australia. 2007a. “Budget Estimates.” In *Standing Committee on Foreign Affairs, Defence and Trade Hansard*, 31 May 2007. Canberra: Commonwealth of Australia. Accessed on 13 February 2011, http://www.aph.gov.au/~/media/Estimates/Live/commttee/S10269.ashx.

Commonwealth of Australia. 2008a. “Budget Estimates.” In *Standing Committee on Foreign Affairs, Defence and Trade Official Committee Hansard*, 4 June 2008. Canberra: Commonwealth of Australia. Accessed on 12 June 2012, http://www.aph.gov.au/Parliamentary_Business/Senate_Estimates/fadtctte/estimates/bud0809/index.

Commonwealth of Australia. 2009a. *Climate Change Risks to Australia's Coast: A First Pass National Assessment*. Department of Climate Change. Canberra: Commonwealth of Australia. Accessed on 4 July 2012, https://www.environment.gov.au/system/files/resources/fa553e97-2ead-47bb-ac80-c12adffea944/files/cc-risks-full-report.pdf.

Commonwealth of Australia. 2009b. “Estimates—Additional Estimates.” In *Standing Committee on Foreign Affairs, Defence and Trade Hansard*, 25 February 2009. Canberra: Commonwealth of Australia. Accessed on 12 May 2011, http://www.aph.gov.au/Parliamentary_Business/Senate_Estimates/fadtctte/estimates/add0809/index.

Commonwealth of Australia. 2012a. “Senate Budget Estimates Questions On Notice—Committees.” In *Senate Standing Committee on Foreign Affairs, Defence and Trade Hansard*, 28 & 29 May 2012. Canberra: Commonwealth of Australia. Accessed on 17 July 2013, http://www.aph.gov.au/Parliamentary_Business/Senate_Estimates/fadtctte/estimates/bud1213/def/Department-of-Defence/defenceqonsindex.

Commonwealth of Australia. 2012b. “Senate Supplementary Budget Estimates Hearing.” In *Senate Standing Committee on Foreign Affairs, Defence and Trade Hansard*, 17 October 2012. Canberra: Commonwealth of Australia.

Commonwealth of Australia. 2013a. *Energy Use in the Australian Government's Operations 2010–11*. Energy and Tourism Department of Resources. Canberra: Commonwealth of Australia. (Available at: http://content.webarchive.nla.gov.au/gov/wayback/20160615122731/http://www.industry.gov.au/Energy/EnergyEfficiency/Non-residentialBuildings/GovernmentBuildings/Documents/report-2011.pdf, accessed 13 June 2017).

Commonwealth of Australia. 2013c. “Energy Use in the Australian Government's Operations (Reports 2001–2011).” Energy and Tourism Department of Resources. Canberra: Commonwealth of Australia. (Available at: http://webarchive.nla.gov.au/gov/20160615065736/http://www.industry.gov.au/Energy/EnergyEfficiency/Non-residentialBuildings/GovernmentBuildings/EnergyUseOperations/Pages/PreviousEditions.aspx, accessed 13 June 2013).

Commonwealth of Australia. 2015a. *Foreign Affairs, Defence and Trade Legislation Committee: Estimates*. Canberra: Commonwealth of Australia. Accessed on 20 May 2017, http://parlinfo.aph.gov.au/parlInfo/search/display/display.w3p;query=Id%3A%22committees%2Festimate%2F6662254f-7408-419c-b81e-4a2e7981ccea%2F0001%22.

Defence, Department of. 2005a. *Australia's National Security: A Defence Update 2005*. Canberra: Commonwealth of Australia.

Defence, Department of. 2005b. *Defence Energy Management Strategy*. Canberra: Department of Defence. Accessed on 20 May 2017, http://www.uat.defence.gov.au/ips/parliament/qons/41st/slc/0506_budget_may05/responses/Q09_ems.pdf.

Defence, Department of. 2007b. *Joint Operations for the 21st Century*. Canberra: Department of Defence. Accessed on 11 September 2011, http://www.defence.gov.au/publications/docs/FJOC.pdf.

Defence, Department of. 2008a. *Combat Climate Change*. Canberra: Department of Defence. Accessed on 7 July 2011, http://www.defence.gov.au/environment/climate_change/. [Author note: Website archived, document in author's possession].

Defence, Department of. 2008b. *Looking over the Horizon: Australians Consider Defence*. Canberra: Department of Defence. Accessed on 7 August 2011, http://www.defence.gov.au/whitepaper/2009/docs/Looking_Over_The_Horizon.pdf.
Defence, Department of. 2009a. *Defence Annual Report 2008–09*. Department of Defence. Canberra: Department of Defence. Accessed on 10 July 2012, http://www.defence.gov.au/AnnualReports/08-09/.
Defence, Department of. 2009b. *Defending Australia in the Asia Pacific Century: Force 2030*. Canberra: Commonwealth of Australia.
Defence, Department of. 2010a. *Defence Annual Report 2009–10*. Department of Defence. Canberra: Department of Defence. Accessed on 1 March 2011, http://www.defence.gov.au/AnnualReports/09-10/.
Defence, Department of. 2010b. *Defence Environmental Strategic Plan 2010–2014*. Canberra: Defence Support Group for Department of Defence.
Defence, Department of. 2011. *Global Change and Energy Sustainability Initiative*. Canberra: Defence Preparedness Branch for Department of Defence.
Defence, Department of. 2012. "Answers to Questions on Notice—Committees." In *Senate Standing Committee on Foreign Affairs, Defence and Trade Senate Budget Estimates Hansard*, 29 May 2012. Canberra: Commonwealth of Australia. Accessed on 10 May 2013, http://www.aph.gov.au/~/media/Estimates/Live/fadt_ctte/estimates/bud_1213/def/Department-of-Defence/Defence_QONs.ashx.
Defence, Department of. 2013b. *Defence White Paper*. Canberra: Commonwealth of Australia. Accessed on 10 January 2014, http://www.defence.gov.au/whitepaper/2013/docs/WP_2013_web.pdf.
Defence, Department of. 2014. *Defence Procurement Policy Manual*. Canberra: Department of Defence.
Defence, Department of. 2016a. *Additional Budget Estimates 2015–2016: Responses to Questions on Notice*. Canberra: Commonwealth of Australia. Accessed on 8 January 2017, http://www.aph.gov.au/Parliamentary_Business/Senate_Estimates/fadtctte/estimates/add1516/def/defenceqonsindex.
Department of Industry. 2011. *Low Emission Technology Demonstration Fund (LETDF): Funded Projects*. Canberra: Department of Industry for Commonwealth of Australia.
DSTO. 2008. "Bubbles to bust a bio-fouling problem." *Australian Defence Science*, vol. 16, no. 4:4–5.
Evans, Ann. 2012. *Australian Survey of Social Attitudes, 2011*. The Australian Survey of Social Attitudes: Canberra: Australian Data Archive.
Field, Chris. 2011. "Operation Queenslander: Ten Ideas for Australian Defence Force Support to Disaster Relief Operations." *Australian Army Journal For The Profession of Arms*, vol. 8, no. 3:121–140.
Fitzgibbon, Joel. 2008a. "Defence Supports Earth Hour 2008." In *Transcripts of the Minister for Defence*, 29 March 2008. Canberra: Department of Defence.
Fitzgibbon, Joel. 2008b. "Labor's Defence Reform Project—Meeting the Strategic Challenges of the 21st Century." In *Transcripts of the Minister for Defence*, 30 July 2008. Canberra: Commonwealth of Australia.
Fitzgibbon, Joel. 2008c. "Speech by the Minister for Defence to the Brookings Institution, Washington DC." In *Transcripts of the Minister for Defence*, 15 July 2008. Canberra: Commonwealth of Australia.
Fitzgibbon, Joel. 2008d. "Speech by the Minister for Defence to the Shangri-La Conference Singapore, 31 May 2008, 'Making Defence Policy in Uncertain Times'." In *Transcripts of the Minister for Defence*, 31 May 2008. Canberra: Commonwealth of Australia.
Fitzgibbon, Joel. 2008e. "Speech to the National Resilience Institute (Lemhannas), Jakarta, Indonesia." In *Transcripts of the Minister for Defence*, 18 September 2008. Canberra: Department of Defence.
Fitzgibbon, Joel. 2014. "Hunter Electorate." Federal Member for Hunter—Joel Fitzgibbon MP. Accessed on 20 February 2014, http://www.joelfitzgibbon.com/about.

Gillard, Julia E. 2011a. "Address to the Carbon Expo Australasia 2011, Melbourne." In *Transcripts of the Prime Ministers of Australia*, 09 November 2011. Canberra: Commonwealth of Australia. Accessed on 12 March 2012, http://pmtranscripts.dpmc.gov.au/release/transcript-18253.

Gillard, Julia E. 2011b. "A great clean energy future, a great reform agenda: Address to the National Press Club, Canberra." In *Transcripts from the Prime Ministers of Australia*, 14 July 2011. Canberra: Commonwealth of Australia. Accessed on 12 March 2012, http://pmtranscripts.dpmc.gov.au/release/transcript-18004.

Gillespie, Ken J. 2008. "Chief of Army Speech." In *Address to Australian Command and Staff College*. Canberra. (Website archived. Document in transcript format in possession of author).

Gillespie, Ken J. 2009. "The Adaptive Army Initiative." *Australian Army Journal For The Profession of Arms*, vol. 6, no. 3:7–19.

Gurran, Nicole, Elisabeth Hamin, and Barbara Norman. 2008. *Planning for climate change: Leading Practice Principles and Models for Sea Change Communities in Coastal Australia.* National Sea Change Taskforce: Sydney: The University of Sydney. Accessed on 19 April 2012, http://stokes2013.files.wordpress.com/2014/10/planningforclimatechange.pdf.

Houston, Angus. 2007. "The ADF of the Future." *Australian Defence Force Journal: Journal of the Profession of Arms*, no. 173.

Houston, Angus. 2010. "Future security prospects, future capability plans and how the ADF makes a contribution to a 'whole of government' approach to national security." In *Address to National Security Lecture Series*, 29 October 2010. Canberra: University of Canberra.

Howard, John. 2010. *Lazarus Rising: A Personal and Political Autobiography*. Sydney: Harper Collins.

Howard, John W. 2004. "Address to National Press Club Canberra, Securing Australia's Energy Future." In *Transcripts from the Prime Ministers of Australia*, 15 June 2004. Canberra: Department of the Prime Minister and Cabinet. Accessed on 13 March 2012, http://pandora.nla.gov.au/pan/10052/20040821-0000/www.pm.gov.au/news/speeches/speech912.html.

Howard, John W. 2006a. "Address at the Business Council of Australia Annual Dinner Sofitel Wentworth, Sydney." In *Transcripts from the Prime Ministers of Australia* 13 November 2006. Canberra: Department of Prime Minister and Cabinet. Accessed on 13 March 2013, http://pmtranscripts.dpmc.gov.au/release/transcript-22579.

Howard, John W. 2006b. "Address at the Launch of the Asia-Pacific Partnership on Clean Development and Climate 'Partnership for Action 2006' CSIRO, Yarralumla." In *Transcripts from the Prime Ministers of Australia*, 1 November 2006. Department of Prime Minister and Cabinet. Accessed on 11 March 2013, http://pmtranscripts.dpmc.gov.au/release/transcript-22552.

Howard, John W. 2006c. "Address at the Liberal Party of Australia (Queensland Division) Annual State Convention." In *Transcripts from the Prime Ministers of Australia*, 4 November 2006. Canberra: Department of the Prime Minister and Cabinet. Accessed on 13 March 2013, http://pmtranscripts.dpmc.gov.au/release/transcript-22559.

Howard, John W. 2006d. "Address to the APEC CEO Summit National Convention Centre, Hanoi." In *Transcripts of the Prime Ministers of Australia*, 18 November 2006. Canberra: Department of Prime Minister and Cabinet. Accessed on 15 March 2013, http://pmtranscripts.dpmc.gov.au/release/transcript-22590.

Howard, John W. 2006e. "Address to the Asia-Pacific Partnership on Clean Development and Climate Inaugural Ministerial Meeting Government House, Sydney." In *Transcripts from the Prime Ministers of Australia*, 12 January 2006. Canberra: Department of Prime Minister and Cabinet. Accessed on 8 March 2013, http://pmtranscripts.dpmc.gov.au/release/transcript-22099.

Howard, John W. 2006f. "Address to the Australia/UK Leadership Forum Parliament House, Canberra." In *PM Transcripts: Transcripts from the Prime Ministers of Australia*, 27 March

2006. Canberra: Department of the Prime Minister and Cabinet. Accessed on 10 March 2013, http://pmtranscripts.dpmc.gov.au/release/transcript-22190.

Howard, John W. 2007a. "Address to the National Press Club Great Hall, Parliament House." In *Transcripts from the Prime Ministers of Australia*, 25 January 2007. Canberra: Department of Prime Minister and Cabinet. Accessed on 13 March 2013, http://pmtranscripts.dpmc.gov.au/release/transcript-15149.

Howard, John W. 2007b. "APEC Leaders' Sydney Declaration on Climate Change, Energy Security and Clean Development." In *Transcripts from the Prime Ministers of Australia*, 8 September 2007. Canberra: Department of Prime Minister and Cabinet. Accessed on 5 April 2013, http://pmtranscripts.dpmc.gov.au/release/transcript-15401.

Howard, John W. 2007c. "Climate Change Announcement." In *Transcripts from the Prime Ministers of Australia*, 17 July 2007. Canberra: Department of Prime Minister and Cabinet. Accessed on 12 April 2013, http://pmtranscripts.dpmc.gov.au/release/transcript-15296.

Howard, John W. 2007d. "Interview with Simon Beaumont, Radio 6PR, Perth." In *Transcripts from the Prime Ministers of Australia*, 19 February 2007. Canberra: Department of Prime Minister and Cabinet. Accessed on 19 April 2013, http://pmtranscripts.dpmc.gov.au/release/transcript-15212.

Howard, John W. 2007e. "Meeting of Major Economies of Energy Security and Climate Change." In *Transcripts from the Prime Ministers of Australia*, 4 August 2007. Canberra: Department of Prime Minister and Cabinet. Accessed on 29 March 2013, http://pmtranscripts.dpmc.gov.au/release/transcript-15377.

Howard, John W. 2007f. "National Plan for Water Security; Qantas; Australia-Japan relations." In *Transcripts from the Prime Ministers of Australia*, 5 February 2007. Canberra: Department of Prime Minister and Cabinet. Accessed on 3 March 2013, http://pmtranscripts.dpmc.gov.au/release/transcript-15193.

Howard, John W. 2007g. "Nuclear Power Station Plebiscites." In *Transcripts of the Prime Ministers of Australia*, 23 August 2007. Canberra: Department of Prime Minister and Cabinet. Accessed on 8 April 2013, http://pmtranscripts.dpmc.gov.au/release/transcript-15491.

Howard, John W. 2007h. "Weekly Radio Message—Climate Change." In *Transcripts from the Prime Ministers of Australia*, 16 July 2007. Canberra: Department of Prime Minister and Cabinet. Accessed on 18 April 2013, http://pmtranscripts.dpmc.gov.au/release/transcript-15292.

Howard, John W. 2013. "One Religion is Enough." London: The Global Warming Policy Foundation. Accessed on 7 February 2014, http://www.thegwpf.org/john-howard-religion/.

Jotzo, Frank. 2013. "Australia needs climate institutions, whoever is in power." *The Conversation*, no.

Kelly, Mike. 2008. "Regional Security and Middle Power Diplomacy." In *John Gee Memorial Lecture* 29 August 2008. Canberra: Department of Defence. Accessed on 20 May 2017, https://www.lowyinstitute.org/sites/default/files/pubfiles/Kelly%2C_Regional_security_and_middle_power_diplomacy_1.pdf.

Kelly, Mike. 2012. "PNG: Securing a Prosperous Future, Domestic and Regional Security (Prosperous Future Conference, Deakin University Geelong)." In *Transcripts Parliamentary Secretary for Defence*, 13 April 2012. Canberra: Department of Defence.

Lawson, Robert. 2007. "Climate Change in the Asia-Pacific Region: Security Implications for Australia." In *Centre for Defence and Strategic Studies Australian Defence College Shedden Papers*. Canberra: Department of Defence. Accessed on 5 May 2012, http://www.defence.gov.au/ADC/Publications/Shedden/2008/Publctns_ShedPaper_050310_ClimateChangeintheAsiaPacificRegion.pdf.

Leahy, Peter. 2004. "The Australian Army Reserve: Relevant and Ready." *Australian Army Journal For the Profession of Arms*, vol. 2, no. 1.

Leahy, Peter. 2008. "Introduction By The Chief of Army." *Australian Army Journal For The Profession of Arms*, vol. 5, no. 2.

Macintosh, Andrew, Deb Wilkinson, and Richard Denniss. 2010. "Climate Change." In *The Rudd Government: Australian Commonwealth Administration 2007–2010*, edited by Chris Aulich and Mark Evans, pp. 596–652. Canberra: The Australian National University E Press. Accessed on 3 March 2014, http://press.anu.edu.au/apps/bookworm/view/The+Rudd+Government%3A+Australian+Commonwealth+Administration+%092007+-+2010/5091/ch11.xhtml#toc-anchor.

Mackellar, Dorothea. 1908. "My Country." Pippa Masson. Published online: The Estate of Dorothea Mackellar. Published online 5 September 1908, http://www.dorotheamackellar.com.au/archive/mycountry.htm.

Morrison, David. 2012. "Speech to Australian Strategic Policy Institute." In *Chief of Army Speeches and Transcripts*, 11 April 2012. Canberra: Department of Defence.

Nelson, Brendan. 2007a. "Address to the CEDA State of the Nation Conference 2007, 'Constructing a Regional Security Community'." In *Transcripts of the Minister for Defence* 14 June 2007. Canberra: Department of Defence.

Nelson, Brendan. 2007b. "Opening Address to D+I Conference Adelaide." In *Transcripts of the Minister for Defence* 22 August 2007. Canberra: Department of Defence.

Palazzo, Albert. 2012. "The Future of War Debate in Australia: Why has there not been one? Has the need for one now arrived?" In *Land Warfare Studies Centre Working Paper Series*, Working Paper 140. Canberra: Land Warfare Studies Centre. Accessed on 20 May 2017, https://www.army.gov.au/sites/g/files/net1846/f/wp140-the-future-of-war-debate-in-australia.pdf.

Palmer, Dr Jennifer L. 2009. "Addressing Energy as a Military Cost." *Australian Army Journal For The Profession of Arms*, no. 178:6–31.

Personal Communication. 2013. "Email correspondence." Unpublished. 17 July 2013.

Press, Anthony, Anthony Bergin, and Eliza Garnsey. 2013. *Heavy Weather: Climate and the Australian Defence Force*. Special Report 49. Canberra: Australian Strategic Policy Institute. Accessed on 12 March 2014, https://www.aspi.org.au/publications/special-report-issue-49-heavy-weather-climate-and-the-australian-defence-force.

Reich, Jenny, John Hearps, Andrew Cohn, Jeromy Temple, and Peter McDonald. 2006. *Defence Personnel Environment Scan 2025*. Canberra: Director Strategic Personnel Planning and Research for Department of Defence. Accessed on 13 February 2012, http://trove.nla.gov.au/work/26715104?selectedversion=NBD41326420.

Royal Australian Navy. 2007. *Future Maritime Operating Concept—2025: Maritime Force Projection and Control*. Canberra: Department of Defence. Accessed on 12 June 2013, http://www.navy.gov.au/media-room/publications/future-maritime-operating-concept-2025.

Rudd, Kevin. 2010a. "Prime Minister Speech at the opening of the National Security College Australian National University, Canberra: The establishment of the National Security College of Australia." In *PM Transcripts: Transcripts from the Prime Ministers of Australia*, 24 April 2010. Canberra: Department of the Prime Minister and Cabinet. Accessed on 3 March 2014, http://pmtranscripts.pmc.gov.au/release/transcript-17245.

Rudd, Kevin M. 2008a. "Address to the East Asia Forum in conjunction with the Australian National University, Advancing Australia's Global and Regional Economic Interests." In *PM Transcripts: Transcripts by the Prime Ministers of Australia*, 26 March 2008. Canberra: Department of Prime Minister and Cabinet. Accessed on 11 February 2014, http://pmtranscripts.pmc.gov.au/release/transcript-15823.

Rudd, Kevin M. 2008b. "Address to the London School of Economics: Australia and the UK—Global Partners in shaping the future Global Order, London." In *PM Transcripts: Transcripts from the Prime Ministers of Australia*, 7 April 2008. Canberra: Department of Prime Minister and Cabinet. Accessed on 18 February 2014, http://pmtranscripts.pmc.gov.au/release/transcript-15847.

Rudd, Kevin M. 2008c. "Australia 2020—Setting our Nation's Sights for the Future, Address to the Sydney Institute Annual Dinner 2008, Grand Habour Ballroom, Star City, Sydney." In *PM Transcripts: Transcripts from the Prime Ministers of Australia*, 16 April 2008. Canberra:

Department of Prime Minister and Cabinet. Accessed on 5 March 2014, http://pmtranscripts.pmc.gov.au/release/transcript-15873.

Rudd, Kevin M. 2008d. "Building a Better World Together, Speech at Kyoto University." In *PM Transcripts: Transcripts from the Prime Ministers of Australia*, 09 June 2008. Canberra: Department of Prime Minister and Cabinet. Accessed on 2 April 2013, http://pmtranscripts.pmc.gov.au/release/transcript-15955.

Rudd, Kevin M. 2008e. *Building on ASEANs success: towards an Asia-Pacific Community, Singapore Lecture Series*. Singapore: Institute of South East Asian Studies Publishing.

Rudd, Kevin M. 2008f. "Climate Change and Water." Canberra: National Library of Australia Accessed on 10 February 2014, http://pandora.nla.gov.au/pan/79983/20080812-0001/www.pm.gov.au/topics/climate.html.

Rudd, Kevin M. 2008g. "The First National Security Statement to the Parliament Address by the Prime Minister of Australia The Hon. Kevin Rudd MP." In *PM Transcripts: Transcripts of the Prime Ministers of Australia*, 4 December 2008. Canberra: Department of Prime Minister and Cabinet. Accessed on 5 March 2013, http://pmtranscripts.pmc.gov.au/release/transcript-16289.

Rudd, Kevin M. 2009a. "Address to the 64th session of the United Nations General Assembly, New York." In *PM Transcripts: Transcripts from the Prime Ministers of Australia*, 23 September 2009. Canberra: Department of Prime Minister and Cabinet. Accessed on 12 March 2014, https://pmtranscripts.pmc.gov.au/release/transcript-16828.

Rudd, Kevin M. 2009b. "Address to the Australia-Israel Chamber of Commerce and General Sir John Monash Foundation Leadership Luncheon Docklands, Melbourne." In *PM Transcripts: Transcripts by the Prime Ministers of Australia*, 24 April 2009. Canberra: Department of Prime Minister and Cabinet. Accessed on 12 March 2014, https://pmtranscripts.pmc.gov.au/release/transcript-16515.

Rudd, Kevin M. 2009c. "Address to the Lowy Institute." In *PM Transcripts: Transcripts from the Prime Ministers of Australia*, 06 November 2009. Canberra: Department of Prime Minister and Cabinet. Accessed on 12 March 2014, http://pmtranscripts.pmc.gov.au/release/transcript-16904.

Rudd, Kevin M. 2009d. "Focus on Climate Change—Prime Minister Blog." Canberra: National Library of Australia. Accessed on 11 February 2014, http://pandora.nla.gov.au/pan/79983/20091030-1529/www.pm.gov.au/PM_Connect/PMs_Blog/Climate_Change_Blog.html.

Rudd, Kevin M. 2009e. "Partners in the Global Economy—Australia and the United States, Address to the US Chamber of Commerce, Washington." In *PM Transcripts: Transcripts from the Prime Ministers of Australia*, 31 March 2009. Canberra: Department of Prime Minister and Cabinet. Accessed on 3 March 2014, https://pmtranscripts.pmc.gov.au/release/transcript-15832.

Rudd, Kevin M. 2009f. "Transcript of address at the close of the Pacific Islands Forum Cairns." In *PM Transcripts: Transcripts from the Prime Ministers of Australia*, 7 August 2009. Canberra: Department of Prime Minister and Cabinet. Accessed on 4 March 2013, https://pmtranscripts.pmc.gov.au/release/transcript-16737.

Rudd, Kevin M. 2009g. "Transcript of joint doorstop at the United Nations with Minister Wong New York." In *PM Transcripts: Transcripts by the Prime Ministers of Australia*, 23 September 2009. Canberra: Department of Prime Minister and Cabinet. Accessed on 2 March 2014, https://pmtranscripts.pmc.gov.au/release/transcript-16829.

Sales, Leigh. 2013. "The toxic political debate of climate change." Canberra: Australian Broadcasting Corporation. Accessed on 11 February 2014, http://www.abc.net.au/7.30/content/2013/s3831850.htm.

Shanahan, Rodger. 2013. "Editorial." *Australian Army Journal for The Profession of Arms*, vol. 10, no. 4:5–7.

Smith, Stephen. 2012. "Minister for Defence and Minister for Defence Materiel—Joint Media Release—Ocean Shield arrives in Australia." In *Transcripts of the Minister for Defence* 30 June 2012. Canberra: Department of Defence.

Snowden, Warren. 2008. "Australia's Strategic Imperatives Presented to the ASPI Global Forces Conference." In *Defence Minister: Ministerial Releases*, 2 July 2008. Canberra: Department of Defence.

Steffen, Will, Andrew A. Burbidge, Lesley Hughes, Roger Kitching, David Lindenmayer, Warren Musgrave, Mark Stafford Smith, and Patricia A. Werner. 2009. *Australia's Biodiversity and Climate Change: A Strategic Assessment of the vulnerability of Australia's biodiversity to climate change*. Canberra: Commonwealth of Australia. http://www.environment.gov.au/system/files/resources/eab369d6-76f9-46c8-beb4-aaae8ece112e/files/biodiversity-vulnerability-assessment.pdf.

Sturrock, Robert, and Peter Ferguson. 2015. *The Longest Conflict: Australia's Climate Security Challenge*. Melbourne: Centre for Policy Development.

Thomson, Mark. 2013. *The Cost of Defence: ASPI Defence Budget Brief 2013–14*. Canberra: Australian Strategic Policy Institute. Accessed on 15 May 2014, https://www.aspi.org.au/publications/the-cost-of-defence-aspi-defence-budget-brief-2013-2014/ASPI-CostDefence2013.pdf.

Watt, David. 2012. "Budget 2011–12: Australian Defence Force (ADF) operations." In *Parliamentary Library Research Publications, Australian Defence Force (ADF) Operations*. Canberra: Commonwealth of Australia. Accessed on 10 May 2013, http://www.aph.gov.au/About_Parliament/Parliamentary_Departments/Parliamentary_Library/pubs/rp/BudgetReview201112/ADF.

Wesley, Michael. 2012. "Climate views have moved on." *The Australian*, Lowy Institute for International Policy, 2 July 2012. Accessed on 23 March 2013, http://www.lowyinstitute.org/publications/climate-views-have-moved.

White, Alexander. 2014. "The ten worst environment decisions in Abbott's first year." *The Guardian*, The Guardian, 6 September 2014. Accessed on 9 December 2016, https://www.theguardian.com/environment/southern-crossroads/2014/sep/05/abbott-first-year-environment-climate-ten-worst-decisions.

White, Ben. 2010. "Sustainable Defence Capability: Australia's national security and the role of defence industry." *Journal of the Australian Profession of Arms*, no. 183:87–96.

White, Martin. 2008. "The Compelling Requirement to Energy-Proof the Australian Defence Force." *Journal of the Australian Profession of Arms*, no. 175:12–21.

Wikipedia. 2017c. "Chief of the Defence Force (Australia)." Online: Wikipedia. Accessed on 15 January 2017, https://en.wikipedia.org/wiki/Chief_of_the_Defence_Force_(Australia)).

Wikipedia. 2017d. "Department of Defence (Australia)." Online: Wikipedia. Accessed on 20 May 2017, https://en.wikipedia.org/wiki/Department_of_Defence_(Australia)).

Wikipedia. 2017f. "Minister for Defence (Australia)." Online: Wikipedia. Accessed on 20 May 2017, https://en.wikipedia.org/wiki/Minister_for_Defence_(Australia)).

Woodward, Peter. 2011. "The Reserve Response Force and Public Safety." *Australian Army Journal For The Profession of Arms*, vol. 8, no. 3:101–120.

Zhang, Chun. 2009. "Regional Reactions to the Defence White Paper 2009: Rebuilding middle power leadership for Australia." *Australian Defence Force Journal: Journal of the Australian Profession of Arms*, no. 180:7–9.

Chapter 6
Climate Security Case Study: United States

Abstract This chapter examines the response by the United States political-military sector to climate change through the lens of securitisation theory. Research undertaken used mixed content analysis techniques to systematically examine more than 2,000 speech-acts, policies, doctrinal articles and other publications between 2003 and 2013. Particular attention was paid to the speech-acts of prominent persons in command and leadership appointments as well as major climate programme and initiatives of the United States Navy (USN). The case study reveals that climate security first appeared as a concern during the George W. Bush Administration, though it was not a major factor. In early 2008 the Bush Administration released the *National Defense Authorisation Act* (NDAA) that directed the US military to undertake a sweeping examination of climate impacts on its operations. A key component of this was the inclusion of climate change in the 2010 National Security Strategy and Quadrennial Defense Review. It also laid wider foundations, bringing climate security to greater prominence in the US military at a time that segued with the ascension of the Obama Administration and its more activist climate agenda. The US Navy—the service most at risk from sea-level rise, natural disasters and with increased operational opportunities in the Arctic—adopted a proactive approach to climate change that included prominence in the speech-acts of USN leaders and via publication of several adaptation and mitigation strategies.

Keywords Climate security · Case study · Climate change · US military · Climate securitisation

6.1 Introduction

This case study examined how the US Navy and selected areas of the US political establishment responded to climate change through the prism of securitisation theory between 2003 and 2013.[1] In total, 2,093 documents were examined that

[1]Cut-off date for documents included in this case study was March/April 2013.

M.D. Thomas, *The Securitization of Climate Change: Australian and United States' Military Responses (2003–2013)*, The Anthropocene: Politik—Economics—Society—Science 10, DOI 10.1007/978-3-319-49658-0_6

Table 6.1 US case study: sources, domains and numbers of documents examined

	Programme	Document type					Total
		SPR	DTN	S&A	MDA	JNP	
Strategic	President	2	0	107	82	0	191
	Secretary of Defence	4	0	807	0	0	811
	Chairman of Joint Chiefs	2	0	609	0	20	631
							(1633)
Tactical	Secretary of the Navy	0	0	214	0	0	214
	Chief of Naval Operations	3	4	191	2	1	201
	US Oceanographer	0	0	5	0	2	7
	Commander of Pacific Fleet	0	0	38	0	0	38
							(460)
Total		11	4	1971	84	23	2093

Strategic and operational represent the two different domains of national security. Each domain consisted of various programme from where each document originated. (Document Type: *SPR* Strategic Policy, *DTN* Doctrine, *S&A* Speeches and Announcements, *MDA* Media, *JNP* Journals, Newspapers and Other publications). *Source* The author

consisted mostly of speech-acts but also included media publications, journal articles, military doctrine and departmental policy. In a similar fashion to the Australian case study, the research boundaries for this study were grouped into three separate areas: strategic, operational and tactical. Table 6.1 shows how the case study was structured and also annotates how many documents, by type, were examined for each area.

The strategic levels for this case study included examination of documents from the President, the Secretary of Defense (SECDEF) and the Chairmen of the Joint Chiefs of Staff (CJCS). The operational and tactical levels are largely combined in this case study and included examination of programme areas from the Secretary of the Navy (SECNAV), the Chief of Naval Operations (CNO), and—on a somewhat limited basis—US Navy Oceanographer and the Commander of the Pacific Fleet. The case study is presented in two main sections. Section 6.2 provides the results and discussion of the US Strategic Programmes while Sect. 6.3 provides the results and discussion of the operational and tactical level programme. Supplementary to the specific findings derived from the content analysis (coding) is an expanded discussion and analysis of selected speeches and policies. The aim here was to build upon the content analysis and to provide context to the particular research outcomes. On this basis, the discussion also provides an overall narrative of how the US political-military establishment responded to the security implications of climate change between 2003–2013.

6.2 Climate Security in US Political and Military Strategic Programme

Three separate programme were analysed under the functional grouping of Strategic Programmes (President, SECDEF and the CJCS). These areas are widely recognised as central decision-makers and securitising actors responsible for US strategic direction and national military strategy. Combining them is useful in identifying major trends on climate change across the higher levels of US strategic policy. Individual results and analysis for each of the Strategic Programmes is subsequently presented.

A word search for *climate change* within all 1,636 artefacts in the Strategic Programmes returned seventy-seven separate sources containing 185 direct references (filtered).[2] Of these 185 direct references, 119 were coded. Thirty-three of these (roughly 30%) were coded as securitised references while the remaining eight-six (70%) were coded as non-securitised (Table 6.2).[3] At the broadest level, it was found that climate change was predominately framed as a non-securitised issue by US strategic actors.

In order to provide context against other important strategic issues, a search of specific words/phrases was conducted for all available strategic artefacts. Figure 6.1, for instance, shows how *climate change* compared in frequency against *terrorism* (and its derivatives *terror*, *terrorist*, *counter-terrorism* and *counter-terrorist*). A search of all 1,636 documents in US Strategic Programmes revealed a total of 4,342 references across 796 different sources (unfiltered). To put this in perspective, of the programme analysed, approximately half of all artefacts contained some kind of reference to terrorism across the period 2003–2013. This clearly shows that climate change was discussed relatively infrequently against one of the more recent defining issues of US strategic policy.

The most common securitised categories for US Strategic Programme are shown at Fig. 6.2. This indicates that when climate change was framed as a security issue within the Strategic Programme, it was predominately done so (in descending order) as an energy security issue, a global security issue and as convergence/multiple security issue and—what this called—an 'Arctic security' issue. (Humanitarian Aid and Disaster Relief frame was included under 'Convergence'.) Table 6.3 shows that when climate change was framed as a security issue it was done so in a neutral manner but tending toward forceful language by presenting it as a challenge or threat. Two passages of text were coded as framing climate security opportunities. These aspects are further analysed under Presidential Programme.

The most common non-securitised categories for Strategic Programmes found it was done so dominantly as (1) a global/international issue; and (2) as an energy issue.

[2]'Filtered' is a term used in this case study to describe search results that have removed extraneous references to the term 'climate change' included in footnotes, endnotes, headings, margins and other inconsequential areas of publications. 'Unfiltered' includes all publication content.

[3]Non-securitised references consist of 'Other Context' and 'Defence Context'.

Table 6.2 Climate change framing in US Strategic Programme, 2003–2013

Frame	Strategic Programmes			
	President	SECDEF	CJCS	Total
Securitised	14	11	8	33
Non securitised	74	4	8	86
Total	88	15	16	119

Source The author

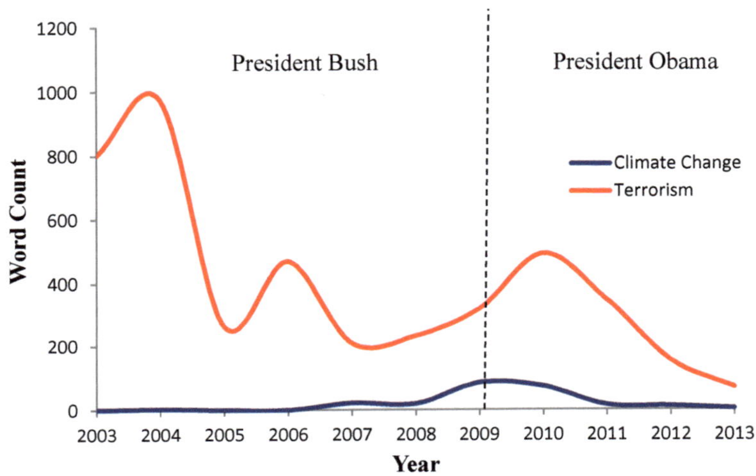

Fig. 6.1 Frequency of *climate change* versus *terrorism* in US Strategic Programmes. *Source* The author

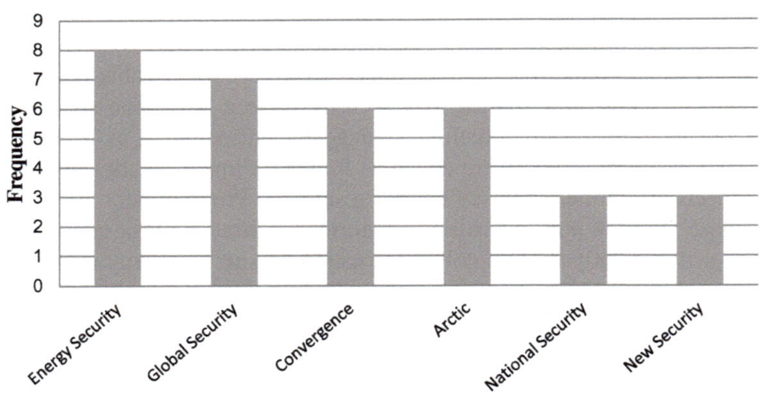

Fig. 6.2 Contextual framing of climate issues (securitised), strategic level (US). *Source* The author

Table 6.3 Language used to frame climate security issues, strategic levels (US)

Frame	President	SECDEF	CJCS	Total
Opportunity	2	0	0	2
Neutral	4	6	4	14
Challenge	3	2	4	9
Threat	5	3	0	8
Total	14	11	8	33

Source The author

Thereafter, climate change was framed across a range of categories, including economic and regulatory issues. Neutral language (seventy-four codes) was dominantly used to frame non-securitised references for all Strategic Programmes (*challenge* or *threat* were coded on twenty-six separate occasions). Climate change as a non-securitised *opportunity* was coded on twelve separate occasions (graphics not shown).

6.2.1 *Presidential Programme Results and Analysis*

The Presidential Programme analysed 191 artefacts, and found fifty separate sources containing 178 references to *climate change*.[4,5] Table 6.4 showed that, for the overall Presidential Programme, fourteen references were coded in a securitised context and seventy-four were coded in a non-securitised context. Broadly, the results indicate that climate change was mainly framed as a non-securitised issue by Presidents George W. Bush (2000–2008) and Barack Obama (2009–2017) across the period 2003–2013. Of these (non-securitised) coded references, Table 6.4 shows that Presidents Bush and Obama predominately framed climate change as an energy issue (with thirty-two coded references), or an international issue (with

[4]This statistic is 'filtered' insofar as it has removed references contained in speech endnotes. All references contained in major policy documents (NSS, QDR, EO) were also removed. Various search results are identified as being either 'filtered' or 'unfiltered' (as previously defined).

[5]The majority of presidential artefacts were speeches (56%) and the remaining were official (White House published) media conferences (42.5%) and strategic policy (1.5%). A 2001 document was included to capture all Inaugural Addresses by both Presidents for each of their two terms. Strategic policy included the 2002, 2006 and 2010 National Security Strategies. Other presidential documents were selected on 'estimated-relevance', and included: (a) All Inaugural Addresses (both terms) for both US Presidents (2001–2013); (b) All US Presidential State of Union addresses (2003–2013); (c) All US Presidential speeches to UN General Assembly (2003–2013); (d) All major addresses by the US President to the Nation (2003–2013); (e) All US Presidential Addresses to Foreign Legislature (2003–2013); (f) All US Presidential News Conferences between 2003–2013 (these were classified as 'Media'); (g) Un-categorised speeches by a US President containing the terms 'climate' or 'energy' in their title (2003–2013) were downloaded; and (h) Selected Executive Orders from US Presidents 2003–2013 (a search by title for 'energy' (all downloaded), 'environment' (all downloaded) and 'climate' (nil recorded)).

Table 6.4 Contextual framing of climate issues (non-securitised), US Presidents Bush and Obama 2003–2013

Non-securitised frame	Total
Energy	32
International/Global	29
Economic	5
Government/Regulatory	3
General reference	2
Agriculture, fisheries, water	1
Forestry	1
Health	1
Total	74

Source The author

twenty-nine coded references). Further analysis of the Presidential Programme revealed differences and similarities to how these presidents framed climate change.

6.2.1.1 President George W. Bush

Americans have never left our destiny to the whims of nature, and we will not start now.

President Bush, Address to the Nation on Hurricane Katrina (2005).

Between 2003 and 2006 climate change was not a major factor in the speech-acts, national security strategies or executive orders of President George W. Bush.[6] Between 2003 and 2006 climate change was mentioned just four times (from two separate sources) in *all* presidential sources examined by this research (resulting in a single coded passage). By contrast, a text search of *terrorism* (and derivative words) during this period (2003–2006) yielded 599 references from fifty-three distinct sources (out of a possible sixty-five sources examined).[7] A search of *Iraq* in the sources yielded 1,887 references between 2003 and 2006 (fifty-five sources). Put another way, 85% of all presidential sources examined between 2003 and 2006 contained some reference to *Iraq*, 80% for *terrorism* and just 3% for *climate change*. This simple example offers a clear insight into the strategic dominance of the Global War on Terror (including operations in

[6]This finding, however, must be placed in context. Specifically, because this research did not code material from 2001–2002, it did not include a number of policies and documents which could have influenced the outcomes. These included President Bush's 2001 initiatives—*Climate Change Technology Program* (CCTP) and the *Climate Change Science Program* (CCSP) (2006)—President Bush's 2002 'Clear Skies Initiative' and the 2002 *National Security Strategy*. Although these documents were not coded as part of content analysis, they were included in the discussion.

[7]Of these sixty-five sources, sixty were either speeches or official media conferences. The remaining five were executive orders. All searches were unfiltered.

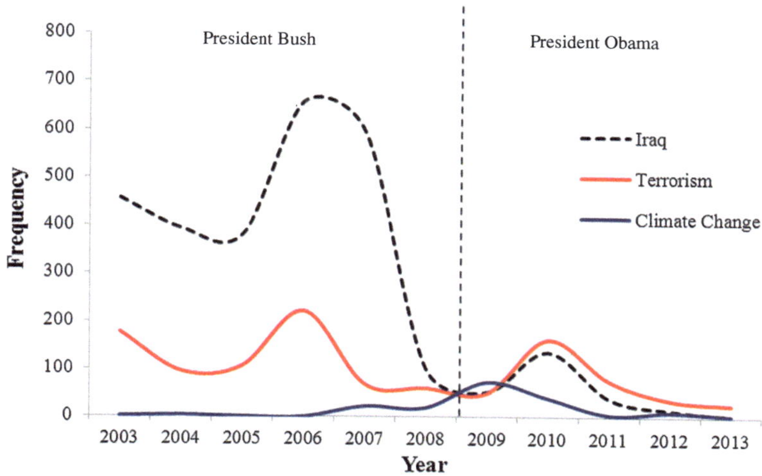

Fig. 6.3 Comparison of Presidential priorities, 2003–2013. Mentions (Frequency) of the words *Iraq*, *Terrorism* and *climate change* in US Presidential sources 2003–2013. *Source* The author

Afghanistan (from 2001) and then Iraq (2003–2011)) compared with climate change throughout most of the Bush presidency. A comparative graphic is at Fig. 6.3.

The 2002 *National Security Strategy* (2002 NSS) highlighted the major security priorities that dominated the Bush presidency. Widely known for articulating a doctrine of pre-emption ('we will not hesitate to act alone, if necessary, to exercise our right of self-defense by acting pre-emptively'), the 2002 NSS envisioned that the US use its 'unprecedented—and unequalled—strength and influence' by implementing a 'distinctly American internationalism' (White House 2002: 1–6). Centred on three goals (political and economic freedom, peaceful relations with other states, and respect for human dignity), the 2002 NSS prioritised its effort to combating terrorism as well as the 'looming threat' of 'aggressive behaviour of rogue states' that see 'weapons of mass destruction as weapons of choice … [to] overcome the conventional superiority of the United States' (White House 2002: 15).

Despite a focus on other security problems, the 2002 NSS *did* include a discussion on climate change. In Chap. 4, under 'A New era of Global Economic Growth through Free Markets and Free Trade', the NSS noted that economic growth 'should be accompanied by global efforts to stabilise greenhouse gas concentrations … [to contain] dangerous human interference with the global climate'. The 2002 NSS asserted that the US's overall objective was to reduce its emissions intensity[8] (18% over the decade), and listed six strategies to support this goal

[8]As distinct from emissions *quantity*.

(including a commitment to the *intent* of the UNFCCC, engaging industry to cut the most potent GHGs, promotion of renewable energy, improved measuring standards, increase research and development on new technologies and provision of assistance to developing countries) (White House 2002: 20). Beyond these aspects, there was no discussion in the NSS of the broader security risks posed by climate change.

By the 2006 US NSS, climate change had largely been removed from the narrative (however tenuously it previously existed). In this document, energy security (diversifying energy markets and ensuring energy independence) had become a supplementary issue in what continued to remain a singular focus on 'fighting and winning the war on terror and promoting freedom' (White House 2006: i).

By 2007 and 2008, however, President Bush increasingly referred to climate change. Across this period, this research identified forty-one direct references to climate change in speech-acts (resulting in twenty-one coded passages of text). Over the total period examined under the Bush Presidency (2003–2008), content analysis returned twenty-two coded passages of text. The majority of these (seventeen) were coded as non-securitised references and in a largely benign context (*neutral*). The five securitised references were all framed as *energy security*.

The language used by Bush to frame climate change was, for the most part, benign. Of those references coded, most were neutral (that is, neither as a threat, challenge or as an opportunity). The research identified only three occasions when Bush framed climate change as a challenge. On at least one of these occasions he identified it as a 'serious' challenge. Of significance, this research did not identify any instances of Bush framing climate change as a threat or crisis. As will be shown, this contrasted sharply with the rhetoric of President Obama, who frequently referred to climate change as a threat (Table 6.5c).

Overall, the content analysis identified three prominent categories within which climate change was framed during the Bush presidency: *energy*, the *economy* and as a *global* issue (Table 6.5a, b). Although the coding identified the general trends, further detail was added through content analysis of the speeches and policies themselves. Thus, using a combination of research methods, the three interlocking propositions that guided President Bush's approach to climate change can be summarised:

(1) Action on climate change was undertaken in the national self-interest. Primarily, it was not to damage economic growth (since economic growth represented the best solution to combatting global warming);
(2) Action on climate change was to promote US energy security (with a focus on technological innovation to produce clean energy as well as concepts of 'energy independence' from foreign oil) and;
(3) Action on climate change was a collective (international) problem and any global agreement was to encompass the major developing nations, particularly China (Bush also 'upheld' the principles of UNFCCC but strongly disagreed with the Kyoto Protocol).

Table 6.5 Climate change framing by President Bush and Obama, 2003–2013

(a) Securitised			
Securitised Frame	**Bush**	**Obama**	**Total**
International/Global	0	5	5
National	0	2	2
Energy	5	2	7
Total	5	9	14
(b) Non-securitised			
Non Securitised Frame	**Bush**	**Obama**	**Total**
International, global	3	26	29
Energy	10	22	32
General reference	1	1	2
Economic	2	3	5
Government regulatory	0	3	3
Ag., fisheries, water	0	1	1
Forestry	1	0	1
Health	0	1	1
Total	17	57	74
(c) Frame			
Securitised Frame	**Bush**	**Obama**	**Total**
Opportunity	0	11	11
Neutral	14	26	40
Challenge	3	4	7
Threat	0	16	16
Total	17	57	74

Source The author

Regarding the economic dimension, Bush spoke adamantly that reducing emissions and meeting the challenge of climate change must not come at the expense of economic growth.[9] Indicative of this was President Bush's first major initiative on climate change (*Clear Skies Initiative*), announced on 14 February 2002 at the National Oceanic and Atmospheric Administration (Bush 2002b). Although this speech occurred outside the time frame considered by this study, it provided an insight into Bush's early mind-set on climate change. Arguably, it was a mind-set that remained throughout his presidency. Specifically, the speech emphasised the centrality of acting in US' economic self-interest and in placing the economy at the forefront of a climate response:

[9]The 'economy' (as would be expected) was one of the highest stated priorities of President Bush. Notwithstanding terrorism and the impending invasion of Iraq, in his 2003 State of the Union address, President Bush stated: 'Our first goal is clear: We must have an economy that grows fast enough to employ every man and woman who seeks a job.'

> We must address the issue of global climate change … *But let's always remember, let's do what is in the interest of the American people.* Today, I'm confident that the environmental path that I announce will benefit the entire world. This new approach is based on this common-sense idea: that *economic growth is the key to environmental progress, because it is growth that provides the resources for investment in clean technologies* … growth is what pays for investments in clean technologies, increased conservation and energy efficiency (Bush 2002b: 233; emphasis added).

In the same speech, Bush further argued:

> Addressing global climate change will require a sustained effort over many generations. My approach recognizes that *economic growth is the solution, not the problem.* Because a nation that grows its economy is a nation that can afford investments and new technologies.
>
> … We can tap the power of economic growth to further protect our environment … that's what we're going to do (Bush 2002b: 236; emphasis added).

Similar sentiments were prevalent across the entire Bush presidency. Late in his presidential term on 3 April 2007, for example, during the *Presidents News Conference* Bush stated:

> In terms of the broader issue [of climate change], first of all, I've taken this issue very seriously. I have said that it is a serious problem. I recognize that man is contributing greenhouse gases, that – but here are the principles by which I think we can get a good deal. *One, anything that happens cannot hurt economic growth.* And I say that because, one, I care about the working people of the country, but also because, in order to solve the greenhouse gas issue over a longer period of time, it's going to require new technologies, which tend to be expensive. And it's easier to afford expensive technologies if you're prosperous (Bush 2007b: 415; emphasis added).

In a speech on 28 September 2007, *Remarks During a Meeting on Energy Security and Climate Change*, President Bush again emphasised the importance of a strong economy in guiding decisions on climate change when he stated:

> Our guiding principle is clear: We must lead the world to produce fewer greenhouse gas emissions, and we must do it in a way that does not undermine economic growth (Bush 2007d: 1261).

Same year, separate speech, Bush (2007c) argued that 'whatever we do' in relation to addressing climate change it must 'not wreck the economy'. He further added:

> If you don't have money, it is really hard to develop new technologies. And so we need to be prosperous for a lot of reasons, primarily so our citizens can have a good life, but also so that we're wealthy enough to make the investments necessary to deal with greenhouse gases (Bush 2007c: 1581).

The second distinguishing aspect of President Bush's approach to climate change was linking the issue with *energy.* First, in a securitised context, all coded passages were done so in the context of—or associated with—energy security (see Table 6.5a). Secondly, in a non-securitised context, ten of the possible seventeen coded passages were in the context of *energy.* The conflation of climate change and energy formed the most dominant narrative on the issue of climate change during

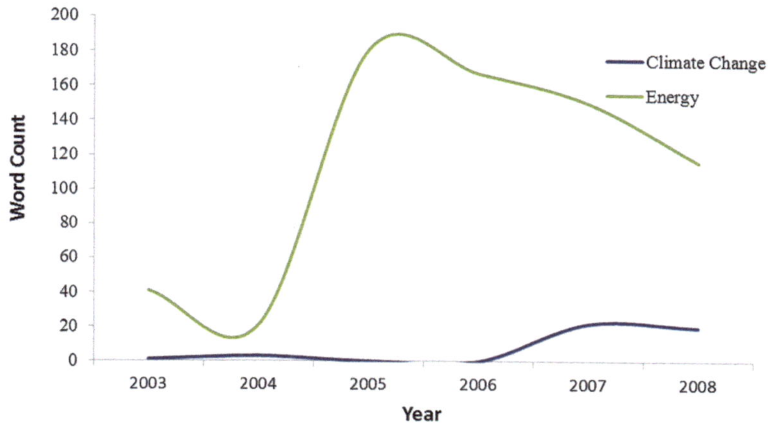

Fig. 6.4 President Bush mentioning *energy* versus *climate change* in speeches. Documents examined also included the US NSSs and selected Executive Orders of President Bush. All data is unfiltered. *Source* The author

the Bush Administration (over the period analysed). The energy-climate nexus also formed a feature within the military's framing of climate change, though this was observed to occur later in the decade.

Within President Bush's energy-climate nexus, nuances were found. Firstly, examination of President Bush's speech-acts suggested that action on climate change formed a supplementary justification in what was a far broader energy agenda dominated by concepts of energy security, energy independence and a belief that technology and innovation would solve (first) the energy 'crisis' and (later) the climate 'problem'. A general indication of the relative importance of *energy* over *climate change* is shown at Fig. 6.4 which compares the frequency of the words in speeches and policy (unfiltered) examined for President Bush between 2003–2008.

The dominance of *energy* issues (over *climate* issues) was also reflected in proposed and approved congressional legislation as well as in Presidential Executive Orders (EO). Prominent legislative examples included: the *Clear Skies Act of 2003* (introduced, not enacted), the *Energy Policy Act of 2005* (EPACT—P. L. 109-58, 8 August 2005), the *Advanced Energy Initiative* (proposed in the 2006 State of Union address) and the *Energy Independence and Security Act of 2007* (EISA—P.L. 110-140, 19 December 2007). Prominent Presidential Executive Orders relating to energy included: *EO 13302—Actions to Expedite Energy-Related Projects* (2003), *EO 13423—Strengthening Federal Environmental, Energy and Transportation Management* (24 January 2007) and *EO 13432—Cooperation Among Agencies in Protecting the Environment With Respect to Greenhouse Gas Emissions From Motor Vehicles, Nonroad Vehicles and Nonroad Engines* (14 May 2007). Overall, these initiatives largely formed the basis of President Bush's broad-based national approach on climate change through a focus on reducing US energy intensity, reducing the relative size of emissions to the overall economy and the development of clean energy.

The focus on energy regulation and legislation (over climate change) by President Bush was also reflected in initiatives within the US DoD. A 2009 report by the Congressional Research Service, *Department of Defense Facilities Energy Conservation Policies and Spending*, surmised that during the Bush era, Congress appropriated $443 million in Defense energy conservation projects with a further $2.8 billion in 'Energy Savings Performance Contracts' between 1999 and 2007. An examination of work by US Congressional Research Service shows that climate change mitigation was *not* a factor when reducing US DoD energy consumption or improving its efficiency (though it was later to become one during the Obama Administration). Undoubtedly, the main driver of the energy programme within the US DoD was the requirement to 'to rein in [Defense] energy consumption and spending' (Andrews 2009: ii).[10]

Not surprisingly, *energy* was accorded a high priority from the outset of the Bush Administration. In just his second week in office, President Bush established the National Energy Policy Development Group (NEPDG, led by the Vice President Dick Cheney) to develop a national energy policy to 'help bring together business, government, local communities and citizens to promote dependable, affordable and environmentally sound energy for the future' (National Energy Policy Development Group 2001). In its opening passages, the report (*National Energy Policy*) gave the staple narrative of President Bush's subsequent speeches on energy. Namely, the fear that the US was facing a 'national energy crisis' that would undermine 'our economy, our standard of living and our national security' (National Energy Policy Development Group 2001: vii). Crucial to this narrative was also a determination to decouple the US from its reliance on foreign oil and achieve energy independence. While these concepts have a long history in US political discourse (particularly since the energy crisis of the 1970s), the speech-acts of President Bush contained prominent examples and were noted by this research for their repetition across both presidential terms.

In his first address to a joint session of congress (27 February 2001), for instance, Bush singled out 'rising energy prices' as a 'warning sign'. Invoking a sense of urgency and the spectre of a national and geopolitical crisis, Bush gravely stated:

> As we meet tonight, many citizens are struggling with the high cost of energy. We have a serious energy problem that demands a national energy policy. The West is confronting a major energy shortage that has resulted in high prices and uncertainty. I've asked … my administration to develop a national energy policy … America must become more energy independent, and we will (Bush 2001: 356).

---

[10]The CRS research highlights how energy conservation programmes in US federal government can be traced to 1970, including the *Federal Energy Management Program* (FEMP) of 1973, the 1978 *National Energy Conservation Policy Act* (NECPA, P.L. 95-619) and the 1985 *Deficit Reduction Act* (P.L. 99-272). Measured against a 1985 baseline, by 2005 the US DoD reported a 28.3% reduction in energy consumption (CRS 2009: 1).

Concepts of energy independence also featured in *all* eight of President Bush's State of the Union addresses. In 2002, Bush argued that he would act to 'increase energy production at home so America is less dependent on foreign oil' (Bush 2002a: 137). Notwithstanding the war on terror, in the 2003 State of Union address 'energy independence' was identified as his third priority (after strengthening the economy and healthcare reform). Many of President Bush's speeches also linked energy independence with different forms of security, namely: energy security, economic security and national security (including the threat from terrorism). Again, a dominant theme in this context was the spectre of foreign oil:

- "One of the greatest results … will be energy independence for this Nation. It's important for our country to understand … we import that oil from countries that don't particularly like us … it jeopardizes our national security to be dependent on sources of energy from countries that don't care for America, what we stand for, what we love." Remarks on Energy Independence (Bush 2003: 162).
- "Extending hope and opportunity depends on a stable supply of energy that keeps America's economy running and America's environment clean. For too long, our Nation has been dependent on foreign oil. And this dependence leaves us more vulnerable to hostile regimes and to terrorists." State of the Union address (Bush 2007a: 59).
- "The dependency upon oil also puts us at the mercy of terrorists. If there's tight supply and demand, all it requires is one terrorist disruption of oil and that price goes even higher. It's in our interests to end our dependency on oil because that dependency presents a challenge to our national security." Remarks at the Washington International Renewable Energy Conference (Bush 2008c: 325).

The underpinning tenet, as Bush viewed it, of freeing the US from foreign oil dependence, was that it would provide stability for the pursuit of economic growth. In his 2005 State of the Union address, Bush argued that 'to keep our economy growing, we also need reliable supplies of affordable, environmentally responsible energy … that makes America more secure and less dependent on foreign oil' (Bush 2005: 115). To break the dependence on foreign oil, Bush constantly spruiked the virtues of the 'technological fix'. In the 2006 State of the Union address Bush argued: 'Keeping America competitive requires affordable energy. And here we have a serious problem: America is addicted to oil … the best way to break this addiction is through technology' (Bush 2006: 151). In 2007, Bush linked the development of energy technologies to solving climate change by stating:

> For many years, those who worried about climate change and those who worried about energy security were on opposite ends of the debate. It was said that we faced a choice between protecting the environment and producing enough energy. Today, we know better. These challenges share a common solution: technology … The key to this effort will be the advance of clean energy technologies (Bush 2007d: 1262).

In this capacity, Bush looked to the use of clean coal technologies ('our challenge is to take advantage of it [coal] while maintaining our commitment to the environment'

(Bush 2007d: 1262); revitalising US domestic nuclear energy industry—'there can be no single solution to climate change, but there can be no solution without nuclear power' (Bush 2007e: 838); as well as new fuel technologies (including renewables for broad base energy generation as well as bio-fuels for transportation). By 2007, Bush began linking many aspects of the energy-climate narrative (foreign oil, energy independence, technology and climate change) in concise form. In the 2007 State of the Union address Bush deployed his grand narrative:

> For too long, our Nation has been dependent on foreign oil. And this dependence leaves us more vulnerable to hostile regimes … [that could] do great harm to our economy. It's in our vital interest to diversify America's energy supply. The way forward is through technology … that will enable us to live our lives less dependent on oil. And these technologies will help us be better stewards of the environment, and they will help us to confront the serious challenge of global climate change (Bush 2007a: 59).

In his final State of the Union address, Bush reinforced these points—particularly noting the deployment of new technologies to enhance energy security and combat climate change:

> Our security, our prosperity, and our environment all require reducing our dependence on oil … The United States is committed to strengthening our energy security and confronting global climate change. And the best way to meet these goals is for America to continue leading the way toward the development of clearer and more energy efficient technology (Bush 2008a: 118).

In relation to the military and climate change, it was not until 2008 that the Bush Administration directed a specific action (though the Department of Defense's Office of Net Assessment did independently commission a report on the abrupt implications of climate change for national security in 2003). In the 2008 *National Defense Authorization Act* (NDAA), the Bush Administration directed the US DoD to examine the potential effects of climate change on facilities, capabilities and missions. Specifically, the 2008 NDAA required the next NSS (2010) to include guidance for military planners, including (NDAA 2008: 122, Section 951):

(1) The risks of projected climate change to current and future missions of the armed forces;
(2) A requirement to update Defense plans based on these assessments, including working with allies and partners to incorporate climate mitigation strategies, capacity building, and relevant research and development; and
(3) Development of the capabilities needed to reduce future impacts.

Similarly, the 2008 NDAA directed the next QDR to use mid-range projections from either the IPCC AR4 report or comparable research to:

> [E]xamine the capabilities of the armed forces to respond to the consequences of climate change, in particular, preparedness for natural disasters from extreme weather events and other missions the armed forces may be asked to support inside the United States and overseas (NDAA 2008: 122, Section 951).

Although this initiative was made late in Bush's term, it came off a growing volume of literature regarding the intersection of climate change and national security, and increased understanding of implications for national security equities (discussed in Chap. 3). In the context of a growing domestic momentum for action, Bush struck a further directive eleven days before he left office, signalling renewed US interest in the Arctic (*National Security Presidential Directive 66/Homeland Security Directive 25*). Partly based on 'the effects of climate change and increasing human activity', the directive also cited a 'growing awareness that the Arctic is … rich in resources' as well as 'several developments' in both national and international approaches as a pretext for a new approach (Bush 2009: 48). In this context, the directive highlighted the US had 'fundamental security interests in the Arctic region' that included 'missile defence and early warning, deployment of sea and air systems for strategic sealift, strategic deterrence, maritime presence and maritime security operations, and ensuring freedom of navigation and over-flight' (ibid). It specifically directed the Secretary of Defense to, amongst other things, '[d]evelop greater capabilities and capacity … increase Arctic maritime domain awareness … preserve [US] global mobility … and project a sovereign United States maritime presence' (ibid: 49). Thus, although there is a perception that the Obama Administration advanced climate change within the military, it was the Bush Administration that initiated the broad scale US military response.

The third—and final—feature distinguishable in the speech-acts of President Bush and his approach to climate change was his opposition to the Kyoto Protocol and the desire to develop an international agreement beyond or outside the existing framework. This aspect was (weakly) reflected in the research results, whereupon relevant text was coded under *global/international*.[11] Throughout his presidency, Bush repeated his opposition to the Kyoto Protocol. Not surprisingly, the main reason Bush refused to ratify the Kyoto Protocol was that it might weaken the US economy. In his first major speech on climate change in 2002, for instance, Bush argued:

> As President of the United States, charged with safeguarding the welfare of the American people and American workers, I will not commit our nation to an unsound international treaty that will throw millions of our citizens out of work (Bush 2002b: 235).

In addition to economic reasons, Bush repeatedly cited the omission of major developing nations, particularly China and India, as a reason not to ratify the Kyoto Protocol. This position was made explicit when he remarked "[o]ne of the main reasons I was against Kyoto was that China wasn't at the table. I mean, we could do

[11]Many passages that were coded under *energy* contained references to international aspects of climate change. However, the method used in this case study largely coded text to a single theme.

all we wanted to do, but it wouldn't affect greenhouse gases over the long run unless a country like China had agreed to participate" (Bush 2007c: 1581).[12]

In his opposition to the Kyoto Protocol, Bush sought a new 'international approach to energy security and climate change' based on the development of national plans that 'reflect[ed] each country's different energy resources, different stages of development, and different economic needs' (Bush 2007d: 1262). Thus, whereas the UNFCCC Kyoto Protocol required a 'common [strategy] but differentiated responsibility' to reducing emissions, Bush adopted the outlook that 'while our strategies may be differentiated, we share a common responsibility to reduce greenhouse gas emissions while keeping our economies growing' (2007d: 1262). Whereas the UN differentiated by 'responsibility', Bush differentiated by 'strategy'. In this context, Bush stressed the importance of developing global emission objectives that included 'concerted effort by *all* our nations' [emphasis added], according to its own specific circumstances in which 'each nation must decide for itself the right mix of tools and technologies to achieve results' (Bush 2007d: 1262). Like his conservative Australian counterpart, Prime Minister Howard, national self-interest was actively promoted to the public as the pre-eminent rationale. Despite this intent, a new agreement failed to materialise. On 4 November 2008, the Republican nominee Senator John McCain (R-Az.) lost the election to Democrat, Barack Obama (D-Il.).

6.2.1.2 President Barack Obama

No nation, however large or small, wealthy or poor, can escape the impact of climate change.

President Obama, Address to United Nations (2009g)

The election of Barack Obama signified a re-prioritisation of climate change by the US. In November 2008, as president-elect, Obama delivered a video-message to the *Governors Global Climate Summit,* in which he declared the issue an urgent priority which he would immediately take up to 'strengthen our security':

[12]Literally dozens of quotes on this point were found during the research. Three further examples are provided here: 'The Kyoto Protocol would have required the United States to drastically reduce greenhouse gas emissions. The impact of this agreement, however, would have been to limit our economic growth and to shift American jobs to other countries while allowing major developing nations to increase their emissions. Countries like China and India are experiencing rapid economic growth [but] this also means they're emitting increasingly large quantities of greenhouse gases, which has consequences for the entire global climate'. 2008, *Remarks on Energy and Climate Change* (Bush 2008d). 'In order for there to be effective international agreement, it must include—these agreements must include—commitments, solid commitments, by every major economy, and no country should get a free ride'. 2008, *Remarks at the Washington International Renewable Energy Conference* (Bush 2008c). 'Look, we can't have an effective agreement unless China and India are part of it. It's as simple as that' (Bush 2008b).

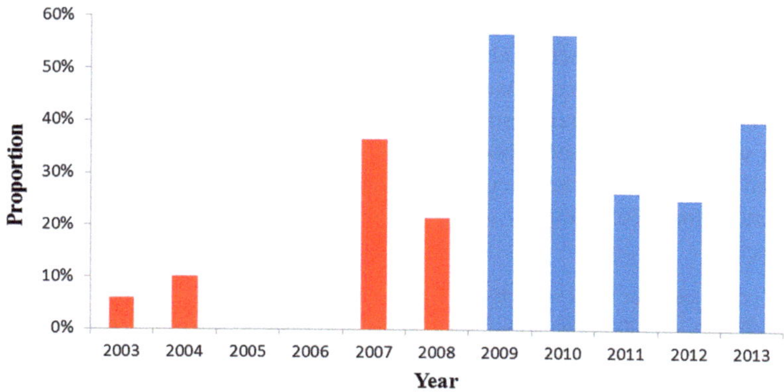

Fig. 6.5 Proportion of *climate change* mentions in examined presidential speeches and policies, 2003–2013. The rise of climate change from 2007 onwards in presidential sources, and particularly during the Obama Administration (*blue*), is evident. *Source* The author

> Few challenges facing America - and the world - are more urgent than combating climate change. The science is beyond dispute and the facts are clear. Sea levels are rising … Now is the time to confront this challenge once and for all. Delay is no longer an option. Denial is no longer an acceptable response. The stakes are too high. The consequences, too serious (Obama 2008).

In a 2009 speech to the United Nations General Assembly, President Obama identified the 'preservation of the planet' as one of his key 'four pillars' (Obama 2009l: 1443). In this context, Obama specifically identified climate change as a 'responsibility that must not be deferred' and declared that 'the days when America dragged its feet on this issue are over' (ibid: 1445). The Obama Administration's new outlook was reflected in this research, which recorded a surge in the number of references to climate change in the speech-acts and policies of President Obama (Fig. 6.3). Indeed, of the Obama sources examined between 2009 and 2010, almost 60% contained one or more references to climate change (Fig. 6.5).[13] This highlights the increased prominence that climate change had in the Obama Administration, particularly the years 2009 and 2010.[14]

The following passages further examine the climate-securitisation issues (and climate-response more broadly) of President Obama. The discussion focuses on differences and similarities to his predecessor.

[13]In making this assessment, it should be noted that some speeches analysed also contained a question and answer session following the major speech. Thus there are instances where climate change may not have been raised in the speech itself, but was then included in discussions thereafter.

[14]The research examined 101 artefacts between 2003 and 2008 for President Bush and 90 artefacts for President Obama (2009–2013). The research also examined fewer artefacts across 2011–2013 compared to the first two years of the Obama Administration (2009—31 artefacts, 2010—22 artefacts, 2011—19 artefacts, 2012—12 artefacts, 2013—5 artefacts).

Similarities. This research identified a number of similarities between President Bush and Obama's approach to climate change. First, like President Bush, President Obama's approach to climate change was dominated by *energy* considerations. This was reflected in the research results (Table 6.5b) which totalled twenty-two coded passages within the non-securitised energy-climate nexus and a further two securitised coded passages under energy security. Similarly, concepts of foreign oil and energy independence featured strongly as did ideas of acting in the national self-interest:

> We're convinced, for our own self-interest, that the way we use energy, changing it to a more efficient fashion, is essential to our national security, because it helps to reduce our dependence on foreign oil and helps us deal with some of the dangers posed by climate change (Obama 2009e: 1821).

Also similar to Bush, Obama often spoke of responding to climate change whilst ensuring the growth of the US economy and protecting jobs. However, unlike Bush, Obama's outlook on climate change and the economy was heavily influenced by the 2008–2009 global financial crisis (GFC) and resultant American Recovery and Reinvestment Act (ARRA) of 2009 (P.L. 111-5) that directed significant investment towards development of US clean energy economy (ARRA 2009).[15] The military was a major beneficiary of this stimulus package, particularly in the areas of clean energy and energy efficiency (Andrews 2009: 14).

Broadly, both Bush and Obama frequently cited the economy, energy security and climate change as interdependent issues. One example by Obama linking these issues was given during a speech on 26 January 2009:

> I want to say a few words about the deepening economic crisis ... and the need for urgent action.
>
> ... We owe it to each of them and to every single American to act with a sense of urgency and common purpose. We can't afford distractions, and we cannot afford delays, and that is why I look forward to signing an American recovery and reinvestment plan that will put millions of Americans to work and lay the foundation for stable growth that our economy needs and that our people demand. *These are extraordinary times, and it calls for swift and extraordinary action.* At a time of such great challenge for America, no single issue is as fundamental to our future as energy ... [but] ... [t]hese urgent dangers to our national and economic security are compounded by the long-term threat of climate change, which if left unchecked could result in violent conflict, terrible storms, shrinking coastlines, and irreversible catastrophe (Obama 2009i: 1; emphasis added).

Implicit within this speech, and others like it, was the primacy of the economy over any other issue, including the climate response. It also showed that 'extraordinary action'—words that approximate closely with Copenhagen's emergency measures—were only raised in the context of the GFC. Climate change was a part

[15]The American Recovery and Reinvestment Act (ARRA) of 2009 delivered $68.4 billion spending on energy efficiency measures and renewable programmes as well as a range of tax incentives.

of this narrative, but it was not the defining part. Thus, when Obama spoke on the importance of combatting climate change by introducing a federal cap-and-trade scheme via the *Waxman-Markey Bill* he argued it on economic grounds: '[m]ake no mistake: this is a jobs bill' (Obama 2009d).

Like Bush, Obama constantly noted 'my highest priority as President is creating jobs and putting Americans back to work' (Obama 2011: 1). Moreover, when climate change was included in discussions on the economy, both Bush and Obama spoke of the need to grow the economy (first) whilst mitigating or adapting to climate change (second). This was true before the GFC, but was particularly the case after it. In noting economic primacy, both Obama and Bush also regularly argued to achieve both (i.e., jobs and climate action), simultaneously. In November 2012 at the *President's News Conference*, Obama made the case for such a 'win-win' strategy:

> [I]f the message is somehow we're going to ignore jobs and growth simply to address climate change, I don't think anybody is going to go for that. I won't go for that.
>
> If, on the other hand, we can shape an agenda that says we can create jobs, advance growth, and make a serious dent in climate change and be an international leader, I think that's something that the American people would support (Obama 2012b).

But in undertaking an 'all-of-the-above' strategy, Obama presented a realist outlook that also embraced the exploitation of fossil fuels and 'cleaner' sources of energy. It belied his rhetoric on climate change and it reflected a more pragmatic political outlook that was reminiscent of Bush. Obama's 2012 State of the Union address made this apparent where he proudly noted his Administration had 'opened millions of new acres for oil and gas exploration, and tonight I'm directing my administration to open more than 75% of our potential offshore oil and gas resources … American oil production is the highest that it's been in 8 years' (Obama 2012a).

Although at a very broad scale, another similarity between Bush and Obama was that both tended to frame climate change in a non-securitised manner (Table 6.5a, b). Despite this, the research did identify significant differences, and it is to these this book now turns.

Differences in Approaches to Climate Change between President Bush and Obama. Having identified some similarities between the two presidents, the research also identified major divergences. A key point summary includes:

(1) Climate change was heavily dominated by global considerations during Obama's presidency. In particular, COP 15 marked a major moment in international climate negotiations that resulted in the Copenhagen Accord and a commitment by states to limit planetary warming to no more than 2 °C. This aspect was reflected in the research results whereupon more than half of all coded references (non-securitised) framed climate change as an international issue (Table 6.5b).

(2) President Obama spoke more frequently on the issue of climate change, with an emphasis on the opportunities presented through the advent of a clean energy economy. President Obama, for instance, discussed climate change in all but one of his State of Union addresses between 2009 and 2013. In 2011, although there was no direct reference to it, he discussed his Administration's clean energy goals. Table 6.5c shows this research identified climate change was framed as an opportunity on (at least) eleven occasions. Arguably, the case for climate change as an opportunity was used to support clean energy initiatives—and jobs growth more generally—in ARRA of 2009.

(3) President Obama often framed and spoke about climate change as a security threat and had a broader interpretation of the threats posed by climate change that Bush. This was partially born out in the research results (Table 6.5a), but was elaborately evident in his speech-acts (discussed below).

(4) President Obama used a mixture of short-term and long-term outlooks to discuss climate change. The research coded five occasions where Obama spoke about climate change as an urgent, crisis-like issue requiring immediate action. Bush was never found to have spoken about climate change in this manner.

(5) President Obama sought an international binding agreement within the framework of the UN but accepted this would take some time to achieve on account of the limitations of US Congress as well as international 'deadlock'. In a speech in 2009 he argued these points, adding, 'I actually think that it's necessary for us, ultimately, to get such treaty, and I am supportive of such efforts' (Obama 2009c: 1826).

(6) President Obama undertook expanded regulatory action to address climate change, particularly through *Executive Order 13514—Federal Government Leadership in Environmental, Energy and Economic Performance* (2009).[16] This directed all federal agencies (including US DoD) to reduce emissions and undertake organisational risk assessments posed by climate change.

(7) President Obama spoke on a requirement for legislating a 'market-based cap on carbon pollution'. This differed from President Bush who argued that there was a 'wrong way' and a 'right way' (Bush 2008d: 526).

(8) President Obama issued a specific plan relating solely to addressing climate change, *The President's Climate Action Plan* (13 June 2013). Although the plan was not considered during research coding (since it fell outside the time frame), it was found to represent policy continuation with similar rhetoric from previous speeches and policy acts. In particular, the *Climate Action Plan* placed significant emphasis on regulatory initiatives or government-led initiatives.

Returning to the central aims of this book, the following discussion will expand on the key differences each president had in terms of climate securitisation. In the

[16]This was superseded in November 2013 by *Executive Order 13653—Preparing the United States for the Impacts of Climate Change.*

period examined, President Obama discussed the security dimension of climate change in far broader terms than his predecessor. For instance, whereas Bush had a narrow interpretation of climate security (centred on energy security), Obama framed climate change as an international, national and energy security threat (Table 6.5). In addition, Obama often went into some detail to describe the actual physical, social, economic and security threats posed by climate change.

Obama also made a point of emphasising the magnitude of the threats by arguing that it represented a planetary-wide threat in which no state (or people) would be immune. Furthermore, Obama also spoke on how climate change threats were already occurring and that they were not solely to be viewed in long timescales. These aspects were not evident during President Bush's speeches. Selected quotes by Obama highlight these points:

- "No nation, however large or small, wealthy or poor, can escape the impact of climate change. Rising sea levels threaten every coastline. More powerful storms and floods threaten every continent. More frequent droughts and crop failures breed hunger and conflict in places where hunger and conflict already thrive. On shrinking islands, families are already being forced to flee their homes as climate refugees. The security and stability of each nation and all people—our prosperity, our health, and our safety—are in jeopardy, and the time we have to reverse this tide is running out", Presidential remarks at United Nations Secretary General Ban Ki-Moon's Climate Change Summit (Obama 2009g).
- "The danger posed by climate change cannot be denied. Our responsibility to meet it must not be deferred … wars over refugees and resources … devastated by drought and famine. Land that human beings have lived on for millennia will disappear", Presidential remarks to the United Nations General Assembly (Obama 2009h).
- "A warming planet will spread disease, shrink water resources, and deplete crops, creating conditions that produce more famine and more conflict", Remarks to the Ghanian Parliament in Accra (Obama 2009k: 1090).
- "Every nation on this planet is at risk", Remarks on the Declaration by the Major Economies Forum on Energy and Climate Change in L'Aquila, Italy (Obama 2009j: 1073).

Obama also constantly referred to climate change as an urgent problem requiring both short- and long-term responses. In many respects, the language used by Obama came close to approximating the qualifying language of existential threat requiring emergency measures under the Copenhagen School securitisation framework. Again at the 2009 UN General Assembly:

> [T]he threat from climate change is serious, it is urgent, and it is growing. Our generation's response to this challenge will be judged by history, for if we fail to meet it boldly, swiftly, and together, we risk consigning future generations to an irreversible catastrophe (Obama 2009h).

Again, in the 2009 declaration during the Major Economies Forum on Energy and Climate:

> Climate change is one of the greatest challenges of our time. As leaders of the world's major economies, both developed and developing, we intend to respond vigorously to this challenge, *being convinced that climate change poses a clear danger requiring an extraordinary global response* (Obama 2009j: 1075; emphasis added).

Similar rhetoric was also evident at COP 15 in the context of establishing a new international climate agreement. During a speech at a Plenary Session of the UN Climate Change Conference in Copenhagen, Obama argued:

> The time for talk is over … there is no time to waste. America has made our choice … We are ready to get this done today, but there has to be movement on all sides to recognize that it is better for us to act than to talk; it's better for us to choose action over inaction, the future over the past. And with courage and faith, I believe that we can meet our responsibilities to our people and the future of our planet (Obama 2009e: 1822).

Building on President Bush's 2008 *National Defense Authorization Act (NDAA)*, the new policy directions of the Obama Administration on climate change had a measured impact on the US DoD. At the broadest level, Obama's 2008 presidential victory set the political conditions necessary for a robust military response to climate change. Part of this meant mainstreaming climate change as a legitimate national security issue, from which detailed policies could be designed and implemented to better prepare the military for the actual risks. This was done—as a continuation of the directives established late in the Bush term—in the 2010 US NSS (and also the 2010 QDR discussed in the military section below). Furthermore, Obama commissioned a number of separate initiatives in addition to the 2010 NSS, including the requirement for the US DoD to develop a departmental level climate change adaptation plan (addressed in detail below, under SECDEF Programme). Lastly, Obama also cited the US DoD in speeches (and policy) designed to communicate that the US DoD was actively contributing to broader national climate change mitigation efforts. Similarly, Obama also cited the military leadership as advocating for action on climate change on the basis of its risk to security. In his 2009 Nobel Lecture, Obama noted:

> It is not merely scientists and environmental activists who call for swift and forceful action – it's military leaders in my own country and others who understand our common security hangs in the balance (Obama 2009b).

Highlighting the military's concerns regarding climate change, and ensuring a substantive military response, represented a vastly different approach from Bush who—in the speeches analysed by this book—never sought to link the US military to climate change. Three examples from Obama's term highlight this difference.

First was *EO 13514*, that specifically called upon the Federal Government to 'lead by example' through creation of a 'clean energy economy that will increase our Nation's prosperity, promote energy security, protect the interests of taxpayers, and safeguard the health of the environment' (Executive Order 13514 2009).

Within this framework, *EO 13514* mandated the US DoD take specific action on climate change that included (but was not limited to): (1) a requirement to produce an annual strategic sustainability performance plan that included a climate change adaptation plan; (2) a requirement to evaluate climate change risks and vulnerabilities to manage the effects of climate change on agency operations and missions in both the short and long-term; and (3) a requirement to participate in interagency Climate Change Adaptation Task Force. Moreover, while 2008 NDAA initiated *general* guidance for military planners on climate change (born out by the 2010 NSS and 2010 QDR), it was *EO 13514* that effectively bought forward *specific* time frames, detailed requirements and an understanding that US DoD leaders would be held accountable.

Secondly, climate change was 'mainstreamed' as a security issue in the 2010 NSS. It represented the first time that climate change had been dealt with in detail by the pre-eminent US national security document. In total, there were twenty-three direct references to climate change, with three specific paragraphs that addressed the issue under the heading 'key global challenge'. The amplified rhetoric in relation to climate change within the 2010 NSS marked a further significant departure from the Bush Administration (the 2006 NSS did not refer to climate change at all). Highlighting the scale and urgency, the 2010 NSS argued:

> The danger from climate change is real, urgent and severe. The change wrought by a warming planet will lead to new conflicts over refugees and resources; new suffering from drought and famine; catastrophic natural disasters; and the degradation of land across the globe. The United States will therefore confront climate change based upon clear guidance from the science, and in cooperation with all nations (White House 2010: 47).

A third aspect was to link the US DoD as part of broader national climate change mitigation and adaptation plans. In his 2012 State of the Union address, Obama noted the inability of Congress to pass a 'comprehensive plan to flight climate change' led him to direct the dėvelopment of clean energy on federal land. He specifically singled out the US DoD and US Navy as making 'one of the largest commitments to clean energy in history' (Obama 2012a). This was further reinforced in the *President's Climate Action Plan* that identified the US DoD as contributing to a broad range of mitigation and adaptation measures.

6.2.2 *Secretary of Defence (SECDEF) Programme*

Analysis of the SECDEF Programme was undertaken using 811 sources published between 2003–2013 (Table 6.5 lists the SECDEFs analysed and Table 6.7 lists the source breakdown).[17] A text search for *climate change* and *global warming*

[17]Artefacts for the US Secretary of Defence consisted of 807 speeches and 4 strategic policy documents. The strategic policy documents included the 2001, 2006 and 2010 Quadrennial Defense Reviews.

Table 6.6 List of US Secretary of Defense included in this research

Secretary	Took office	Left office	Serving President
Donald Rumsfeld	January 20, 2001	December 18, 2006	George W. Bush
Robert M. Gates	December 18, 2006	July 1, 2011	George W. Bush Barack Obama
Leon Panetta	July 1, 2011	February 27, 2013	Barack Obama
Chuck Hagel	February 27, 2013	February 17, 2015	Barack Obama

Note This research ended early in the tenure of Secretary Hagel and did not include all additional speeches in which he addressed the security implications of climate change
Source Wikipedia (2017e)

Table 6.7 US Secretary of Defence sources and references to *climate change*, 2003–2013

Secretaryy	Speech-acts	Policies	References to *climate change* in speech-acts	References to *climate change* in other documents	Total sources
Donald Rumsfeld	298	2	0	0	300
Robert M. Gates	365	2	15	20	367
Leon Panetta	122	0	2	0	122
Chuck Hagel	22	0	0	0	22
Total	807	4	17	20	811

Source The author

returned eleven separate sources containing a total of thirty-seven references. From this, eleven passages of text were coded within a securitised context and four passages of text were coded in a non-securitised context (Table 6.6).

The majority of references came from official government policy documents, notably half of all references to *climate change* (nineteen) came from the 2010 *Quadrennial Defense Review* (QDR), with a single reference in the 2008 *National Defense Strategy* (NDS). The remaining references came from nine separate speech-acts; all of these occurred from 2009 onwards. The majority of speeches that referenced *climate change* were given during the incumbency of Secretary Gates (five speeches in 2009 containing eleven references and two speeches in 2010 containing four references). Closer examination of these speeches shows that most of these references were attributable to the Deputy Secretary of Defense, William J. Lynn III (eleven of the fifteen). In the 122 speeches examined by this case study and delivered by Leon Panetta, climate change was referenced climate change in two of those speeches in 2012 and 2013. Climate change was not mentioned in any speech-acts or policy documents examined by this book under Secretary's Rumsfeld or Hagel emphasizing that only 22 speeches were examined for Chuck Hagel on the basis that the case study concluded at the start of 2013. If this case study had been extended, additional mentions of climate change under Chuck Hagel on multiple occasions are evident and certainly consistent with the

Table 6.8 Climate change framing by the office of US Secretary of Defense, 2003–2013

Securitised frame	Rumsfeld	Gates	Panetta	Hagel	Total
International/Global	0	2	0	0	2
National	0	0	1	0	1
Convergence	0	3	0	0	3
New security	0	2	0	0	2
Arctic	0	2	1	0	3
Total	0	9	2	0	11
Non Securitised Frame	**Rumsfeld**	**Gates**	**Panetta**	**Hagel**	**Total**
International/Global	0	1	0	0	1
Energy	0	3	0	0	3
Total	0	4	0	0	4

Source The author

trajectory of climate security emphasis under the Obama Administration, see for example Hagel (2013, 2014a, b).

As noted, a total of fifteen passages of text (from thirty-seven references) were coded. All texts coded came from sources published after 2008. Eleven passages were coded as securitised across five categories that included global and regional security, national security, convergence, new security and the Arctic (Table 6.8). No particular securitised category was considered dominant. Four passages of text were coded from sources published in 2009 and 2010 under two non-securitised categories (*global* and *energy*).

Discussion on climate change in SECDEF sources peaked in 2009 and 2010 (noting that the study only included documents until early 2013). The issue was extensively addressed in the 2010 QDR and received partial description in the 2008 NDS. Beyond this, climate change was not a major consideration in the speeches of any Secretaries, with the exception of the Under Secretary of Defense (William Lynn) who, on occasion, stood in for an absent Secretary Gates (of course, this precludes examination of Chuck Hagel's speeches beyond 2013). Furthermore, climate change, or matters pertaining to the environment more broadly, were not addressed in either the 2005 *National Defense Strategy* (NDS) or the 2006 QDR. This period, particularly under President Bush and Secretary Rumsfeld, was characterised by US strategic efforts toward fighting the 'long war' against terrorism.

The language used by Secretary Gates (including Deputy Secretary Lynn) and Panetta to describe climate change was predominately neutral (six codes), with a tendency to also frame it as a *challenge* (two codes) or *threat* (three codes). Two passages of text were coded for temporal framing. The first was the portrayal of climate change as requiring a short-term response by the US Department of Defence (outlined by the 2010 QDR). The second was the portrayal of climate change as a long-term pressure (over the next twenty years) that would create greater uncertainty (outlined in the 2008 NDS).

A broad deduction from these results is that climate change was *not* a major consideration in the speech-acts or major strategic policy documents (namely the *National Defence Strategy* and *Quadrennial Defence Review*) of the US Secretary of Defense until at least 2008. From this point onwards, and particularly between 2009 and 2010, climate change was framed as a global and national security issue/challenge/threat that roused military attention, justifying the military to consider its implications. At no point, was 'emergency measures' spoken of, or undertaken, in response to climate change. Nor was climate change framed as an urgent matter requiring immediate military response.[18] To build a more complete picture of US climate securitisation, the following paragraphs will describe its trajectory under each US Secretary of Defense (2003–2013) in more detail.

6.2.2.1 Secretary Rumsfeld

The period under Secretary Rumsfeld was dominated by a focus on transforming the US DoD (away from Cold War structures, capability and culture) and fighting the Global War on Terror (including the proliferation of WMD by 'rogue' nations). A simple indicator of this point was a word search of *Iraq*, *terrorism*, *transformation* and (for comparison) *climate change* in 811 speeches and policies of the Secretary of Defense between 2003 and 2013 (Fig. 6.5). Further evidence is provided by an examination of speeches and policies under Secretary Rumsfeld (Fig. 6.6).

The 2001 QDR (developed prior to September 11, but published shortly thereafter) was built around four policy goals: (1) assurance to allies and friends; (2) dissuading future military competition; (3) deterring threats and coercion against US interests; and (4) acting decisively to defeat any adversary. To achieve this, Rumsfeld set out to 'transform' the US Armed Forces by exploiting new approaches to operational concepts, capabilities, structures and technology. The focus on transformation, Global War on Terror and general priorities under Rumsfeld were recorded in a budget speech to the House Armed Services Committee in 2003:

> When our nation was attacked, there was a great deal of pressure to put off transformation – people cautioned, you can't fight the global war on terrorism and simultaneously transform this institution. The opposite is the case. The global war on terror has made transforming an even more urgent priority (Rumsfeld 2003).

Both major strategic documents (the 2005 *National Defence Strategy* and the 2006 *Quadrennial Defense Review*) represented a continuation of these goals. The 2005 NDS highlighted that the security threats of the twenty-first century would be dominated by hard security threats; key drivers being globalisation and the potential

[18]Noting that later iterations of US Department of Defense documents began to describe climate change in much more urgent terms. For example, the 2014 DoD Climate Change Adaptation Roadmap describes climate change as posing "immediate risks to US national security" (US DoD 2014a: 1).

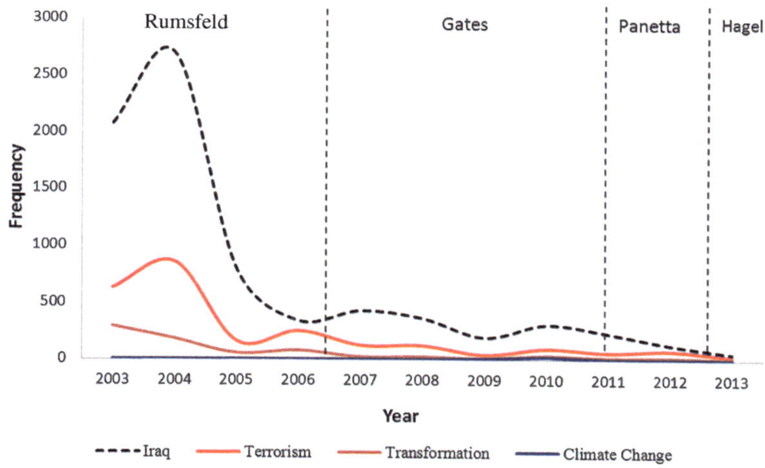

Fig. 6.6 Comparative importance of security issues, Secretary of Defense 2003–2013. 'Transformation' included data from the 2001 QDR (counted in the year 2003). *Source* The author

proliferation of weapons of mass destruction (US DoD 2005: 1). In this context, the NDS foresaw the main threats to US interests coming from traditional state (military) challenges, irregular challenges (asymmetry), catastrophic challenges (WMD) and disruptive challenges (adversaries who develop breakthrough technologies). Both the 2005 NDS and 2006 QDR also emphasised the US as a nation at war. In the opening line, the 2006 QDR states: 'The United States is a nation engaged in what will be a long war' (US DoD 2006: v). Rather than represent a 'new beginning', the 2006 QDR reinforced that the US military would continue 'transforming along a continuum that reflects our best understanding of [the] world' (ibid). Transformation and a focus on hard security threats remained paramount in the 2006 QDR.

A further aspect of the Rumsfeld period was that climate change was not identified in any speech or policy document as an opportunity, issue, challenge or threat. Indeed, environmental issues more generally appear to have been given short shrift. Neither the 2005 NDS nor 2006 QDR, for instance, use the term 'environmental security' or similar terms. To be sure, of the limited environmental references, Rumsfeld noted in a 2003 speech to the House Armed Services Committee on the need to 'clarify environmental statutes which restrict access to, and sustainment of, training and test ranges essential for the readiness of our troops … in the global war on terror' (Rumsfeld 2003). For Rumsfeld, environmental controls would need to be amended, lest they prohibited military readiness in the war on terror.[19] This outlook was firmly intended to be put into practice across the US

[19]These findings correspond with other scholarly works on the subject. See, for example, Floyd (2010).

DoD, as evidenced by a leaked copy of the 2002 Sustainable Defence Readiness and Environmental Protection Act which read:

> Federal departments and agencies shall not place the conservation of public lands, or the preservation or recovery of endangered, threatened, or other protected species found on military lands, above the need to ensure that soldiers, sailors, airmen and Marines receive the greatest possible preparation for … the hazards and rigor of combat (quoted in Floyd 2010: 134).

As noted by the scholar Rita Floyd, under Rumsfeld's leadership 'environmental laws suddenly constituted a risk to, or an obstacle in, the provision of national security' (Floyd 2010: 141). Sidestepping environmental rules and regulations in the name of military readiness became Rumsfeld's policy *du jour*.

6.2.2.2 Secretary Gates

During Secretary Gates' term, climate change became an issue of some importance. The 2008 *National Defence Strategy* (NDS) was the first official policy document within the SECDEF programme to discuss climate change (over the period examined by this research). Under the sub-heading of 'The Strategic Environment', climate change was identified as one of five major trends that 'could combine with rapid social, cultural, technological and geopolitical change to create greater uncertainty' (US DoD 2008: 4).

The 2008 NDS also noted how converging trends (demographics, resource, environmental and climate pressures) may generate new security challenges. The importance of the energy-climate nexus was also highlighted, in which it stated that the 'risks will require managing the divergent needs of massively increasing energy demand to maintain economic development and the need to tackle climate change' (US DoD 2008: 5). The 2008 NDS also made reference to the emerging security aspects of climate change, but highlighted the uncertainty surrounding how this would unfold. Overall, while these descriptions represented a small concern in the broader strategic picture, they were a significant departure from the 2005 NDS. Arguably they were at the vanguard of mainstreaming climate change as a legitimate security issue in (formal) US strategic literature following previous attempts by the Clinton Administration in the late 1990s (Brauch 2011b).

One of the most prominent speech-acts addressing climate change by a SECDEF (or a representative) was made by Under Secretary William Lynn in 2009. The speech was made in the context of President Obama's trip to Copenhagen and the anticipated release of the 2010 QDR. Lynn identified climate change as an 'instability accelerant' with the potential to exacerbate existing tensions through 'food and water shortages, increases in the spread of disease, and … migration' (Lynn 2009). The main focus, however, was the energy-climate nexus. In this context, Lynn highlighted the mitigation efforts of the US DoD through 'increasing the energy efficiency of our forces, both in theater [tactical] and at fixed installations [strategic]' (ibid).

By far the most important strategic policy document in the SECDEF Programme to address climate change (over the period examined) was the 2010 QDR. In total, the 2010 QDR contained nineteen direct references to climate change. As previously noted, the assessment was initiated during the final stages of the Bush Administration via 2008 *National Defense Authorization Act for Fiscal Year 2008*.

The 2010 QDR identified that climate change would impact the US DoD in two broad ways. Firstly, climate change 'will shape the operating environment, roles, and missions that we undertake' (US DoD 2010: 84). It further noted that 'climate-related changes'—such as heavy downpours, rising temperature and sea-level, retreating glaciers, thawing permafrost, lengthening growing seasons, lengthening ice-free seasons, earlier snow melt and altered river flows—were already occurring 'in every region of the world' (US DoD 2010: 84). Backed by estimates from the US intelligence community and shedding previous strategic uncertainty on the subject, QDR (2010: 85) declared 'climate change *will* contribute to food and water scarcity, *will* increase the spread of disease, and may spur or exacerbate mass migration' acting as an 'accelerant of instability or conflict' (emphasis added). In this context, the 2010 QDR noted the likely increase in demand for HADR operations both domestically and internationally as well as the opportunity to use the moment as a means to undertake 'environmental security cooperative initiatives with foreign militaries' as a 'non-threatening way of building trust, sharing best practices' (ibid). Further operational focus was directed on the opening of Arctic waters and a requirement to 'address gaps in Arctic communications, domain awareness, search and rescue, and environmental observation and forecasting to support [Arctic] operations' (US DoD 2010: 86). These aspects became prominent justifications in US Navy Arctic-climate efforts (see below).

The second way the 2010 QDR noted climate change would impact the US DoD was on facilities and military capabilities. It highlighted the risk to coastal infrastructure, citing the 2008 *National Intelligence Council* report that judged more than 30 coastal installations at risk from rising sea-levels (US DoD 2010: 85). In view of such risks, the 2010 QDR called for a 'comprehensive assessment of all installations to assess potential impacts of climate change', noting a requirement to engage with a range of other government departments to 'assess, adapt to, and mitigate [its] impacts' (US DoD 2010: 85 and 86). Thus, the risk that climate change presented to US military global infrastructure cannot be underestimated. Taken in their totality, the underlying thrust of these assessments spoke of a requirement for the US DoD to begin the process of adapting to the coming impacts posed by climate change. Not surprisingly, these were not for altruistic or 'feel-good' reasons, they were commenced on the central premise of ensuring the US DoD could 'operate into the future without decline' (US DoD 2012a: 3).

Beyond these assessments, the 2010 QDR focused on the energy-climate nexus which it described as 'inextricably linked'. In this context, the 2010 QDR focused on energy efficiency and energy reduction strategies that would 'improve operational effectiveness, reduce greenhouse gas emissions in support of US climate change initiatives and protect the Department from energy price fluctuations' (ibid: 87). To do this, the QDR identified a range of tactical initiatives (e.g., inclusion of fully burden cost of fuel measures and energy efficiency targets) as well as strategic initiatives (e.g., the development of large scale on-base renewable energy).

The final observation on the 2010 QDR is that while intimations were made that the US DoD would contribute to national mitigation efforts, the focus remained on ensuring 'mission assurance', improved base 'resilience' and greater 'readiness'. Nonetheless, that the military was even linked to national mitigation efforts represented a major departure from the Bush era.

6.2.2.3 Secretaries Panetta and Hagel

Limited references to climate change were made in the speeches of Secretary Panetta and none were identified by Secretary Hagel (emphasising that examination of documents under Secretary Hagel stopped in early 2013, and that he discussed climate change in detail throughout his tenure but outside the timeframe considered for this case study). In 2012, Secretary Panetta noted that climate change posed an 'environmental threat' that constituted 'threats to our national security' (Panetta 2012a). Panetta recited rising sea-levels, severe droughts, melting polar ice caps and more frequent and devastating natural disasters as threats that would 'raise demand for humanitarian and disaster relief' (ibid). The Arctic, energy security and US DoD energy efficiency strategies were all highlighted in the same speech. Although limited in the overall context, the speech did represent something of an escalation in language by a SECDEF to describe climate change. Beyond this, there were no new or additional insights.

Moreover, by late 2012, this research detected in the literature that climate change had peaked as an issue in the minds of senior military leadership and—having already established initial programme in the 2010 QDR—had not much else to contribute to programme under way. The 2012 *Strategic Sustainability Performance Plan* (SSPP) and associated *Climate Change Adaptation Plan* are a good case in point. The SSPP detailed how Departmental energy, climate and sustainability programme were now delegated to the Under Secretary of Defense Acquisition, Technology and Logistics (also badged as the US DoD Senior Sustainability Office). Although the SSPP was not coded by this research, several aspects deserve comment. First, the SSPP set departmental goals to reduce its GHG emissions with the accompanying objective of being 'a US Government leader in reducing Greenhouse Gas Emissions'. Notably, the SSPP excluded the military's operational energy and rather focused on reducing GHG at fixed military facilities and commercial fleets (US DoD 2012b: I-15).

Secondly, the *Climate Change Adaptation Plan* represented the first of its type in US military history. Although it was somewhat hidden (attached as an appendix to the SSPP), the *Climate Change Adaptation Plan* rehashed many of the broad aspects already cited in 2010 QDR and NSS, but also elaborated on a range of lower level efforts (both under way and planned). The following specific examples highlight how climate change was, by 2012, being progressively became mainstream within and across the US Department of Defense (US DoD 2012a: 10):

- Incorporating climate considerations into installation-level planning and training plans, including formal military training and education.
- United Facilities Criteria (UFC) 2-100-01, paragraph 3-5.6.2.3 required master planners to consider climatic changes (for example, changes in land use and population density in the vicinity of installations; changes in climatic conditions such as temperature, rainfall patterns, storm frequency and intensity and water levels) when crafting long-range installation infrastructure master plans.
- UFC 2-100-01, paragraph 3-5.6.2.3 specifically identified the US *National Climate Assessment* as a source for climate change scenarios.
- The Department's Natural Resources Conservation Program Instruction (DoDI 4715.03) required installation natural resources management plans to assess the potential impacts of climate change on natural resources and to adaptively manage such resources to minimise adverse mission impacts.

Thus, as climate change considerations became mainstream across lower levels of the US military, the attention of some senior leaders of the military—for the time being at least—had moved on. At this point in the narrative, it is worth providing a strategic perspective on why this may have been the case. Largely, two major drivers were reshaping US grand strategy. First was the on-going impact of the 2008 global financial crisis. Although the military had received an injection of funding via ARRA 2009, it was largely directed toward creating US jobs and stimulating the economy rather than substantively growing the military per se. Moreover, funding via ARRA was counter to the larger trend under way from 2010: a transitional *decrease* in US DoD funding.

Figure 6.7, from the *National Defense Budget Estimates for FY 2014*, shows the growth and then contraction of the US Defense Budget 2000–2014 (2013 current $US dollars).[20] The broader point to be made here, for the purpose of this study, is that framing climate change as a national security threat did not yield an *increase* in US Defense spending. On the contrary, at the point in time that climate change was identified as a threat, US Defense spending *decreased.* This was in sharp contrast to 'emergency measures' enacted during the war on terror, where in FY 2003 US Defense spending alone grew at more than 22%. Suffice to say, climate change could hardly be considered an immediate threat in the sense that it required or justified a sizeable increase in military budget.

The second factor was a strategic rebalance by the US away from the Middle East (end of Iraq War, drawdown in Afghanistan) and towards the Asia-Pacific. As it was put by Secretary Panetta in a speech delivered in Singapore:

> The purpose of this trip, and of my remarks today, is to explain a new defense strategy that the United States has put in place … and how the United States military supports that goal by rebalancing towards this [Asia-Pacific] region … Before I detail these specific efforts, let

[20]Not included in these estimates are additional measures brought about by the *Budget Control Act* of 2011 (BCA, Public Law 112-25), which, among other actions, set caps on discretionary appropriations through 2021, and included automatic enforcement procedures if pre-defined limits were not met (the automatic cuts were generally referred to as 'sequestration').

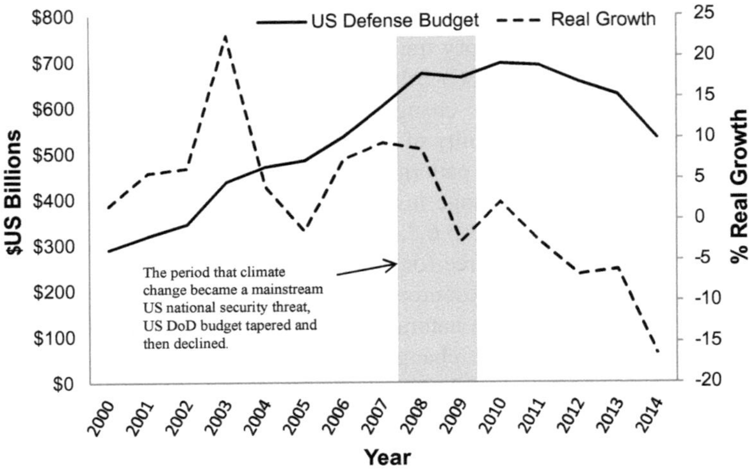

Fig. 6.7 US Defense spending FY 2000–2014. *Source* US DoD (2013b: 146–148)

> me provide some context for our broader defense strategy in the 21st century. *The United States is at a strategic turning point after a decade of war …*
>
> At the same time, the United States, like many other nations, is dealing with large debt and large deficits, which has required the Department of Defense to reduce the planning budget by nearly half a trillion dollars … by the Congress in the Budget Control Act over the next decade. But this new fiscal reality, a challenge that many nations confront these days, has given us an opportunity to design a new defense strategy for the 21st century that both confronts the threats that we face and maintains the strongest military in the world (Panetta 2012b; emphasis added).

And so, placed in this broader strategic context, the noted reduction in references to climate change in SECDEF speeches (and, other leaders examined in this case study) reflected the new strategic realities and the fact that substantive action on climate change by the US military appeared to be already underway.[21]

6.2.3 Chairman of the Joint Chiefs of Staff

Analysis of the Chairman of the Joint Chiefs of Staff (CJCS) Programme was undertaken on 631 official government sources published between 2003–2013

[21]Notwithstanding this point, the rebalance to the Asia-Pacific may also portend a stimulating effect on US military climate response on account of the fact that the Asia-Pacific region is one of the most at high-risk from climate change. This facet of US military pivot and climate security discourse was considered to be beyond the scope of this research.

Table 6.9 List of CJCS examined by this research

Chairman	Service	Appointed	Left office	Serving President
Gen. Richard B. Myers	USAF	1 Oct 2001	30 Sep 2005	Bush
Gen. Peter Pace	USMC	1 Oct 2005	30 Sep 2007	Bush
Adm. Mike Mullen	USN	1 Oct 2007	30 Sep 2011	Bush/Obama
Gen. Martin Dempsey	US Army	1 Oct 2011	25 Sep 2015	Obama

Source Wikipedia (2017a)

Table 6.10 Breakdown of CJCS sources and references to climate change (filtered)

Chairman	Source type			References to 'climate change' in speech-acts	References to 'climate change' in other documents
	Speech-acts	Other	Total		
Myers	0	8	8	0	0
Pace	0	0	0	0	0
Mullen	499	7	506	19	3
Dempsey	110	7	117	2	0
Total	609	22	631	21	3

'Other' consisted of policy documents, journals and media. *Source* The author

(Table 6.9 lists the CJCS and Table 6.10 lists source breakdown). The majority of the sources were from 2007 onwards (98%).

A text search for *climate change* returned seventeen separate sources, containing a total of twenty-four references (filtered). From this, eight passages of text were coded within a securitised context and eight passages of text were coded in a non-securitised context. The majority of references to climate change were attributed to Admiral Mike Mullen (twenty-two references from thirteen sources). These all occurred between 2009 and 2011.

Of the references that were framed, the language tended to be moderate (e.g., either in a neutral frame or as a challenge). The research did not find an escalation of language (e.g., 'threat') as typified by President Obama's language.

Of the securitised passages coded, it was found that CJCS tended to frame climate change across a number of securitised areas; including global security, energy security, as a new security issue, Arctic security and as a convergence of security issues. No one particular category dominated. In those passages coded as non-securitised, climate change tended to be framed as a global issue above any other. Only one passage of text was coded regarding the timeframe to address climate change (coded under long-term response).

Admiral Mullen provided the most comprehensive views by a Chairman on the subject of climate change. In *Joint Force Quarterly*, Mullen argued:

> Across the Nation, there is a growing recognition of the interconnection between energy, national security, and America's future … For our military, enhancing energy security carries even greater benefit – doing so will reduce risk, improve efficiencies, and preserve freedom of action (Mullen 2011: 2).

In the same article, Mullen added that by enhancing energy security, the military might even 'help stem the tide of strategic security issues' related to climate change. On this, Mullen cited 'far-reaching' consequences, including:

> Near the polar cap, waterways are opening that we could not have imagined a few years ago, rewriting the geopolitical map of the world. Rising sea levels could lead to mass migrations similar to what we have seen in Pakistan's recent flooding.
>
> Climate shifts could drastically reduce the arable land needed to feed a burgeoning population as we have seen in parts of Africa. As glaciers melt and shrink at a faster rate, crucial water supplies may diminish further in parts of Asia.
>
> This impending scarcity of resources compounded by an influx of refugees if coastal lands disappear not only could produce a humanitarian crisis, but also could generate conditions that could lead to failed states and make populations more vulnerable to radicalization. These troubling challenges highlight the systemic implications – and multiple-order effects - inherent in energy security and climate change (Mullen 2011: 3).

Despite this assessment, the 2011 *National Military Strategy* (NMS) included a short reference to climate change within a broad description on the strategic environment. On consideration of the certainty of other strategic documents, the 2011 NMS presented a more cautious tone:

> The *uncertain* impact of global climate change combined with increased population centers in or near coastal environments *may* challenge the ability of weak or developing states to respond to natural disasters (US DoD 2011: 2; emphasis added).

Given the increased prominence of climate change within US strategic guidance during the Obama Administration, the lack of attention on the issue in the 2011 NMS appeared incongruent. However, although it was more than the 2004 NMS (which did not address climate change in any aspect), it reinforced a broader point that by 2011 climate change within the most senior levels of the US military, particularly for the Joint Force Chiefs, did not outwardly register as a priority issue. Moreover, climate change did not appear as a major strategic consideration for the Chairman of the Joint Chiefs of Staff (CJCS). Apart from the article in the *Joint Force Quarterly*, cited above, no major speeches or policy initiatives on climate change were made. The attitude that climate change represented a long-term issue, undeserved of overreach, was characterised by CJCS General Dempsey when he chided:

> You can always get people to talk about 2030, 2040, 2050 – assuming we get past, you know, the Mayan prophecy of 2012 in the first place. (Laughter). But you can always get people to talk about 2050, you know, whether it's demographic shifts, climate change. You know, people are willing to talk about that (Dempsey 2011; parenthesis in original).

In a separate but similarly flippant speech, Dempsey added:

> Go to a cocktail party sometime … see if you can get somebody in conversation about 2050. No problem. Talk about global warming … life on other planets … But ask somebody about, hey, what do you think about 2020? Crickets (Dempsey 2012).

Striking a more serious tone, Dempsey then identified the 'mid-term' as being the most difficult since present-day office holders would actually be present and accountable for any successes or, more importantly, any failures. In his own words: 'It's intimidating because we actually have the opportunity to shape it, and we're going to own it' (Dempsey 2012). Not surprisingly, outside of these references, this research did not identify any descriptions by Dempsey on the security, economic, social or other implications of climate change.[22]

6.3 Climate Security in US Operational and Tactical Programme

This section presents results from the qualitative content analysis of 460 US Navy documents between 2003 and 2013. Specifically, the research analysed four areas/individuals within the US Navy that included the Secretary of the Navy (SECNAV), the Chief of Naval Operations (CNO), the US Navy Oceanographer and the Commander of the Pacific Fleet. The majority of documents analysed were speeches made by the Secretary of the Navy, Chief of Naval Operations and the Commander of the Pacific Fleet. To provide a complete picture and to build a narrative prior to discussion in Chap. 7, inclusion of selected quotes from sources beyond these units-of-analysis are highlighted.

6.3.1 Secretary of the Navy

Analysis of the SECNAV Programme was undertaken on 214 departmental sources published between 2003 and 2013 (Table 6.11 lists the SECNAVs considered by this research and Table 6.12 lists source breakdown). The majority of documents were sourced from 2007 onwards (84%). All SECNAV documents analysed were speech-acts.

A text search of *climate change* and *global warming* returned ten separate sources containing a total of thirty-five references (filtered) from 214 speeches. All references to climate change are attributed to Secretary Mabus in the years 2009, 2010 and 2012.

These results are consistent with findings in other US programmes analysed, in which climate change was not a major point of discussion until at least 2007, but mainly from 2009 through 2010. In the eighty-six speeches by Secretaries England and Gordon between 2003 and 2009 there was no reference to climate change or global warming. The appointment of Secretary Mabus in May 2009 and the associated uptake of climate change in his speeches, as well as Navy policy and

---

[22]The identification by Dempsey of the inherent challenge to 'own' the 'mid-term' is particularly relevant for climate change since to avoid a 2 °C world, many scientists have identified the period 2010–2020 as the *critical decade* to commence emissions reductions.

Table 6.11 List of US Secretary of Navy considered by this research

Secretary	Took office	Left office	Serving President
Gordon R. England	1 October 2003	29 December 2005	George W. Bush
Donald C. Winter	3 January 2006	13 March 2009	George W. Bush Barack Obama
Ray E. Mabus	19 May 2009	Incumbent	Barack Obama

The list does not include Secretary Dionel M. Aviles (served 29 December 2005–3 January 2006) and Secretary B.J. Penn (served 13 March 2009–19 May 2009) since none of their speeches were sourced for analysis. *Source* Wikipedia (2017g)

Table 6.12 Total references to climate change in speeches of US Secretary of Navy, 2003–2013 (filtered)

Secretary	Number of speech-acts examined	References to 'climate change' in speech-acts
Gordon R. England	25	0
Donald C. Winter	61	0
Ray E. Mabus	128	35
Total	214	35

Source The author

doctrine, reflected the emerging importance placed on climate change by the Obama Administration. As a case in point, within the first two years of Secretary Mabus' appointment, the US Navy published four key documents outlining its organisational stance on climate change and energy: 2009 US Navy *Task Force Climate Change*, 2009 *Navy Arctic Roadmap*, 2010 *Navy Climate Change Roadmap* and 2010 *Navy Energy Vision.*[23]

Of the thirty-five references to climate change attributed to Secretary Mabus, nine passages of text were coded within a securitised context and thirteen passages of text were coded in a non-securitised context. Of the securitised references, this research identified three areas of focus by Secretary Mabus: the Arctic; as a threat multiplier with a range of other precipitating security factors (convergence); and as an energy security issue. Of the non-securitised references, the majority were coded under matters pertaining to *energy*. In its totality, the energy-climate nexus tended to be the dominant narrative.

The language used by Secretary Mabus to frame climate change was predominately neutral but on multiple occasions he framed climate change as a *challenge*. No instances were identified where Secretary Mabus framed climate change as a *threat*. Furthermore, although no instances were identified where Secretary Mabus directly framed climate change using the word *opportunity*, there is broad evidence that he used the issue to justify the strategic opportunities of advancing his 'energy agenda' that included improving energy efficiency, reducing US Navy reliance on fossil fuels and the development of renewable energy to supply US Naval bases.

[23]Each of these documents is dealt in further detail under CNO.

Lastly, few instances were identified regarding temporal framing of climate change. In the most telling quote on the topic, Mabus saw it as a long-term issue: 'we established TFCC to develop policy and investment and force structure recommendations regarding long-term climate change' (Mabus 2012a: 6).

In June 2009, in his first major speech as Secretary of Navy to the *Current Strategy Forum* at the US Naval War College, Secretary Mabus highlighted the need to combat both 'traditional security challenges posed by the military forces of other states' and also 'new, non-traditional, but very dangerous threats' arising from non-state actors as well as '[i]rregular warfare, insurgency, criminal activity, social unrest and low intensity civil conflict' (Mabus 2009a: 4). Secretary Mabus also identified 'trends in demographics, climate change, globalization, immigration, and resource availability' as factors that 'will intensify the strain' on nations, particularly those 'least equipped to deal with them' (ibid: 4). In this strategic context, Mabus argued for a more 'powerful … inventive [and] agile' maritime force capable of 'meeting the challenges' that lay 'beyond the horizons' (2009a: 2). He then cited the forthcoming 2010 QDR as justifying a 'major examination of our strategy, [and] how we design and deploy our forces' (ibid: 6). Central to this 'major examination' was a focus on energy:

> We must move towards greater energy independence ashore and afloat … Since our operational flexibility and sustainability are directly linked to our energy supplies, energy reliability is a strategic concern for our force.
>
> The potential disruption of our Nation's fuel supplies threatens our ability to perform our missions, and threatens the security of America.
>
> We will move toward alternative energy for two reasons: security *and* diminishing our impact on the environment. … We will have a comprehensive energy strategy to increase conservation, develop alternative energy options and provide secure energy (Mabus 2009a: 6; emphasis added).

In his next major speech on 7 October 2009 to the *International Seapower Symposium*, Secretary Mabus expanded on the 'challenges' of climate change with particular reference to 'a polar region free of [summer] ice' in which 'the security, economic and environmental implications of a Northwest Passage that is open' would increasingly factor (Mabus 2009b). Mabus then proceeded to specifically link the security implications of climate change and energy security concerns as justification to transform the US Navy's approach to energy:

> As a result, the global security implications of climate change and energy use will become one of the great challenges for our successors, as they are becoming for us, and *I am committed to placing the US Navy on a path to do something about it.*
>
> I am committed to looking at the way in which our Navy uses power, and committed to taking leadership inside the United States in looking for ways to make our energy use more efficient. I want to … shift to alternative sources of energy in our infrastructure, in our shore management, and in our fleet (Mabus 2009b: 4; emphasis added).

Two days later, in a speech at Jackson State University, Mabus (2009e) emphatically declared, 'I am committed as Secretary to addressing climate change and energy consumption'. On 14 October 2009, Secretary Mabus delivered a pivotal speech in which he prefaced the scale of his solution to addressing these challenges by invoking President Kennedy's 1961 moon aspirations and noting:

> Bold steps are in our nature as Americans and what make us great as a nation; no one has ever gotten anything done by being timid. I'm here to commit the Navy and Marine Corps to meet bold and ambitious goals in energy. I mean this about being bold and ambitious (Mabus 2009d: 7).

Mabus proceeded to announce five energy goals that formed a cornerstone of his tenure as US Secretary of Navy. Describing it as a 'strategic imperative' (2009d: 1), Mabus' energy targets were formalised in the 2010 Navy policy document *Energy Program for Security and Independence* (US Navy 2010a: 3).[24] In November 2009, Mabus identified energy as 'one of the defining issues in my time as Secretary of the Navy' and reinforced climate change as one of the key drivers for energy transformation:

> I see the significant challenges and global security implications of climate change and energy use as one of the great challenges for us, and it's for these reasons that last month I put the United States Navy and Marine Corps on a path to … energy reform (Mabus 2009c: 3).

Noting the US military's significant energy usage ('the DoD uses more than 90% of all energy used by the federal government and 2% of all the energy used in the United States') as well as its contribution to global warming, Mabus looked 'beyond the military' to:

> … second and third order effects on the environment. The carbon that's emitted from our ships, aircraft, and vehicles is a contributor to global warming and climate change. According to the projections endorsed by our own Task Force on Climate Change, global warming could result in an Arctic Ocean free of summer ice within twenty-five years. The security implications of this are dramatic. In short, we have not acted as very responsible stewards of our environment (Mabus 2009d: 3).

In addition to 'societal and global obligation to promote a better environment', Mabus (2009c: 3) also considered that the Navy had a requirement to lead the nation on energy transformation, improve its war-fighting capabilities and strengthen its energy security by reducing reliance on external sources. Concerned that expanding demand and continued reliance on 'finite' fossil fuel reserves would

[24]Five energy goals: (1) Energy Efficient Acquisition. Inclusion of specific energy targets and efficiency standards as mandatory requirements in US Naval and Marine corps acquisition contracts. (2) Sail the 'Great Green Fleet'. By 2016, the deployment of a Carrier Strike Group ('Great Green Fleet') composed of vessels powered by nuclear, biofuels and hybrid-electric drives. (3) Reduce Non-Tactical Petroleum use. By 2015, a 50% reduction of petroleum use in its commercial fleet. (4) Increase Alternative Energy Ashore. By 2020, to produce 50% of shore based energy requirements from alternative energy sources; 50% of DON installations to be 'net zero'. (5) Increase Alternative Energy Use DON-Wide. By 2020, to have 50% of *total* US Navy energy consumption (including operational fleet of ships, aircraft, tanks, vehicles as well as shore installations) to come from alternative energy sources.

see 'costs ... certainly continue to go up', Mabus (2009c: 2) argued 'military organisations rely too much on fossil fuels'.

In targeting the military's own reliance on fossil fuels, Mabus' energy goals also extended to getting the US Defense industry to deliver more energy efficient capability. On this, Mabus (2010a) also declared the Navy would begin to assess 'how the manufacturers uses energy ... we have to do this because of where this energy is coming from'. For Mabus (2010a: 6): 'The smaller the carbon footprint, the better that manufacturer is going to be in those competitions'. Nothing of this sort had been found in the Australian case study.

In a telling acknowledgement on the criticality of political leadership in legitimising and spurring military action on energy and climate change, Mabus directly credited President Obama as being a:

> [S]pringboard for the Navy and Marines to do more, to go farther, and to take up leadership across the DoD, across the federal government and across the broader United States in developing and using alternative sources of fuels (Mabus 2009d: 4).

He repeatedly derided critics of the programme, arguing that the US Navy had historically been at the forefront of previous energy transformations:

> The Navy has always led in the change to new types of energy. In the 1850s from sail to coal, in the early 20th century from coal to oil, and in the 1950s embracing nuclear power as a way to propel our submarines and carriers. Every single time that this was done, there have been naysayers and there have been doubters ... Every time, every single time, these naysayers have been wrong, and we have been a better Navy and better war fighters because of it. And I'm absolutely confident that's going to be the case now (Mabus 2010a: 7).

Despite this, as criticism mounted and as the US Defense budget tightened, US strategy shifted and following political failure at Copenhagen, Mabus progressively changed emphasis away from justifying energy transformation on environmental and climate change grounds. Instead, Mabus began to focus his reasons on more practical (traditional) grounds by stressing the strategic and tactical (war-fighting) benefits. A turning point was reached in April 2010 when, at the Center for Naval Analysis, Mabus conceded that using climate change as part justification for energy transformation tended to be too distracting:

> I think as this debate unfolds about things like climate change, things like the effect of fossil fuels on America and on the world, that if we make this debate one of energy independence for America, and national security for America and for our armed forces, it is a debate that we are going to win. It's a debate that the American people can unify behind.
>
> *I think our chances are less certain if we base it on what to most people are a little more nebulous ideas like climate change – harder to see, harder to understand and harder to qualify* (Mabus 2010a: 3; emphasis added).

In the context of events, these comments represented an important change in tactics that, as will be seen, filtered downwards throughout the US Navy. With or without this change of justification, Mabus faced heavy criticism. In one withering critique, Senator John McCain (R-Az.) on July 27, 2012 wrote:

> Your misrepresentations of the provisions threatens the credibility of the senior Navy leaders you have ordered to advocate on behalf of a speculative program that does not address the core needs of the Navy or the Marine Corps … You are the Secretary of the Navy, not the Secretary of Energy (McCain 2012).

In 2012, a senior member of the Senate Armed Services Committee and prominent climate change denialist, Senator Jim Inhofe (R-Okla.) accused Mabus of risking military readiness, citing 'grave concerns about the cost of 'greening' our military' (Inhofe 2012c) and—more broadly—the 'Obama Administration's attempt to force its liberal green agenda through the Department of Defense' (Inhofe 2012a, b). In 2011, RAND Corporation published a report (*Alternative Fuels for Military Applications*) which concluded that 'the use of alternative fuels offers the armed services no direct military benefit' and that efforts should rather be directed towards 'using energy more efficiently in weapon systems and at military installations' (Bartin/Bibber 2011: xix). RAND argued that these findings were consistent with other high level reports previously published by the Defense Science Board and the JASON Defense Advisory Group.[25]

In both realisation and response, Mabus' justification for making US Navy energy transformation began to increasingly coalesce around 'hard' strategic imperatives (energy independence, national security, reduced operating costs, aiding US agricultural sector through promotion of biofuel markets) and tactical imperatives (decrease fuel demand at forward operating bases, offering war-fighters greater endurance and reducing fuel convoys). This facet was reflected in *Energy Program*, which directly cited two priorities for Naval energy reform as 'Energy Security and Energy Independence' (US Navy 2010a: 2).

The trend was also reflected in speech-acts made by Secretary Mabus. This research found *no* direct references to climate change or global warming in Secretary Mabus speech-acts throughout 2011 and just four references in 2012.[26] Gradually, climate change and environment considerations were airbrushed out of the US Navy's energy transformation narrative. Initially it depended on which audience Mabus addressed, but, increasingly, climate change was framed as a bi-product or secondary side-effect. In May 2012, during a speech at the *Global Conference on Oceans, Climate and Security: Making the Connections*, Mabus argued:

> We're doing it all for one underlying reason. We're doing it to be a better military. We're doing it to be better war fighters. And that has some great side-effects. It makes us better stewards of the environment at the same time, makes us better stewards of the ocean (Mabus 2012a: 4).

---

[25]The US military has a long history of seeking to reduce its energy requirements (as described in CRS 2009). For other examples, see *Report of the Defense Science Board Task force on DoD Energy Strategy: More Fight—Less Fuel*, February 2008.

[26]Although in 2012 at the *Global Conference on Oceans, Climate and Security: Making the Connections*, Mabus detailed the risks of climate change, but without actually uttering the words 'climate change'.

By October 2012, at the *Naval Energy Forum*, Mabus had removed any reference to climate change or the environment in a lengthy speech outlining his energy initiatives[27]:

> We're doing what we're doing, and we're leading in what we're doing because it increases national security and, I would argue, increases international stability. It reduces the impact of price spikes on military readiness and on our procurement, and it can save lives (Mabus 2012b: 7).

Ultimately, (environmental) idealism appeared to give way to (national security) realism:

> The Great Green Fleet is not about some environmental agenda. *It is about maintaining America's military and economic leadership across the globe in the 21st century* (Mabus 2012c: 6; emphasis added).

Briefly, three other points in relation to Secretary Mabus' energy-climate strategy warrant attention. Firstly, as US military budget cuts began to take effect, Mabus increasingly justified his energy transformation as ultimately saving taxpayers money for reinvestment into alternate areas of capability.[28] Along these lines, he also regularly cited that the US Navy would contribute to national economic aims by helping to establish a domestic, commercially viable biofuel industry ('as we build demand, supply will come'), which would 'provide energy independence and American jobs' (Mabus 2012c: 6). Secondly, the organisational penetration of Mabus' energy strategy became increasingly evident through a range of initiatives aimed at changing the Navy's cultural attitude toward energy use. Though many of these initiatives already existed, Mabus sought to extend these far deeper into the organisation. The direction that 2014 officer promotions would be tied to performance in contributing to energy resource management measures provided an example of this (Mabus 2012b).

Lastly, analysis of Mabus' speeches indicated that climate change *was* invoked as a near-term driver of US Navy capability and force structure. This was particularly evident in relation to the Arctic. Clear indications of climate change as a force structure/capability determinant is evident in the US Navy *Task Force Climate Change*, *Arctic Roadmap* and *Climate Change Roadmap* (analysed under 'CNO', see below) as well as Mabus' speech-acts. As Mabus described it:

---

[27]The general trend in the US to downplay climate security matters was also evident through the closure of the CIAs 'Center on Climate Change and National Security' in late 2012. See: http://www.judicialwatch.org/blog/2012/11/cia-quietly-closes-global-warming-unit/ (13 May 16).

[28]Mabus regularly cited a number of 'practical' examples of how his energy transformation would save money, resources and lives. One example was USS MAKIN ISLAND, which was described as the 'Prius of the seas', a 'hybrid ship' operating on electric-diesel engines which saved US$15 million in fuel on its first deployment. Other Naval examples included the use of 'stern flaps' and different hull coatings both used to limit drag. Other examples included the naval fighter 'green hornet', Marine LED combat lighting, solar blankets and solar generators at forward operating bases. The references may be found in many of Secretary Mabus' speeches across the period 2009–2013.

> … the short answer is, we need a bigger fleet to do what we do today, but if you have things like an ice-free Arctic, you will need additional ships to make sure that you do that correctly (Mabus 2010b: 6).

In summary, although these latter examples indicate forms of organisational adaptation to climate change, the vast concentration of effort under Secretary Mabus can be best described as mitigation through reducing energy consumption and improving energy efficiency. More specifically, while climate change was initially framed as one of the reasons to reduce emissions, it was progressively removed from the justifying narrative on the grounds that it exposed Secretary Mabus to critics ideologically and politically opposed to climate change. It also roughly coincided with Obama's failure to get his climate change initiatives through Congress and failure at Copenhagen. Justification for US Navy energy transformation thus came to rest on more practical, hard-power reasons associated with advancing the national interest and war-fighting capabilities. The framing and justification of climate change as a force structure and capability determinant for the US Navy will be explored in further detail in the next section.

6.3.2 *Chief of Naval Operations (CNO)*

Analysis of the Chief of Naval Operations (CNO) Programme was undertaken on 201 departmental sources published between 2003–2013 (Table 6.13 lists the CNOs considered by this research and Table 6.14 lists source breakdown). All documents (except one) were sourced from 2005 onwards.[29] The majority of documents analysed were speech-acts (95%) with the remainder ('Other') sourced from a combination of doctrine, media interviews, policy and journal articles.

A text search of *climate change* and *global warming* returned twenty-six separate sources containing a total of 213 references (unfiltered). All references to these terms (with one exception) occurred between 2009 and 2011. The exception included two references contained in a 2007 naval doctrine publication *A Cooperative Strategy for the 21st Century Seapower* (CSP21). All references to climate change in this programme occurred during the appointment of Admiral Roughead. As expected, the majority of references occurred in Naval policy documents, including the 2009 US Navy *Task Force Climate Change* (thirty-four references) and the 2010 US *Navy Climate Change Roadmap* (145 references). Beyond this, the majority of references (thirty-two in total) occurred in twenty-one separate speeches made by ADM Roughead across 2009–2011. Broadly, about one-fifth of all speeches analysed for ADM Roughead in his appointment as CNO contained one or more references to climate change.

---

[29]CNO Programme, Year (number of sources): 2005 (17), 2006 (29), 2007 (22), 2008 (13), 2009 (39), 2010 (29), 2011 (22), 2012 (26) and 2013 (3)).

Table 6.13 List of US Chief of Naval Operations (CNO) 2000–2013

CNO	Appointed	Left office	President
ADM Vern Clark	21 July 2000	22 July 2005	Bush
ADM Michael Mullen	22 July 2005	29 September 2007	Bush
ADM Gary Roughead	29 September 2007	23 September 2011	Bush/Obama
ADM Jonathon Greenert	23 September 2011	18 September 2015	Obama

Source Wikipedia (2017b)

Table 6.14 Total references to climate change in speeches and policy of US CNO, 2003–2013 (unfiltered)

Chairman	Source type			References to *climate change* in speech-acts	References to *climate change* in other documents
	Speech-acts	Other	Total		
Clark	0	1	1	0	0
Mullen	64	1	65	0	0
Roughead	97	7	104	32	181
Greenert	30	1	31	0	0
Total	191	10	201	32	181

Source The author

These results are largely consistent with previous findings that show climate change was not a major point of discussion until at least 2007, but mainly from 2009 through 2011. On this point, no references to climate change were found in the speeches analysed during the appointment of ADM Greenert (although subsequent to the data collection phase a speech was found in which he discussed it at length in 2011 as the Vice Chief of Naval Operations).[30]

Of the 213 references to climate change, twenty-five passages of text were coded within a securitised context and seven passages of text were coded in a non-securitised context.[31] Of the securitised references, this research identified three areas of focus by ADM Roughead: the Arctic; as a threat multiplier with a range of other precipitating security factors (i.e., convergence); and as a general/national security issue. Of the non-securitised references, climate change

[30]This aspect raised a limitation with the research method. Moreover, by 2011 the US military had commenced its response to climate change, evident through the various adaptation/mitigation plans. Thus, the most senior leaders in US Defense had delegated responsibility. It is highly plausible that while senior leaders examined in this study were no longer 'talking' about climate change, lower level commanders and administrators were (i.e., those who now had responsibilities to see the plans through).

[31]Although there were 213 distinct references to the term 'climate change'/'global warming', coding rules were strictly applied so that the vast majority of references in the two major climate policy documents did not skew results.

was found to be mainly framed as a generalised issue or under matters pertaining to energy.

Language used to frame climate change was predominately neutral but on several occasions climate change was cast as a challenge. No instances in the CNO Programme were identified where climate change was framed as a threat. Only three passages of text were coded that identified the temporal framing of climate change (one was urgent while the remaining two were long-term). A broad observation from these findings is the shift, or rather de-escalation, in rhetoric used to frame climate change compared with President Obama. It pointed to a divergence between the emergency language of the president on climate change and the benign language of operational military commanders.

The earliest identified discussion of climate change under the CNO Programme occurred in CSP21. This document is noteworthy for two reasons. First, it anticipated the Arctic as a region of opportunity such that climate change would 'gradually [open] up the waters of the Arctic … to new resource development, but also to new shipping routes' (US Navy 2007: 6). Conversely, the strategy noted a potential negative risk to Arctic warming such that there existed 'potential sources of competition and conflict for access and natural resources' (US Navy 2007: 6). The dual nature of climate change acting as either an opportunity or as a challenge/threat in relation to the Arctic became an increasing feature of the US Navy narrative, built upon on by the annual *Arctic Symposium* (since 2001) and culminating in the publication of US Navy *Arctic Roadmap* in 2010. The second observation is that CSP21 linked climate change to concepts of human security by declaring:

> The effects of climate change may also amplify human suffering through catastrophic storms, loss of arable lands, and coastal flooding, could lead to loss of life, involuntary migration, social instability, and regional crisis (US Navy 2007: 7).

This represented an early articulation of the breath and scale of the threats posed by climate change and was considered a rare occurrence under the auspices of the Bush presidency and its attendant focus on the War on Terror.

Speeches made by ADM Roughead on climate change also highlighted the scale of threats posed by climate change. He framed climate change as both a current and long-term issue. On one occasion at least, Roughead indicated it required a 'sense of urgency' (Roughead 2010d). In 2009, addressing the *Current Strategy Forum*, ADM Roughead further noted that climate change was already 'upon us' and affecting 'everything from weather pattern shifts to changes in the ice caps … water densities and salinity to transit routes' (Roughead 2009). As was found during coding, ADM Roughead regularly framed climate change as part of a broader convergence of issues that would impact security. In many respects, he adopted a neo-Malthusian outlook: 'The limited access to fresh water, dwindling agricultural yields, the overfishing, mass migrations, climate change will continue to stress the global order' (Roughead 2010d). In this context, Roughead was touching on a frame that also dominated broader US thinking on climate change by presenting it as a 'threat multiplier'.

ADM Roughead (and the US Navy more broadly) also worked to proactively position the US Navy as the leading service in its response to climate change. This was manifest in five major initiatives spearheaded by the office of the CNO (ADM Roughead) between 2008 and 2011: firstly, a 2008 request to the National Academy of Sciences to conduct a comprehensive study on the national security implications of climate change for US naval forces (delivered in 2011); secondly, the 2009 *US Navy Task Force Climate Change*; thirdly, the 2010 *Arctic Roadmap*; fourthly, the 2010 *Energy Roadmap*; and lastly, the 2010 *Navy Climate Change Roadmap*.[32]

What emerged from coding and analysis of these policies and through the speech-acts of ADM Roughead was an overarching strategy to leverage the threats and opportunities presented by climate change to position and strengthen US Naval capabilities in pursuit of the US Navy maritime agenda, as articulated in CSP21, and also in support of US national interests (as presented in the 2010 QDR). Furthermore, there is some evidence to suggest that the US Navy sought to promote climate change (amongst other strategic issues) as an opportunity to bulwark—indeed, strengthen and grow—itself during a period of US military austerity. From a political perspective, the timing of the US Navy climate policy initiated by the CNO is important, since its demand for increased resources to meet threats and opportunities posed by climate change was likely to be viewed more favourably by an Administration and Secretary sympathetic to the need for climate action and ideologically aligned to the cause more generally. Confidence in this assessment—and an endorsement of US Navy climate strategy vis-à-vis other agencies—was strengthened when President Obama specifically singled out the US Navy for additional funding on renewable energy initiatives in his 2012 State of the Union Address (Obama 2012a).

Nonetheless, the profusion of climate policy activity across 2009–2011 by the US Navy can only be truly assessed against broader US national-strategic and global orchestrations. Described earlier: cessation of US military operations in Iraq (2010), drawdown in Afghanistan (2013), 'rebalance' by the Obama Administration of a US strategic posture orientated toward Asia-Pacific (2012) and—prominently—an overall reduction in current and planned US military expenditure (circa 2010). Thus, although ADM Roughead was a parochial advocate for an increase in budget, the larger picture presented a sharp decrease in real growth Naval funding (US DoD 2013b: 161–162).[33]

Deeper examination of the CNO Programme revealed a US Navy climate policy that can be assessed under the two broad areas of adaptation and mitigation. In terms of adaptation, the US Navy (via the CNO) developed climate policies primarily focused on adapting its force structure and capability as a consequence of physical environmental changes being wrought by climate change. Although the

---

[32]The Naval Studies Board Report was not included in the source document coding, but was included in the book discussion. The report was not considered for coding since it existed outside any of the programmes analysed in this research.

[33]For instance, US Navy real growth in parenthesis: 2011 (−1.3), 2012 (−3.2), 2013 (−3.4), 2014 (−10.6).

majority of this effort was directed toward enhancing US Navy Arctic capabilities it was also extended to other regions. In terms of mitigation, the CNO from 2009 onwards was principally responsible for delivering the Secretary of Navy energy goals. Arguably, although climate change formed part-justification for these mitigation efforts, other key drivers remained dominant (cost cutting and savings measures, opportunity for re-investment, opportunities for new capability, opportunities to enhance war-fighting).

Two months prior to the election of President Obama, ADM Roughead wrote to the President of the National Academy of Sciences requesting that the National Research Council's (NRC) Naval Studies Board (NSB) conduct a comprehensive study on the national security implications of climate change for US naval forces. From the moment of request to point of delivery, the report spanned four years (2008–2011). By this stage, however, ADM Roughead had already been active in directing internal Navy resources along similar terms of references. On 15 May 2009, for instance, ADM Roughead directed the US Navy Oceanographer to establish and lead a Task Force Climate Change (TFCC) that would 'recommend policy, strategy, roadmaps, force structure and investments for the Navy regarding the Arctic and Climate Change' (US Navy 2009a: 2). Although the strategy focused on the Arctic, it extended its remit to examine broader implications concerning how the Navy would adapt, mitigate and assess the 'extent, timing and impacts' of climate change across its organisation. The breadth of the directive was significant: every functional area of the Navy ('N-code') was required to deliver against a range of specified tasks (US Navy 2009a). Likewise, was the penetration of TFCC. As a case in point, the directive identified the formation of an executive level Naval Climate Change Coordination Office (NCCCO) responsible for coordinating working groups across middle management (O5 and O6). Far from being a 'nebulous' issue, climate change had in fact become a very real, day-to-day, practical requirement for US Naval staff officers.

Of special significance in TFCC was a requirement for Navy staff to be on the lookout for new opportunities that might advance US Naval capabilities. A case in point was a request by the TFCC for the TFCC Executive Steering Committee to expressly 'advocate for Navy Arctic and climate change goals, objectives and funding' (US Navy 2009a: 4). Working groups were likewise directed to consistently look for new opportunities in areas of capability, training, operations, missions, strategy, research and engagement/cooperation. TFCC concluded by re-assuring those involved that Naval climate policy would be backed by a 'strategic communications outreach programme' that included a campaign in which the Navy's senior leadership group would 'develop and maintain talking points … regarding the changing Arctic and Earth's climate' (US Navy 2009a: 7–8). Placed in context, the TFCC represented a stepping stone toward the development of the US Navy *Arctic Roadmap* and then *Climate Change Roadmap*.

On the Arctic, ADM Roughead consistently framed climate change as *the* major issue re-shaping its geography, economy, international relations and security. In a 2011 speech, *Active In the Arctic Seminar*, Roughead enthused:

> Well in my mind, there is a phenomenal event taking place on the planet today, and that is what I call the opening of the Fifth Ocean; that's the Arctic Ocean. We haven't had an ocean open on this planet since the end of the Ice Age. So if this is not a significant change that requires new, and I would submit, brave thinking on the topic, I don't know what other sort of physical event could produce that (Roughead 2011a: 1).

On the specific challenges and opportunities of climate change in the Arctic, ADM Roughead listed a progressive and increasing requirement in the near, medium and long-terms:

> [T]he first push up into the Arctic will be for fishing, as the fishing stocks migrate with the cooler water, or they'll start to move up toward the Arctic. So I think you'll see increased fishing activity up there, probably see some eco-tourism …
>
> Then I believe you've going to see a push into the extraction of resources from the bottom, or even, you know, beneath the bottom. That probably will be taking place in the next 10 years or so. And then … profitable trade routes is about maybe two decades out (Roughead 2011a: 20).

In a separate speech, Roughead depicted findings from an earlier 2008 US Geological Survey Report, *Circum-Arctic Resource Appraisal*, which brought the Arctic's significant resource potential into national prominence. Emphasising a requirement for the US Navy to position itself to advance US Arctic resource interests, Roughead proceeded to argue:

> … efforts are already underway, and not just among the Arctic nations, to seize the economic potential of a region estimated to contain 22% of the world's oil and natural gas resource base, and trillions in economic potential (Roughead 2011b: 2).

In this context, Roughead sought to invoke concepts of strategic competition, intimating:

> I don't find it coincidental that China, a nation with previously no claim on Arctic waters, has called for universal access rights to increasingly navigable arctic waters (Roughead 2010c).

At large, Roughead argued that climate change was causing a strategic re-appraisal of the US Navy role in the Arctic. He envisioned it as an area of distinct opportunity for the US Navy:

> Because at a fundamental level, the trends point undeniably towards a new venue of operations and responsibility for our global Navy – for preserving American interests in free and fair access there – and in light of this we remain committed to preparing exhaustively for the challenges and especially for the opportunities that are going to exist in an ice-diminishing Arctic (Roughead 2011b: 2).

In framing climate change as one of several new factors shaping the strategic environment, Roughead (2010b) was 'blatantly and shamelessly parochial' in consistently arguing for a larger navy to advance US national interests through the maintenance of an orderly and stable maritime commons. Lamenting US 'sea blindness', Roughead argued that the US Navy had failed to grow its fleet during the decade-long upsurge in Defence expenditure experienced under the Bush Administration's War on Terror. Thus, juxtaposed against the dawning era of 'compressed defense budget[s]', Roughead argued that the emerging new order required '*more* naval power' (Roughead 2010b; emphasis added). Though climate change was cited as one of many reasons for this requirement it did, on occasion, become tenuous. In one instance, Roughead (2010b) even went so far as to link naval expansion plans on the somewhat incredulous idea that 'there is a lot of area we need to cover and when you consider the changes in the Arctic the oceans will only increase in size'.

In advocating for more naval resources ('because I'm the CNO I can say things like that'), Roughead spoke with conviction on the requirement for US Navy climate change adaptation (Roughead 2010b). In one such speech, he identified several areas including in national policy (specifically US ratification of UN Convention on the Law of the Sea Treaty), naval communications architecture, naval ship design, military exercises as well as in areas of international maritime cooperation.

US Naval climate adaptation efforts were overtly addressed by ADM Roughead in 2009 and 2010 when he commissioned the *Arctic Roadmap* and *Climate Change Roadmap*. Noteworthy for their scale, the *Arctic Roadmap* alone identified thirty-five distinct action items across five functional areas of US Navy bureaucracy, including (1) Strategic policy, Missions and Plans; (2) Operations and Training; (3) Capability Investments; (4) Strategic Communications and Outreach; and (5) Environmental Assessment and Prediction.

An indication of the programme's scale was revealed under 'Capability Investments'. Listing five specific sub-action items, 'Action item 3.2' directed a 'capabilities based assessment for Naval Arctic capabilities' that included but was not limited to, 'current and required capability to execute undersea warfare, expeditionary warfare, strike warfare, strategic sealift, regional security cooperation, HA/DR, and DSCA … C4ISR capability … current and required infrastructure, installations, and facilities in the region' (US Navy 2009b: 14). Put simply, in this single example, the US Navy sought to review its entire Arctic capability portfolio in response to the rapidly changing Arctic climate.

Although the outcomes of this analysis are not accessible to the public, other action items are more readily traceable. What is important at this point is to appreciate how strategic guidance translated into policy response which generated 'action items' and 'metrics' that delivered operational and tactical outcomes. It was the bureaucratisation of strategic direction; *ends* articulated into operational policy (*ways*) to becoming implemented by tactical *means*. Also, as has already been shown, since there was no additional funding allocation to the US Navy, resources for these initiatives came from existing budgets. This indicated a level of prioritisation that climate change had previously not held.

Under the area of 'Environmental Assessment and Prediction', for example, action item 5.7 directs the biennial production of an *Arctic Environmental Assessment and Outlook Report* to 'inform Navy policy, strategy and investment decisions' (US Navy 2009b: 24). Published in August 2011, the report noted that the IPCC reports are insufficient since the 'refresh rate is too long to meet the budget POM cycle' (US Navy 2011: v). The report proceeds to compile a detailed description of the state of the Arctic environment, the changes under way as a consequence of climate change and concluded with a detailed list of the operational challenges and opportunities for the US Navy. Despite the fervour, the assessment also portended a more tempered approach by US Navy on climate change by warning 'the Arctic will still remain a difficult and dangerous operating theater … it is doubtful that the Arctic will shift much of Navy's attention from current deployment patterns' (US Navy 2011: 22).

This degree of caution, however, was not evident in the 2010 *Climate Change Roadmap*. Considered an 'extension to the Navy Arctic Roadmap' to be superseded by the 2014 QDR, the opening statement is noteworthy for the both the currency and confidence of climate change impacts:

> Climate change *is* a national security challenge … [that] … *will* lead to increased tensions in nations … *It will affect the type, scope, and location of future Navy missions* (US Navy 2010b: 3; emphasis added).

Similar in format to the *Arctic Roadmap*, the *Climate Change Roadmap* specified three phases across four years in areas of strategy, operations, capability, communications and environmental assessment. Again, the broad conclusion from the *Climate Roadmap* was the scale of organisational response by the US Navy regarding climate change. This aspect was well captured by the Roadmap objectives, which sought the following effects (US Navy 2010b: 3):

- The Navy is fully mission-capable through changing climatic conditions while actively contributing to national requirements for addressing climate change.
- Naval force structure and infrastructure are capable of meeting combatant commander requirements in all probable climatic conditions over the next 30 years.
- The Navy understands the timing, severity, and impact of current and projected changes in the global environment.
- The media, public, government, joint, interagency, and international community understand how and why the Navy is effectively addressing climate change.
- The Navy is recognised as a valuable joint, interagency, and international partner in responding to climate change.

Moreover, the *Climate Change Roadmap* related not just to improvements in the Navy's understanding on the long-term impacts of climate change (such as force structure and capability), but also more short term (intangible) aspects such as how it is perceived by, and related with, a broad selection of stakeholders. These facets underline how the US Navy sought to leverage climate change as an opportunity to improve its institutional standing and advance cooperative measures enshrined in CSP21. This was recognisable throughout the *Climate Change Roadmap* wherein the Navy sought opportunities to 'formalise cooperative relationships', initiate 'intergovernmental, multilateral and bi-lateral activities', 'form new and expand existing cooperative agreements' as well as 'develop and strengthen partnerships' with a range of other public agencies, private entities, foreign militaries and NGOs on a range of climate change matters (US Navy 2010b).

Less prominent, but worthy of comment, was the identification of a number of operational risks posed by climate change, including the opening of the Arctic and sea-level rise, reducing port availability for refuelling and resupply, increasing requirement for earth-moving projects and changes in ability of critical resource storage. Somewhat rare in public military discussion on climate change was also a direction to monitor areas of uncertainty, including ocean acidification, abrupt climate change and geo-engineering.

The final point to be made on both the *Climate Change* and *Arctic Roadmaps* is that both were scheduled to be incorporated into the Navy's *Program Objective Memorandum* for FY-14 (POM-14). This is a significant point, since it signified the time frame when the US Navy would actually commence funding the recommendations made as a consequence of both roadmaps. Furthermore, that climate change considerations were factored into the US Navy Planning-Programing-Budgeting-Execution (PPBE) process at all indicated a level of legitimacy regards the challenges posed by climate change, but also that these challenges were *not* considered exceptional or urgent. Moreover, viewed in the broader US DoD acquisition model, the roadmaps represented a starting point (Planning Phase) that was designed to identify and prioritise the risks posed by climate change for progressive remediation through *normal* military budgetary cycles commencing in FY-14.

Finally, both the strategic and tactical implications of climate change on the US Navy were expanded by the 2011 Naval Studies Board report, *National Security Implications of Climate Change for US Naval Forces*. This report represented the most comprehensive publicly available assessment of the impact of climate change on the US Navy. On the basis of the extensive terms of reference, it is reasonable to suggest that the findings made in this report were unlikely to differ significantly from those found through internal Navy processes as articulated in the roadmaps. It is, therefore, worthwhile to highlight the report's major findings, and an abbreviated version is included at Appendix 3-3.

Although *Climate Change Roadmap* can broadly be described as articulating US Navy adaptation, its strategic direction regarding climate mitigation was specifically addressed in the Navy's energy strategy (*A Navy Energy Vision for the 21st Century*), developed by *Task Force Energy* established under ADM Roughead in 2009. *Energy Vision* represented the start point for Navy's operationalisation of Secretary Mabus' energy goals (*Energy Program for Security and Independence*). Even though climate change mitigation was cited as a part-justification for improving the Navy's use of energy, the main focus undoubtedly remained enhancing US Naval energy security. Not surprisingly, this correlated strongly with views espoused by Secretary Mabus. In the opening statement in *Energy Vision*, ADM Roughead cited record oil prices in 2008 as a 'glimpse of an energy future … [in which] ever-rising costs and [cause of] strategic vulnerability' to the Navy (Roughead quoted in US Navy 2010c: 1). The bottom line became apparent when it surmised: '[f]or the Navy, high [oil] prices and price volatility pressurize budgets that could otherwise go to increased capability' (US Navy 2010c: 2).

Moreover, climate change was *a* factor, but it was not *the* factor. Thus, despite the apparent fervour surrounding Navy's efforts on climate *adaptation*, its rhetoric on contributing toward national climate change *mitigation* efforts was more subdued. Leaving no doubt, *Energy Vision* reinforced this point by declaring:

> Virtually all investments the Navy makes in energy efficiency and alternative energy [are] for the primary purpose of energy security and enhanced combat capability [but] will also serve to reduce greenhouse gas emissions (US Navy 2010c: 2).

Once again, this line reflected a change in rhetoric that drew justification of energy transformation away from 'nebulous' climate change toward 'unifying' concepts of energy security and enhanced war-fighting.

The reduced role played by climate change in US Naval leadership rhetoric was reflected in research results that showed climate change was not cited by ADM Roughead's replacement, ADM Greenert, in thirty speeches given from late 2011 through mid-2013 (though instances were found where he discussed it at length as *Vice* Chief of Naval Operations, prior to his appointment as CNO). Furthermore, while numerous examples were found supporting energy reform being progressed downwards through the Navy's bureaucratic structure, few instances were found regarding equivalent climate change initiatives. As noted by ADM Greenert, the US Navy 'energy strategy is much more mature than our climate strategy' (Greenert 2011: 24). Selective examples of energy reform proliferating through Navy structures included:

- OPNAV Energy Instruction (22 Jun 2012). Directs administrative action from strategic through tactical commanders for US Navy shore establishments regarding Navy energy goals.
- 2014 Officer Selection boards to include consideration of those officers who have excelled at energy resource management.
- Great Green Fleet lead-in training and exercises in anticipation of sailing the green fleet in a major military exercise in 2016.
- Partnerships with other government agencies and industries in the development of bio-fuel supply.

6.3.2.1 Summary of CNO Programme

Having analysed the speeches and policy of the CNO Programme, a number of conclusions can be made. Firstly, while the CNO Programme initiated a range of actions on climate change between 2009 and 2011, it became less prominent in the years 2012 and 2013. The change in focus may be broadly attributed to three reasons: (1) a revision of national strategic and military priorities following operational drawdown from the Middle East, subsequent rebalance toward Asia as well as reductions in US Defense spending; (2) the US Navy's internal prioritisation of energy security over climate change; and (3) that the majority of studies on the impact of climate change on the US Navy by 2012 were either already under way or had already been completed (from an increasing volume of literature generated by a broad cross section of academic, military, other agency, think-tank and NGO sources).

Secondly, the CNO Programme showed that the US Navy response to climate change was largely driven during the appointment of ADM Roughead. The drive to investigate the consequences of climate change on the US Navy pre-dated the election of President Obama and the appointment of Secretary Mabus. While this indicated a level of autonomy, the decisions must be viewed in the context that climate change at that moment in time was viewed as a significant international issue and that moves to address climate change were likely to be viewed favourably by the increasing likelihood of a new Administration. At any rate, it was largely already flagged as a burgeoning requirement by President Bush's 2008 *National Defense Authorization Act*. Thus, from a scientific, national security and political perspective, ADM Roughead confronted few risks, political or otherwise, in commencing a wide scale response to climate change. In some respects, he could have hoped to gain from being recognised as a leader by a favourable Administration in responding to climate change, particularly through opportunities presented by the opening of the Arctic.

Predominately, ADM Roughead's response favoured an examination of how the US Navy should adapt to climate change. While this focused on the Arctic, other geographical areas were to be considered. Broadly, the period between 2009 and 2011 can largely be considered as representing a scoping phase during which the US Navy commissioned a number of information-gathering (fact-finding) studies to assess the strategic implications of climate change. While the studies were extensive in their reach, extending across and down through the Navy's organisational structure, this research did *not* identify any immediate shifts in capability as a consequence of climate change. Furthermore, climate change adaptation was scarcely promoted as a major initiative (in the timeframe considered by this research) in the speech-acts of the CNO nor was it found in any lower-level initiatives.[34] This contrasted sharply with concepts of energy security. Moreover, the scoping studies represented more of a start-point than an end-point. In this sense, the major capability decisions regarding US Navy climate change considerations were scheduled to occur from 2014 onwards via standard PPBE processes. Their very embedding within this structure itself, however, indicated a level of maturity previously unseen in relation to the military and climate change. This was notable given the substantive decline in US Navy budget from around 2010.

Furthermore, US Navy mitigation efforts were largely framed around debates on energy security with an overarching requirement to free up funding in order to (re)invest in other capability priorities. In this context, US Navy energy transformation was framed as an auxiliary benefit to broader US national climate mitigation efforts.

[34]Noting here that the US Navy (in the years after 2013) did participate in Climate Change Adaptation Working Groups.

The third major conclusion on the CNO Programme is that climate change was not considered an urgent issue sufficient to warrant any emergency measures. The rhetoric on climate change by senior naval leadership was considerably muted compared with President Obama, although it was considered to be greater than the Secretary of the Navy and Secretary of Defense. In addition, the CNO Programme revealed that climate change was predominately framed by the CNO as a securitised issue. As a security issue, the US Navy focused on how it would adapt so that it might advance US national interests, centred on the Arctic. Ultimately, the US Navy climate response was one that placed paramount importance on ensuring the protection of US national security interests.

The underlying conclusion from this analysis supports a realist interpretation of US Navy response to climate change. Namely, the US Navy response was driven primarily to advance its own and its nation's self-interest by ensuring the preservation of its capabilities through the gradual adaptation of its doctrine, organisation, training, materiel, leadership, personnel and facilities in the face of a changing climate.

6.3.3 US Navy Oceanographer and Commander Pacific Fleet

Analysis of the Oceanographer Programme was undertaken on seven documents published across 2009 and 2010. Analysis of the Commander of Pacific Fleet was undertaken on thirty-eight documents published across 2012–2013. An overview of the documents and their authors is in Table 6.15.

The US Navy Oceanographer, Rear Admiral (RADM) David Titley was found to have predominately framed climate change as a securitised issue with a focus on the Arctic. In addition, speeches made by RADM Titley make clear that he considered climate change a long-term issue which would gradually impact the US Navy. This steady outlook was reflected in his rhetoric, in which he never framed climate change as a threat but rather in neutral terms. On several occasions he did, however, describe it as a strategic challenge.

Table 6.15 References to climate change in documents of US Oceanographer and Commander Pacific Fleet (N/A indicates that the five speech-acts were specifically about climate change.)

Title	Name	Speech-acts	Other	References to 'climate change'
US Oceanographer	RADM David Titley	5	2	N/A
Comdr. Pacific Fleet	ADM Cecil Haney	38	0	3
Total	214	43	2	3

Source The author

During his tenure as US Navy Oceanographer (2009–2012), RADM Titley was appointed to lead the US Navy response to climate change, including oversight of the *Arctic* and *Climate Change Roadmaps*. He was also variously used as a climate change envoy, in a role similar to that of his British counterpart RADM Neil Morisetti. In this capacity, RADM Titley addressed the US Congress on several occasions, was a regular public speaker on the subject and also spoke at COP 15 on the strategic risks of climate change. Speeches made by RADM Titley during his tenure as US Navy Oceanographer validate findings from the CNO Programme, providing further clarity on US Navy intentions. The following passages summarise these aspects, providing some unique examples that build on those already provided.

RADM Titley regularly articulated a detailed portrayal (the most by any of the officials assessed in this research) of the strategic, operational and tactical risks of climate change on the US Navy and on US national security more generally. Holding a PhD in meteorology, Titley regularly engaged the scientific community and sought a response that was 'science led'. Consistently throughout the speeches examined, Titley was careful not to present climate change as a crisis (or threat) that required emergency measures. On the contrary, Titley regularly provided a sense of orderliness. In a 2010 speech, Titley (2010c: 33) declared: 'The first thing I would say is that, from a US Navy perspective, the Arctic is a challenge and not a crisis'. More broadly, Titley extended this steady-hand approach to climate change in general. Asked by a Congressional Sub-Committee if he considered climate change a crisis, Titley (2010a) responded, 'I am not sure I would call it a crisis. It is a strategic challenge'.

This outlook was manifest in Titley's strategic, cautious and long-term approach to climate change more broadly. In an aptly named 2010 roundtable, *Climate Change Investments in a Fiscally Restrained Environment*, Titley portended a requirement to 'pace the threat' such that 'we don't get ourselves like in a tail chase with climate change or with changes in the Arctic' (Titley 2010b: 3). RADM Titley reiterated this approach in a 2010 Hearing to the House of Representatives, Subcommittee on Energy and Environment, *A Rational Discussion of Climate Change: the Science, The Evidence, The Response* in which he argued 'we are looking strategically out. So not just three, four or five days, but what are the next 20, 30, 50 years going to look like?' (Titley 2010d: 180).

Strictly aware that 'we're going to have to fold these challenges into a tight fiscal budget', Titley (2010b: 3) advanced the deliberate nature of the *Climate Change* and *Arctic Roadmaps* on the grounds that 'one thing I do not want to do is spend a lot of money and then find out we didn't spend it for the right thing' (ibid: 3). In this context:

> We are beginning to conduct the assessments necessary to inform future investments and are initiating adaptation activities in areas where we have enough certainty with which to proceed (Titley 2010d: 155).

Even where Titley saw opportunity, he advanced with caution:

> … the time is ripe to use the change in climate as an opportunity [but we need] to … get this right and make sure … we get best return on our collective investment (Titley 2010b: 7).

This outlook was also evident in downplaying ideas of strategic competition:

> In recent years … a lot of media reports about the Arctic framed the issues in way that made the Arctic sound like the Wild West. This was going to be the last great race for resources. When everybody converged at 90° North, we would all need to watch out. As you may recall, the Russians got there first and, by planting their flag on the North Pole, probably did us a big favour … [by] get[ting] our leaders to focus on the Arctic (Titley 2010c: 33).

So as to leave no doubt, Titley then added, 'we have to understand what it is we are going to do before we start spending money or before we even study how we are going to spend money' (Titley 2010c: 34).

In the interim, Titley advocated a method of building partnerships across a range of public, private, military, government and non-government areas. In 2013 Titley noted TFCC had 'interacted with over 220 governmental, non-governmental, academic, private-sector and international organizations in 14 countries' (Titley quoted in Pearce 2013: 47). Thus, for Titley (2010b) 'there is a real opportunity, too, that this climate change can almost be viewed as a common enemy … that can lead to partnerships'. It was within this framework that Titley advocated a range of measures that could occur in the near term, including military-to-military exercises concerning HADR, Search and Rescue as well as collaboration on climate research. Nonetheless, and as was true of the broader Navy, climate change was 'more than humanitarian disaster relief' (Titley 2010c: 27).

Examination of thirty-eight speeches between 2012 and 2013 made by the Commander of Pacific Fleet returned three general references to climate change. This examination was considered of limited value since it did not capture speeches made during the main period of focus climate change by the US Navy (2009–2011). Despite this, references to climate change were framed as a *challenge* and mostly as an environmental issue with a focus on rising sea-levels threatening Pacific island nations. The seeming low importance of climate change in these speeches somewhat validate the finding that the intense periods of examination by the US Navy had largely passed by 2012.[35]

[35]This summary should be placed in the context of the period for which this book examined documents (i.e., 2003–2013). Specifically, it did not include an assessment of comments made by the US Pacific Commander Admiral Samuel J. Locklear III from around early-mid 2013 onwards in which he identified climate change as a major security threat. Moreover, this book did not analyse speeches of the various US Combatant Commanders (PACOM included). A good summary of US Combat Commanders outlook (2015) on climate change as a security issue can be found at: https://climateandsecurity.files.wordpress.com/2014/01/15_07_24-dod_gcc_congressional-report-on-national-security-implications-of-climate-change.pdf (4 March 2016).

6.4 Chapter Conclusion

Chapter 3 identified three distinct phases of US climate security discourse; this chapter has examined the second phase in detail between 2003–2013.[36] Under the presidency of George W. Bush, climate change was conceived of as a national security risk by administration officials and the US DoD, though in a far smaller number of documents relative to the output of its successor. From the White House lens, President Bush's climate priorities were dominated by actions that would enhance US energy security and adopting measures for the primary purpose of boosting the US economy. He viewed climate change as a collective international problem, was sceptical of UN agreements that could bind the US into a disadvantageous position, and never framed it as a security 'threat' or 'crisis'. Though the US DoD did highlight the need to tackle climate change during his tenure, President Bush himself stopped short of securitising the issue, or raising emergency measures in response, An examination of almost 300 speeches by Donald Rumsfeld, as Secretary of Defence until late 2006, revealed not a single mention of climate change. Wars in Iraq and Afghanistan, as well as the wider war on terrorism, dominated the US DoD during this era. Despite this, the wider US national security establishment increasingly began to examine the threat posed by climate change to US national security. From at least 2006 extensive hearings were held in the US Congress on the matter and reports by the US intelligence and think-tank community (including by senior retired US military officials) began to press for greater action. The idea that climate change was a 'threat multiplier' for instability and a significant threat to US interests emerged in US climate security discourse from 2007. In 2008, the National Defense Strategy included a reference to 'the need to tackle climate change' (US DoD 2008: 5) and President Bush signed the 2008 *National Defense Authorization Act* which called for climate change to be addressed by the US DoD in the 2010 US NSS and QDR.

The 2008 NDAA, followed by the election of Barack Obama in January 2009, marked a significant acceleration of the US climate security discourse. President Obama routinely characterised climate change as a significant national and global security threat, as did key cabinet members responsible for national security, including his secretaries of defense, and the broader intelligence community. Looking to advance his wider agenda on climate change, but accepting the limitations of a divided Congress, Obama pursued mitigation and adaptation measures via executive and regulatory power invested in the president. From 2009 he released a series of Executive Orders that directed greater US DoD action. Less tangible, but equally real, was the favourable political climate that allowed the US DoD and wider US national security establishment to take the issue seriously and to advance more robust measures for addressing climate change impacts on key national security equities. From 2009 the US DoD—particularly the US Navy—

---

[36]For an update of key documents in US climate security discourse from 2013 until 2017, see Chap. 3.

undertook a significant response to climate change through a series of roadmaps, strategies, plans and administrative orders that included both mitigation and adaptation measures. While these documents commenced life as departmental-wide or service-wide proclamations (and therefore quite generic or 'strategic' in nature), they were progressively built upon and released with greater depth (downwards through the military structure) and breadth (to include all services and commands). Broadly, the US DoD came to accept climate change as a mainstream national security issue, though it was not one viewed as so urgent as to require emergency or extraordinary measures until at least 2014 when it characterised as posing 'immediate risks to US national security' (US DoD 2014a: 1). In the US Navy, the most active of the services, climate responses centred on enhancing its energy security, reducing the cost and burden of fossil fuels, and favourably positioning the service in terms of adapting its force structure, training, and capability for the possibility of increased operations in the Arctic. More broadly, the other critical area (partially addressed in this chapter and Chap. 3) was the requirement for the US military to understand the impact that climate change would have on its global and homeland base infrastructure.

References

Andrews, Anthony. 2009. *Department of Defence Facilities Energy Conservation Policies and Spending*. Congressional Research Service. Washington, DC: US Government Printing Office. Accessed on 11 November 2013, http://www.fas.org/sgp/crs/natsec/R40111.pdf.

An act making supplemental appropriations for job preservation and creation, infrastructure investment, energy efficiency and science, assistance to the unemployed, and State and local fiscal stabilization, for the fiscal year ending September 30, 2009, and for other purposes. American Recovery and Reinvestment Act (ARRA). 111th US Congress, H.R. 1, Public Law 111-5, 17 February 2009. Accessed on 3 March 2013, http://www.gpo.gov/fdsys/pkg/PLAW-111publ5/html/PLAW-111publ5.htm.

Bartin, James T., and Larence Van Bibber. 2011. *Alternative Fuels for Military Applications*. Santa Monica: RAND Corporation. Accessed on 17 March 2014, http://www.rand.org/content/dam/rand/pubs/monographs/2011/RAND_MG969.pdf.

Brauch, Hans Günter. 2011b. "Security Threats, Challenges, Vulnerabilities and Risks in US National Security Documents (1990–2010)." In *Coping with Global Environmental Change, Disasters and Security: Threats, Challenges, Vulnerabilities and Risks*, edited by Hans Günter Brauch, Ursula Oswald Spring, Czeslaw Mesjasz, John Grin, Patricia Kameri-Mbote, Bechir Chourou, Pal Dunay and Jorn Birkmann, vol. 5, pp. 249–274. Berlin: Springer.

Bush, George W. 2001. "Address before a Joint Session of the Congress on Administration Goals." In *Compilation of Presidential Documents*, 5 March 2001, vol. 37, no. 9, pp. 351–357. Washington, DC: US Government Printing Office. Accessed on 12 June 2013, http://www.gpo.gov/fdsys/granule/WCPD-2001-03-05/WCPD-2001-03-05-Pg351-2/content-detail.html.

Bush, George W. 2002a. "Address before a Joint Session of the Congress on the State of the Union." In *Compilation of Presidential Documents*, 4 February 2002, vol. 38, no. 5 pp. 133–139. Washington, DC: US Government Printing Office. Accessed on 14 July 2013, http://www.gpo.gov/fdsys/granule/WCPD-2002-02-04/WCPD-2002-02-04-Pg133-3/content-detail.html.

Bush, George W. 2002b. "Remarks announcing the Clear Skies and Global Climate Change Initiatives in Silver Spring, Maryland." In *Compilation of Presidential Documents*, 18 February 2002, vol. 38, no. 7, pp. 232–236. Washington, DC: US Government Printing Office. Accessed on 3 September 2013, http://www.gpo.gov/fdsys/granule/WCPD-2002-02-18/WCPD-2002-02-18-Pg232/content-detail.html.

Bush, George W. 2003. "Remarks On Energy Independence." In *Compilation of Presidential Documents*, 10 February 2003, vol. 39, no. 6, pp. 161–164. Washington, DC: US Government Printing Office. Accessed on 10 August 2013, http://www.gpo.gov/fdsys/granule/WCPD-2003-02-10/WCPD-2003-02-10-Pg161/content-detail.html.

Bush, George W. 2005. "Address before a Joint Session of the Congress on the State of the Union." In *Public Papers of the Presidents of the United States: George W. Bush* 2 February 2005, vol. (Book) I, pp. 113–121. Washington, DC: US Government Printing Office. Accessed on 4 August 2013, http://www.gpo.gov/fdsys/granule/PPP-2005-book1/PPP-2005-book1-doc-pg113/content-detail.html.

Bush, George W. 2006. "Address before a Joint Session of the Congress on the State of the Union." In *Public Papers of the Presidents of the United States: George W. Bush* 31 January 2006, vol. (Book) I, pp. 146–154. Washington, DC: US Government Printing Office. Accessed on 17 August 2013, http://www.gpo.gov/fdsys/granule/PPP-2006-book1/PPP-2006-book1-doc-pg146-2/content-detail.html.

Bush, George W. 2007a. "Address before a Joint Session of the Congress on the State of the Union." In *Compilation of Presidential Documents*, 29 January 2007, vol. 43, no. 4, pp. 57–64. Washington, DC: US Government Printing Office. Accessed on 12 September 2013, http://www.gpo.gov/fdsys/granule/WCPD-2007-01-29/WCPD-2007-01-29-Pg57/content-detail.html.

Bush, George W. 2007b. "The President's News Conference." In *Compilation of Presidential Documents*, 9 April 2007, vol. 43, no. 14, pp. 409–417. Washington, DC: US Government Printing Office. Accessed on 21 September 2013, http://www.gpo.gov/fdsys/granule/WCPD-2007-04-09/WCPD-2007-04-09-Pg409-2/content-detail.html.

Bush, George W. 2007c. "The President's News Conference." In *Public Papers of the Presidents of the United States: George W. Bush* 20 December 2007, vol. (Book) II, pp. 1573–1585. Washington, DC: US Government Printing Office. Accessed on 10 August 2013, http://www.gpo.gov/fdsys/granule/PPP-2007-book2/PPP-2007-book2-doc-pg1573-2/content-detail.html.

Bush, George W. 2007d. "Remarks during a meeting on Energy Security and Climate Change." In *Compilation of Presidential Documents* 1 October 2007, vol. 43, no. 39, pp. 1261–1265. Washington, DC: US Government Printing Office. Accessed on 8 October 2013, http://www.gpo.gov/fdsys/granule/WCPD-2007-10-01/WCPD-2007-10-01-Pg1261/content-detail.html.

Bush, George W. 2007e. "Remarks on Energy in Athens, Alabama." In *Compilation of Presidential Documents* 25 June 2007, vol. 43, no. 25, pp. 837–842. Washington, DC: US Government Printing Office Accessed on 23 August 2013, http://www.gpo.gov/fdsys/granule/WCPD-2007-06-25/WCPD-2007-06-25-Pg837/content-detail.html.

Bush, George W. 2008a. "Address before a Joint Session of the Congress on the State of the Union." In *Public Papers of the Presidents of the United States: George W. Bush*, 28 January 2008, vol. (Book) I, pp. 116–125. Washington, DC: US Government Printing Office. Accessed on 12 August 2013, http://www.gpo.gov/fdsys/granule/PPP-2008-book1/PPP-2008-book1-doc-pg116/content-detail.html.

Bush, George W. 2008b. "The President's News Conference." In *Compilation of Presidential Documents* 7 July 2008, vol. 44, no. 26, pp. 939–943. Washington, DC: US Government Printing Office. Accessed on 12 August 2013, http://www.gpo.gov/fdsys/granule/WCPD-2008-07-07/WCPD-2008-07-07-Pg939/content-detail.html.

Bush, George W. 2008c. "Remarks at the Washington International Renewable Energy Conference." In *Compilation of Presidential Documents* 10 March 2008, vol. 44, no. 9, pp. 324–330. Washington, DC: US Government Printing Office. Accessed on 2 October 2013, http://www.gpo.gov/fdsys/granule/WCPD-2008-03-10/WCPD-2008-03-10-Pg324-2/content-detail.html.

Bush, George W. 2008d. "Remarks on Energy and Climate Change." In *Compilation of Presidential Documents* 21 April 2008, vol. 44, no. 15, pp. 524–527. Washington, DC: US Government Printing Office. Accessed on 12 September 2013, http://www.gpo.gov/fdsys/granule/WCPD-2008-04-21/WCPD-2008-04-21-Pg524/content-detail.html.

Bush, George W. 2009. "Directive on Arctic Region Policy." In *Administration of George W. Bush*, 12 January 2009, pp. 47–54. Washington, DC: US Government Printing Office. Accessed on 5 September 2013, http://www.gpo.gov/fdsys/pkg/WCPD-2009-01-19/pdf/WCPD-2009-01-19-Pg47-2.pdf.

Dempsey, Martin. 2011. "General Dempsey at the Military Reporters and Editors Conference." In *Joint Chiefs of Staff Speeches*, 18 November 2008, vol. 2013. Arlington: Department of Defense. Accessed on 12 July 2012, http://www.jcs.mil/Media/Speeches/tabid/3890/Article/571837/gen-dempseys-remarks-and-qa-at-at-the-military-reporters-and-editors-conference.aspx.

Dempsey, Martin. 2012. "Gen. Dempsey's Remarks at the Ambassador S. Davis Phillips Family International Lecture Series." In *Joint Chiefs of Staff Speeches*, 12 January 2012. Duke University: Department of Defense. Accessed on 12 July 2013, http://www.jcs.mil/Media/Speeches/tabid/3890/Article/571848/gen-dempseys-remarks-and-qa-at-the-ambassador-s-davis-phillips-family-internati.aspx.

Executive Order 13514. 2009. "Executive Order 13514 Federal Leadership in Environmental, Energy, and Economic Performance." In *Presidential Documents, Federal Register* 5 October 2009, vol. 74 no. 194, pp. 52117–52127. Washington, DC: US Government Printing Office Accessed on 10 October 2013, http://www.gpo.gov/fdsys/granule/FR-2009-10-08/E9-24518/content-detail.html.

Floyd, Rita. 2010. *Security and the Environment: Securitisation Theory and US Environmental Security Policy*. Cambridge: Cambridge University Press.

Greenert, Jonathon W. 2011. "Adapting to Climate and Energy Challenges: Options for U.S. Maritime Forces." Adapting to Climate and Energy Challenges Symposium, Laurel, 29 & 30 March 2011. Accessed on 15 June 2013, http://www.jhuapl.edu/ClimateAndEnergy/Book2011/2011CEBook-LR.pdf.

Hagel, Chuck. 2013. "Halifax International Security Forum (DoD Arctic Strategy)." Online: US Department of Defense. Accessed on 15 January 2017, http://archive.defense.gov/speeches/speech.aspx?speechid=1821.

Hagel, Chuck. 2014a. "The Department of Defense Must Plan for the National Security Implications of Climate Change." Online: US Department of Defense. Accessed on 18 January 2018, https://obamawhitehouse.archives.gov/blog/2014/10/13/defense-department-must-plan-national-security-implications-climate-change.

Hagel, Chuck. 2014b. "Realizing the Asia-Pacific Rebalance." Online: Defense One. Accessed on 15 January 2017, http://www.defenseone.com/ideas/2014/04/realizing-asia-pacific-rebalance/81730/.

Inhofe, James M. 2012a. "Inhofe Again Requests Sec. Mabus to disclose full cost of greening fleet demonstration." Washington, DC: US Senate. Published online 2 August 2012, http://www.inhofe.senate.gov/newsroom/press-releases/inhofe-again-requests-sec-mabus-to-disclose-full-cost-of-green-fleet-demonstration.

Inhofe, James M. 2012b. "Inhofe: Jolly "Green Fleet" Throwing Dollars Out to Sea." Washington, DC: US Senate. Published online 19 July 2012, http://www.epw.senate.gov/public/index.cfm/2012/7/post-a145fe86-802a-23ad-4b07-45be3935642e.

Inhofe, James M. 2012c. "Press Release: Inhofe Questions Excessive Cost of 'Greening' Military (25 July)." Washington, DC: US Senate. Accessed on 8 October 2013, http://www.epw.senate.gov/public/index.cfm/press-releases-republican?ID=BFB1C8CE-802A-23AD-4849-5001DD46198D.

Lynn, William J. 2009. "Remarks at the Connect U.S. Fund." In *Deputy Secretary of Defense Speech*, 14 December 2009. Washington, DC: US Department of Defense.

Mabus, Ray. 2009a. "Remarks by the Honourable Ray Mabus Secretary of the Navy at Current Strategy Forum Naval War College, Newport, Rhode Island." In *Transcripts of the Secretary of*

Navy, 16 June 2009. Washington, DC: United States Navy. Accessed on 12 June 2013, http://www.navy.mil/navydata/people/secnav/Mabus/Speech/Current%20Strategy%20Forum.pdf.
Mabus, Ray. 2009b. "Remarks by the Honourable Ray Mabus Secretary of the Navy at International Seapower Symposium Naval War College, Newport, Rhode Island." In *Transcripts of the Secretary of Navy*, 7 October 2009. Washington, DC: United States Navy. Accessed on 8 August 2013, http://www.navy.mil/navydata/people/secnav/Mabus/Speech/International%20Seapower%20Symposium%207%20Oct%2009%20Rel.pdf.
Mabus, Ray. 2009c. "Remarks by the Honourable Ray Mabus Secretary of the Navy, Defense Attache Association, Washington DC." In *Transcripts of the Secretary of Navy*, 18 November 2009. Washington, DC: United States Navy. Accessed on 6 September 2013, http://www.navy.mil/navydata/people/secnav/Mabus/Speech/Defense%20Attache%20Association%20paginated%20-%2018%20Nov%2009.pdf.
Mabus, Ray. 2009d. "Remarks by the Honourable Ray Mabus Secretary of the Navy, Naval Energy Forum, Hilton McLean, Tysons Corner, McLean, Virginia." In *Transcripts of the Secretary of Navy*, 14 October 2009. Washington, DC: United States Navy. Accessed on 8 September 2013, http://www.navy.mil/navydata/people/secnav/Mabus/Speech/SECNAV%20Energy%20Forum%2014%20Oct%2009%20Rel1.pdf.
Mabus, Ray. 2009e. "Remarks by the Honourable Ray Mabus Secretary of the Navy, Hamer Institute Speech, Jackson State University, Jackson, Mississippi." In *Transcripts of the Secretary of Navy*, 9 October 2009. Washington, DC: United States Navy. Accessed on 7 September 2013, http://www.navy.mil/navydata/people/secnav/Mabus/Speech/Hamer%20Institute%20Naming%20of%20Medgar%20Evers%209%20Oct%2009%20Rel.pdf.
Mabus, Ray. 2010a. "Remarks by the Honourable Ray Mabus Secretary of the Navy, Center for Naval Analysis Earth Day Luncheon, Alexandria, VA (29 April)." In *Transcripts of the Secretary of Navy*, 29 April 2010. Washington, DC: United States Navy. Accessed on 8 September 2013, http://www.navy.mil/navydata/people/secnav/Mabus/Speech/Center%20for%20Naval%20Analysis%20Earth%20Day%20Luncheon%2029%20Apr%2010.pdf.
Mabus, Ray. 2010b. "Remarks by the Honourable Ray Mabus Secretary of the Navy, The Commonwealth Club, San Francisco, CA." In *Transcripts of the Secretary of Navy*, 16 August 2010. Washington, DC: United States Navy. Accessed on 9 October 2013, http://www.navy.mil/navydata/people/secnav/Mabus/Speech/commonwealthclub.pdf.
Mabus, Ray. 2012a. "Remarks by the Honourable Ray Mabus Secretary of the Navy, Global Conference on Oceans, Climate and Security: Making the Connections, Boston, Mass." In *Transcripts of the Secretary of Navy*, 22 May 2012. Washington, DC: United States Navy. Accessed on 8 October 2013, http://www.navy.mil/navydata/people/secnav/Mabus/Speech/OceansConferenceMay12v2.pdf.
Mabus, Ray. 2012b. "Remarks by the Honourable Ray Mabus Secretary of the Navy, Naval Energy Forum, Washington DC." In *Transcripts of the Secretary of Navy*, 17 October 2012. Washington, DC: United States Navy. Accessed on 1 October 2013, http://www.navy.mil/navydata/people/secnav/Mabus/Speech/navalenergyforumtranscriptfinal.pdf.
Mabus, Ray. 2012c. "Remarks by the Honourable Ray Mabus Secretary of the Navy, Senate Committee on Energy and Natural Resources, Norfolk, VA." In *Transcripts of the Secretary of Navy*, 12 March 2012. Washington, DC: United States Navy. Accessed on 4 October 2013, http://www.navy.mil/navydata/people/secnav/Mabus/Speech/Senate%20Committee%20on%20Energy%20and%20Natural%20Resources%20oral%20testimonyAsPrepared.pdf.
Major Economies Forum Declaration. 2009. "Declaration by the Leaders of the Major Economies Forum on Energy and Climate." Washington, DC: Office of the Federal Register, National Archives and Records Administration. Accessed on 11 October 2013, http://www.gpo.gov/fdsys/granule/PPP-2009-book2/PPP-2009-book2-Doc-pg1072/content-detail.html.
McCain, John. 2012. "Letter from Senator McCain to Navy Secretary Ray Mabus Regarding Biofuels." In *John McCain Press Releases*, 27 July 2012. Published online: John McCain Accessed on 12 March 2013, http://www.mccain.senate.gov/public/index.cfm/press-releases?ID=ca4414f1-e686-7cb6-820e-995f3289f68a.
Mullen, Michael G. 2011. "From the Chairman." *Joint Force Quarterly*, no. 60:2–3.

National Energy Policy Development Group. 2001. *National Energy Policy: Report of the National Energy Policy Development Group* Washington, DC: US Government Printing Office. Accessed on 13 January 2013, http://www.wtrg.com/EnergyReport/National-Energy-Policy.pdf.

National Defense Authorization Act (NDAA) for Fiscal Year 2008. National Defense Authorization Act for Fiscal Year 2008 (110th US Congress, Public Law 110-181, 28 January 2008. Accessed on 21 September 2013, http://www.gpo.gov/fdsys/pkg/PLAW-110publ181/html/PLAW-110publ181.htm.

Obama, Barack H. 2008. "Videotaped Remarks to the Bi-Partisan Governors Global Climate Summit November 18, 2008." In *Obama-Biden Transition Project Press Releases*. Washington, DC: The Office of the President Elect. Accessed on 20 May 2017, http://www.presidency.ucsb.edu/ws/index.php?pid=84875. (Site archived. Document in possession of author).

Obama, Barack H. 2009b. "A Just and Lasting Peace." In *Nobel Lecture*, 10 December 2009. Oslo: Nobel Media. Accessed on 13 October 2013, http://www.nobelprize.org/nobel_prizes/peace/laureates/2009/obama-lecture_en.html.

Obama, Barack H. 2009c. "The President's News Conference in Copenhagen." In *Public Papers of the Presidents of the United States: Barack Obama* 18 December 2009, vol. (Book) II, pp. 1823–1828. Washington, DC: US Government Printing Office. Accessed on 20 May 2017, https://www.gpo.gov/fdsys/pkg/PPP-2009-book2/pdf/PPP-2009-book2-Doc-pg1823-2.pdf.

Obama, Barack H. 2009d. "President Obama on Historic Energy Bill." In *White House Briefing Room*, 24 June 2009. Washington, DC: White House.

Obama, Barack H. 2009e. "Remarks at a Plenary Session of the United Nations Climate Change Conference in Copenhagen, Denmark." In *Public Papers of the Presidents of the United States: Barack Obama* 18 December 2009, vol. (Book) II, pp. 1820–1822. Washington, DC: US Government Printing Office. Accessed on 13 October 2013, http://www.gpo.gov/fdsys/granule/PPP-2009-book2/PPP-2009-book2-Doc-pg1820/content-detail.html.

Obama, Barack H. 2009g. "Remarks by the President at United Nations Secretary General Ban Ki-Moon's Climate Change Summit (United Nations Headquarters, New York)." In *President Barack Obama, Speeches and Remarks*, 22 September 2009. Washington, DC: White House.

Obama, Barack H. 2009h. "Remarks by the President to the United Nations General Assembly (United Nations Headquarters, New York)." In *President Obama, Speeches and Remarks*, 23 September 2009. Washington, DC: White House.

Obama, Barack H. 2009i. "Remarks on Energy." In *Compilation of Presidential Documents*, (January, 26), 26 January 2009. Washington, DC: US Government Printing Office. Accessed on 3 September 2013, http://www.gpo.gov/fdsys/pkg/DCPD-200900019/content-detail.html.

Obama, Barack H. 2009j. "Remarks on the Declaration by the Major Economies Forum on Energy and Climate Change in L'Aquila, Italy." In *Public Papers of the Presidents of the United States: Barack Obama* 9 July 2009, vol. (Book) II, pp. 1072–1075. Washington, DC: US Government Printing Office. Accessed on 11 October 2013, http://www.gpo.gov/fdsys/granule/PPP-2009-book2/PPP-2009-book2-Doc-pg1072/content-detail.html.

Obama, Barack H. 2009k. "Remarks to the Ghanian Parliament in Accra." In *Public Papers of the Presidents of the United States: Barack Obama*, 11 July 2009, vol. (Book) II, pp. 1087–1092. Washington, DC: US Government Printing Office. Accessed on 8 November 2013, http://www.gpo.gov/fdsys/granule/PPP-2009-book2/PPP-2009-book2-Doc-pg1087/content-detail.html.

Obama, Barack H. 2009l. "Remarks to the United Nations General Assembly in New York City." In *Public Papers of the Presidents of the United States: Barack Obama* 23 September 2009, vol. (Book) II, pp. 1440–1447. Washington, DC: US Government Printing Office. Accessed on 4 May 2013, http://www.gpo.gov/fdsys/granule/PPP-2009-book2/PPP-2009-book2-Doc-pg1440/content-detail.html.

Obama, Barack H. 2011. "The President's news conference in Kapolei, Hawaii." In *Compilation of Presidential Documents*, 13 November 2011. Washington, DC: US Government Printing

Office. Accessed on 13 October 2013, http://www.gpo.gov/fdsys/pkg/DCPD-201100864/content-detail.html.

Obama, Barack H. 2012a. "Address before a Joint Session of Congress on the State of the Union." In *The American Presidency Project*, edited by Gerhard Peters and John T. Woolley. Published online: The American Presidency Project. Accessed on 18 August 2013, http://www.presidency.ucsb.edu/ws/?pid=99000.

Obama, Barack H. 2012b. "Remarks by the President in a News Conference." In *President Barack Obama, Speeches and Remarks*, 14 November 2012. Washington, DC: White House. Accessed on 7 October 2014, http://www.whitehouse.gov//the-press-office/2012/11/14/remarks-president-news-conference. (Site archived. Document in possession of author. For alternate access, see: http://www.presidency.ucsb.edu/ws/index.php?pid=102644, 5 February 2017).

Panetta, Leon E. 2012a. "Environmental Defense Fund Reception (Energy, Security, and the Environment)." In *Secretary of Defense Speech*, 2 May 2012. Washington, DC: Department of Defense.

Panetta, Leon E. 2012b. "Shangri-La Security Dialogue (Singapore)." In *The IISS Shangri-La Dialogue 14th Asia Security Summit*, 2 June 2012. Washington, DC: International Institute for Strategic Studies.

Pearce, Alix. 2013. "The New Security Climate: Rethinking Australian Defence Policy in an Age of Climate Change." Honours in Government and International Relations, School of Social and Political Sciences, University of Sydney.

Roughead, Gary. 2009. "Chief of Naval Operations, ADM Gary Roughead delivers remarks at the Current Strategy Forum, Naval War College, Newport, RI." In *2010 Speeches: ADM Gary Roughead*, 16 June 2009. Washington, DC: US Navy.

Roughead, Gary. 2010a. "Chief of Naval Operations Adm. Gary Roughead deliver remarks at Government Executive Magazine at the National Press Club." In *2010 Speeches: ADM Gary Roughead*, 10 November 2010. Washington, DC: US Navy. http://www.navy.mil/navydata/people/cno/Roughead/Speech/101111-govexec-roughead.pdf.

Roughead, Gary. 2010b. "Chief of Naval Operations Adm. Gary Roughead Delivers Remarks during the Current Strategy Forum Newport, Rhode Island." In *2010 Speeches: ADM Gary Roughead*, 8 June 2013. Washington, DC: US Navy. Accessed on 6 August 2013, http://www.navy.mil/navydata/leadership/mist.asp?x=S&yr=2010.

Roughead, Gary. 2010c. "Chief of Naval Operations Admiral Gary Roughead Remarks as delivered World Affairs Council of Northern California." In *2010 Speeches: ADM Gary Roughead*, 5 August 2010. Washington, DC: US Navy. Accessed on 6 September 2013, http://www.navy.mil/navydata/leadership/mist.asp?x=S&yr=2010.

Roughead, Gary. 2010d. "Chief of Naval Operations, ADM Gary Roughead delivers remarks at the Navy Energy Forum." In *2010 Speeches: ADM Gary Roughead*, 12 October 2010. Washington, DC: US Navy. Accessed on 12 October 2013, http://www.navy.mil/navydata/leadership/mist.asp?x=S&yr=2010.

Roughead, Gary. 2011a. "Chief of Naval Operations (CNO) Adm. Gary Roughead speaks at Active in the Arctic Seminar June 16, 2011." In *2011 Speeches: ADM Gary Roughead*, 16 June 2011. Washington, DC: US Navy. Accessed on 8 June 2013, http://www.navy.mil/navydata/people/cno/Roughead/Speech/110616%20Arctic%20Capitol%20Hill.pdf.

Roughead, Gary. 2011b. "Chief of Naval Operations Adm. Gary Roughead delivers remarks at the 4th Symposium on the Impacts of an Ice-Diminishing Arctic on Naval & Maritime Operations." In *2011 Speeches: ADM Gary Roughead*, 20 June 2011. Washington, DC: US Navy. Accessed on 7 September 2013, http://www.navy.mil/navydata/people/cno/Roughead/Speech/110620%20Arctic%20Symposium.pdf.

Rumsfeld, Donald H. 2003. "FY2004 Defense Budget Testimony (As Prepared)." In *Transcripts of the Secretary of Defense*, 5 February 2003. Washington, DC: Department of Defense.

Titley, David W. 2010a. "Climate Change Finance: Providing assistance for vulnerable countries." In *Hearing Before the Subcommittee on Asia, the Pacific and the Global Environment of the Committee on Foreign Affairs House of Representatives One Hundred Eleventh Congress Second Session*, 111th Congress, Serial No. 111–120, 27 July 2010. Washington, DC:

US Government Printing Office. http://www.gpo.gov/fdsys/pkg/CHRG-111hhrg57687/html/CHRG-111hhrg57687.htm.

Titley, David W. 2010b. "Climate Change Investments in a Fiscally Restrained Environment Time." In *Department of Defense Bloggers Roundtable*, edited by John Ohab, 18 June 2010. Washington, DC: Federal News Service. Accessed on 10 May 2013, http://www.dodlive.mil/files/2010/06/0618_Rear-Adm_Titley.pdf.

Titley, David W. 2010c. "Global Climate Change." Climate and Energy Symposium 2010: Proceedings on Climate and Energy, Imperatives for Future Naval Forces, Baltimore. Accessed on 18 October 2012, http://www.jhuapl.edu/ClimateAndEnergy/Book/2010CEBook.pdf.

Titley, David W. 2010d. "A Rational Discussion of Climate Change: The Science, the Evidence, the Response." In *Hearing before the Sub Committee on Energy and Environment Committee on Science and Technology House of Representatives One Hundred Eleventh Congress Second Session*, 111th Congress, Serial No. 111–114, 17 November 2010. Washington, DC: US Government Printing Office. Accessed on 10 April 2013, http://www.gpo.gov/fdsys/pkg/CHRG-111hhrg62618/pdf/CHRG-111hhrg62618.pdf.

US DoD. 2005. *The National Defense Strategy of the United States of America*. Washington, DC: Department of Defense. Accessed on 6 August 2013, http://www.au.af.mil/au/awc/awcgate/nds/nds2005.pdf.

US DoD. 2006. *Quadrennial Defense Review Report*. Washington, DC: Department of Defense. Accessed on 18 August 2013, http://archive.defense.gov/pubs/pdfs/QDR20060203.pdf.

US DoD. 2008. *National Defense Strategy*. Washington, DC: Department of Defense. Accessed on 2 May 2013, http://www.defense.gov/Portals/1/Documents/pubs/2008NationalDefenseStrategy.pdf.

US DoD. 2010. *Quadrennial Defense Review*. Washington, DC: Department of Defense. Accessed on 6 March 2012, https://acc.dau.mil/adl/en-US/341273/file/48799/Quadrennial%20Defense%20Review%20-%20February%202010.pdf.

US DoD. 2011. *The National Military Strategy of the United States of America 2011: Redefining America's Military Leadership*. Washington, DC: Department of Defense. Accessed on 5 March 2013, http://www.defense.gov/Portals/1/Documents/pubs/2011-National-Military-Strategy.pdf.

US DoD. 2012a. "Department of Defense FY-2012 Climate Change Adaptation Roadmap." In *Department of Defense Strategic Sustainability Performance Plan FY-2012*. Washington, DC: Department of Defence. Accessed on 20 May 2017, http://www.denix.osd.mil/sustainability/dod-sspp/unassigned/dod-sspp-fy-2012/.

US DoD. 2012b. *Department of Defense Strategic Sustainability Performance Plan FY-2012*. Washington, DC: Department of Defense. Accessed on 9 November 2016, http://www.denix.osd.mil/sustainability/dod-sspp/unassigned/dod-sspp-fy-2012/.

US DoD. 2013b. *National Defense Budget Estimates for FY-2014*. Washington, DC: Office of the Under Secretary of Defense. Accessed on 20 May 2017, http://comptroller.defense.gov/Portals/45/Documents/defbudget/fy2014/FY14_Green_Book.pdf.

US DoD. 2014a. "2014 Climate Change Adaptation Roadmap." In *Department of Defense Strategic Sustainability Performance Plan FY-2014*. Washington, DC: Department of Defense. Accessed on 20 May 2017, http://www.denix.osd.mil/sustainability/dod-sspp/unassigned/dod-sspp-fy-2014/.

US House of Representatives. 2006. "Department of Energy's Plan for Climate Change Technology Programs." In *Hearing Before the Subcommittee on Energy and Committee on Science*, 109th Congress, Second session, Serial No. 109-62, 20 September 2006. Washington, DC: US Government Printing Office. Accessed on 16 March 2013, http://www.gpo.gov/fdsys/pkg/CHRG-109hhrg29851/html/CHRG-109hhrg29851.htm.

US Navy. 2007. *A Cooperative Strategy for 21st Century Seapower*. Washington, DC: US Navy. Accessed on 20 May 2017, http://www.navy.mil/local/maritime/150227-CS21R-Final.pdf.

US Navy. 2009a. *Task Force Climate Change Charter*. Washington, DC: US Navy. Accessed on 6 March 2013, http://greenfleet.dodlive.mil/files/2010/09/Task-Force-Climate-Change-Charter.pdf.

US Navy. 2009b. *US Navy Arctic Roadmap*. Washington, DC: US Navy. Accessed on 3 August 2013, http://www.navy.mil/navydata/documents/USN_artic_roadmap.pdf.
US Navy. 2010a. *Energy Program for Security and Independence*. Washington, DC: US Navy. Accessed on 6 August 2013, http://greenfleet.dodlive.mil/files/2010/04/Naval_Energy_Strategic_Roadmap_100710.pdf.
US Navy. 2010b. *Navy Climate Change Roadmap*. Washington, DC: US Navy. Accessed on 6 March 2012, http://www.navy.mil/navydata/documents/CCR.pdf.
US Navy. 2010c. *A Navy Energy Vision for the 21st Century*. Washington, DC: US Navy. Accessed on 6 March 2013, http://greenfleet.dodlive.mil/files/2010/10/Navy-Energy-Vision-Oct-2010.pdf.
US Navy. 2011. *Arctic Environmental Assessment and Outlook Report In support of The Navy Arctic Roadmap Action Item 5.7*. Washington, DC: US Navy. Accessed on 7 June 2013, http://greenfleet.dodlive.mil/files/2011/08/U.S.-Navy-Arctic-Environmental-Assessment.pdf.
White House. 2002. *The National Security Strategy of the United States of America* White House. Washington, DC: Government of the United States of America. Accessed on 4 June 2013, http://www.state.gov/documents/organization/63562.pdf.
White House. 2006. *The National Security Strategy of the United States of America*. White House. Washington, DC: Government of the United States of America. Accessed on 15 September 2013, http://georgewbush-whitehouse.archives.gov/nsc/nss/2006/index.html.
White House. 2010. *The National Security Strategy of the United States of America*. White House. Washington, DC: Government of the United States of America. Accessed on 20 May 2017, http://nssarchive.us/NSSR/2010.pdf.
Wikipedia. 2017a. "Chairman of the Joint Chiefs of Staff." Online: Wikipedia. Accessed on 15 January 2017, https://en.wikipedia.org/wiki/Chairman_of_the_Joint_Chiefs_of_Staff.
Wikipedia. 2017b. "Chief of Naval Operations." Online: Wikipedia. Accessed on 15 January 2017, https://en.wikipedia.org/wiki/Chief_of_Naval_Operations.
Wikipedia. 2017e. "List of United States Secretaries of Defense by time in office." Online: Wikipedia. Accessed on 15 January 2017, https://en.wikipedia.org/wiki/List_of_United_States_Secretaries_of_Defense_by_time_in_office.
Wikipedia. 2017g. "United States Secretary of the Navy." Online: Wikipedia. Accessed on 15 January 2017, https://en.wikipedia.org/wiki/United_States_Secretary_of_the_Navy.

Chapter 7
Climate Securitisation in the US and Australia: Discussion

Abstract This chapter presents a high-level case study comparison of climate security responses by the Australian and US political-military sector between 2003 and 2013. The findings are based on mixed content analysis of more than 3,500 open sourced speech-acts, defence policy documents, military doctrine articles and other such publications emanating from the political, civilian and military leadership of the Australian and US departments of defence. The results indicate that climate change was not framed as a major security issue under the conservative leadership of either Prime Minister Howard or President Bush. Correspondingly, it failed to register as a major issue for the Australian and US militaries during this period. This changed from 2007, when climate change emerged as a centrepiece political issue under the centre-left governments of Kevin Rudd and later, Julia Gillard, in Australia. An equivalent transition was witnessed in the US on the ascension of the Obama Administration in 2009. In response to the more activist political climate, both the Australian and US militaries became progressively engaged. The Australian military however, developed a muted approach to climate change characterised by few programme of consequence and few mentions in the speeches of its senior military leadership. The US Navy, by contrast, developed several high-level mitigation and adaptation programme and it formed a small but notable presence in the speech-acts of its civilian and military leadership.

Keywords Climate security · Case study · Climate change · US military · ADF · Australian military US Department of Defense · Climate securitization · Climate politics

M.D. Thomas, *The Securitization of Climate Change: Australian and United States' Military Responses (2003–2013)*, The Anthropocene: Politik—Economics—Society—Science 10, DOI 10.1007/978-3-319-49658-0_7

7.1 Introduction

This chapter presents a high-level comparative analysis of the proceeding case studies (Chaps. 5 and 6).[1] The chapter opens with a generalised account of the influence of political ideology on military climate change policy in Australia and the US between 2003 and 2013. Attention then shifts to examining the case study findings according to periods of centre-right and then centre-left governments in both the US and Australia. As it happened, these periods largely coincided. In Australia, centre-right leadership under Prime Minister John Howard occurred from 1996 until December 2007. From December 2007 until September 2013 Australia was governed by the centre-left Australian Labor Party. In the US, centre-right leadership under President George W. Bush occurred from 20 January 2001 through until 20 January 2009. US Democrat President Barack Obama was sworn in on 20 January 2009 and was elected for a second term (20 January 2013) which concluded in January 2017. Throughout this chapter, case study findings are contextualised according to the two analytical frameworks, the Copenhagen and Paris Schools. Where applicable, a direct response to the original research lines of inquiry (*contextual framing*, *temporal framing*, response *measures*) are also made.

7.2 General Finding on Political Influence

This research was originally conceived with the idea of exclusively examining military sector approaches to climate change as a security issue. Early on, however, it became apparent that this question could only be fully answered by analysing the *political* context within which climate change was being framed. Copenhagen School securitisation theory adopts this outlook, noting '[t]he military agenda … does not operate in isolation. The entire interplay of military capabilities between states is deeply conditioned by political relations' (Buzan et al. 1998: 52). This proved a valuable approach, as the Australian and US militaries were generally found by this research to act in *response to* and *following the direction of* the (incumbent) political elite. Furthermore, when direction to act on climate change was forthcoming, no instances were identified where either military sought to frame or respond to climate change above or beyond the conditions set by their political masters. In this regard, both militaries appeared finely tuned to the contours and sensitivities surrounding the politics of climate change—acutely aware that the major political parties tended toward contrasting outlooks regarding both the science of climate change, its priority as a public policy issue and the method of policy

[1]In this chapter, 'US military' refers to those military elements examined in the US case study only. This included the Secretary of Defence, Joint Chiefs of Staff and elements of the US Navy. The term US military is therefore used as a generic phrase and it is not intended to infer findings for the entire US military organisation or US Department of Defense.

response. This general finding tended to downplay Paris School ideas on security monoliths (the 'dispositifs'), deliberately positioning themselves to advantage via various institutional and governance mechanisms to control state security agendas (Bigo et al. 2008). Consequently—and somewhat less dramatically than the Paris School might suggest—this research pointed more towards Samuel Huntington's normal theory of political-military relations, in which the military subordinated itself to political direction.[2] Nevertheless, Olaf Corry's (Paris School) ideas on 'riskification' had some resonance, particularly the idea that the militaries adopted risk-based programme, policies and procedures and framed it as an issue to be dealt over the long-term.

Accordingly, this research also identified a general correlation between political ideology and the military's response to climate change. Centre-right governments (US Republican and Australian Liberal-National Coalition) that tended to downplay climate change (compared with their political opposition) resulted in a lack of outward (public policy) activity and seeming minimal internal attention by the military on the subject. Centre-left governments, by contrast, that presented a more proactive suite of climate change policy responses also witnessed an increased military response to climate change. This finding, however, must be placed in context, particularly by the timing of political change (Australia, 2007 and in the US, late 2008/early 2009) which coincided with increased scientific certainty regarding climate change as a consequence of the IPCC AR4 and also the elevation of climate change as a more mainstream social, economic and security concern. The correlation between political ideology and military climate change response is identified as an area of future research.

Lastly, such generalised findings identified above should not be misconstrued with the idea of *political interference* in military decision-making regarding their climate policies. Rather, this chapter concludes that, for the ADF at least, it was most likely a deliberate and active *avoidance* by the military leadership not to become entangled in what became a highly politicised issue. Either way, at the very heart of this finding existed the challenge of an avowedly apolitical institution responding to what emerged in the Australian context as a politically partisan security issue.

7.3 Climate Securitisation in the Centre-Right Governments of US and Australia

Between 2003 and 2008 President Bush and Prime Minister Howard conceived of climate change in terms of energy security, technological solutions to reducing GHG emissions (particularly those that advanced their economies) and a

[2]For a discussion of Samuel Huntington's work on normal theory of political-military relations, see Eliot A. Cohen. *Supreme Command: Soldiers, Statesmen, and Leadership in Wartime*. New York: Free Press, 2002, and for an alternate view of his elite model see Morris Janowitz. *The Professional Soldier, a Social and Political Portrait*. California: Free Press, 1960.

repudiation of international agreements (Kyoto Protocol). Although Bush signalled his intention to examine the effects of climate change on the military (primarily through provisions contained in the 2008 *National Defense Authorization Act for Fiscal Year 2008*), he only rarely framed climate change as being a security issue. For instance, out of more than a hundred speeches and media conferences delivered by President Bush and examined by this research, it was found he only framed climate change as a security issue in the context of energy security on three occasions. This outcome was largely identical to John Howard, who also singularly framed climate change (on nine occasions) as a security issue in terms of its relationship with energy. Significantly, climate change was never framed as being an issue of national or international security by either leader.

The language used by Bush and Howard to frame climate change (either in a securitised or non-securitised manner) also tended towards more benign depictions than the grim outlook used by either President Obama or Prime Ministers Rudd and Gillard. Indeed, of the documents examined, this research found that Bush *never* framed climate change as a threat or crisis (in either securitised or non-securitised contexts). This was similar to John Howard who, in almost a hundred speeches and media released by him (and examined by this research), framed climate change as a threat on just two occasions (even these came very late in his term, when it was arguably done as a means to counter the ascent of his political opposition, led by Kevin Rudd).

In terms of strategic policy, hard-power security issues dominated the conservative era. Climate change was an insignificant or, in the case of Australia, a non-existent factor in these documents. While the 2002 US *National Security Strategy* briefly mentioned climate change, it was done under the heading of 'Economic Growth'. By the 2006 US NSS, climate change had disappeared from the US strategic narrative altogether. In Australia's major strategic policy documents published under John Howard, climate change was not referenced once in either the 2000 White Paper or the 2003, 2005 and 2007 Defence Update(s).

Moreover, although climate change emerged as a *political* issue under the conservative governments of Bush and Howard, it did not seem to be fully consolidated as a *security* issue. More accurately, it was never consolidated as a national security issue by the governments of these countries because it was never viewed as such by the two primary securitising actors in Messrs Howard and Bush. This research found climate change was predominately framed by both Bush and Howard as a non-securitised issue with varying degrees of focus on energy, economics or the establishment (or otherwise) of an international agreement to limit emissions. More than anything else, however, both leaders framed climate change in terms of its relationship around energy. This was reflected in their policy agenda (s) but also in their speeches. For instance, it was found that in the speeches in which President Bush framed climate change (and which were coded by this research), almost 70% of all coded speeches were done so in terms of its relationship with *energy*. For Howard, almost 40% of the speeches coded framed climate change in terms of energy. These findings broadly matched other scholarly works that have examined this subject (Floyd 2010; Brauch 2011b). Lastly, no speeches were identified where either Bush or Howard linked climate change as an issue concerning their military forces (though Bush later instigated the issue by issuing directives in 2008).

Accordingly, the military's response to climate change, based on detailed analysis of speech-acts, under conservative administrations (2003–2008) was marginal (ADF) to limited (US). That is, it largely reflected the outlook and position held by the government of the day where, in the words of Australian Deputy Secretary of (Defence) Strategy, the military was simply being 'consistent with Government policy' (Peter Jennings quoted in Alexander 2011: 31). In this context, both the US and Australian militaries were intensely focused on the Global War on Terror. For the most part (until at least 2008) climate change simply did not register as an issue of any consequence for US or Australian militaries.

In the Australian military, although there was early evidence (from 2006) that the broader military-intelligence community were already conducting internal assessments on climate change, understandings within the ADF appeared mixed. In late 2006 the Departmental Secretary (Mr. Ric Smith) was adamant his Department had undertaken no analysis of the issue (Budget Estimates 2006: 70). However, in 2007 the then Deputy Secretary of Strategy (Mr. Michael Pezzullo), discussed it at length but ultimately conceded it to be a 'non-traditional' security issue that would 'not affect the force … being developed over the current 10 year period' and would only impact the ADF over the 'long-run' of '40–50 year period' (Budget Estimates 2007a: 101–103). Other ADF documents, on the other hand, and admittedly at lower bureaucratic levels, framed climate change as a serious and broad ranging threat capable of degrading ADF capability in the near, medium and longer terms (Reich et al. 2006: 240). The influence of such documents, however, was marginal. In one instance the ADF published a policy outlook document (*Scan 2025*) describing the threat of climate change and its possible impacts on Defence, but then caveated the publication, saying, 'the findings and views expressed in this report … are not to be taken as the official position of the Department' (Reich et al. 2006: cover page). During this period, sources within the Department of Defence also noted the ADF's trepidation surrounding the domestic politics of climate change.

Aversion to climate change on political grounds was publicly confirmed by the Deputy Secretary Strategy in 2007, when he noted during Senate Estimates that Defence would 'steer well clear' of the 'dramatic and active [political] debate' surrounding the *rate* of climate change (Budget Estimates 2007a: 102). Arguably, as an institution designed to assess strategic risk, this is where the ADF needed to locate its thinking. In this context, climate change came to be initially framed as just one of many non-traditional security issues on a par with 'water resources, resource depletion … pandemic threats, HIV-AIDS and the rest' (Michael Pezzullo quoted in Commonwealth of Australia 2007a: 102). Gradually, the topic emerged from 2007 onwards in the speeches of the Australian military elite (such has the Chief of Defence Force and Service Chiefs) as an adjunct and minor security issue. Very little attention, however, was given to understanding the broader implications concerning its impact on infrastructure, workforce health, energy or the possibility of future regulation. The apparent lack of interest in climate change by the ADF hierarchy even pointed towards a 'strategic lag' in which the political establishment was operating well in advance of the military in announcing climate change as a national security threat. Given the lack of connection made by the government regarding climate change, security and the military—and that the ADF was heavily occupied in Iraq and Afghanistan—this was somewhat understandable. But from the perspective of climate change gauged from the scientific community, it was a questionable strategic outlook.

In the US military—and despite the very active debate occurring in Congress on the subject—no single reference to climate change was found in more than 380 speeches given by Secretary of Defence Donald Rumsfeld and US Navy Secretaries Gordon England and Donald Winter between 2003 and 2009. With some exceptions (discussed below), climate change was not as prominent in the major national strategic policy documents published during the Bush era, as it would become later.[3] Although this research covered a much smaller percentage of the US military (as compared against the Australian military), a similar lack of attention (by the military) under the auspices of a conservative government was evident. A telling passage on the priority of environmental issues more generally was delivered by Donald Rumsfeld when he sought to clarify environmental statutes that restricted or limited troop readiness against the Global War on Terror. The imputation was clear: environmental statues would be diluted (at best) or removed (preferred) lest they interfere with US troop readiness. These findings resonated strongly with other scholarly works that have addressed this matter (notably Floyd 2010).

But notable actions on climate security took place late in the Bush presidency. In 2008 President Bush signed *National Defense Authorization Act for Fiscal Year 2008* (NDAA) and *National Security Presidential Directive 66/Homeland Security Directive 25* that directed the US military to examine the national security implications of climate change—the former having a significant impact on DoD efforts to address climate change through until at least 2010. These directives marked a point of divergence between Australian and US military climate policy response. In many respects, that issuance appeared somewhat incongruent, particularly when contextualised against the limited climate policies and obfuscation of the Bush era and sceptical attitudes of US Republican leaders (Revkin 2005; Council on Foreign Relations 2007; Goldenberg 2009). It raised the question: Why did a Republican Administration direct the US military to seriously consider climate change when it had largely ignored it in other areas? Three main factors are identified here. The first was that despite the partisan nature of US climate change politics, that country still had a far deeper, richer and more sustained climate security discourse than had occurred in Australia. This was evident across a number of areas in US society. In politics, for example, the US Congress had held multiple sessions explicitly examining climate security matters (refer to Chap. 3). These often involved the

[3]This assessment excluded the 2002 NSS which *did* include climate change but remained outside the time frame considered by the case study.

call-up of dozens of experts from the scientific and military communities to testify at such proceedings.[4] Notably, a bill was even introduced on 19 April 2007 by Senator Edward Markey to address the risks posed by climate security, the Global Climate Change Security Oversight Act, but was not enacted (Climate Security Bill 2007; Tauberer 2015). These debates (occurring as they had since the 1980s) pointed towards a far deeper institutional understanding of climate change, influenced as it always was by the spectre of 'energy security' but, increasingly, by the potential opportunities created by the opening of the Arctic. Save for some limited probing from minority parties during Senate estimates, Australia never appeared to have had such an equivalent interrogative political discourse on climate security.

The US military (or at least affiliated people and institutions) also had a greater presence in climate security discourse than the Australian military. This was evident not just through congressional testimony by military personnel but also in its early appearance in National Security Strategy (2002), National Defence Strategy (2008) and—most prominently—through the persuasive and powerful think-tank contributions of many former senior US military officers. This latter cohort of retired military officers—and the US think-tanks more broadly—were particularly influential in US climate security discourse from 2007 and in framing climate change as a national security 'threat multiplier'. Once again, this did not occur in Australia where, with the exception of former Chief of Defence Force Admiral Chris Barrie, not one retired high-ranking military officer was identified by this research as a vocal participant in the emerging climate security discourse.[5]

The second possible factor to have prompted President Bush to issue the climate security directives was one of *political necessity*, in which it was important to be seen to do something about climate change which—during a presidential election year—had emerged as a central issue. This possibility related to the Republican Party taking action on climate change in an area of perceived policy strength, namely national security. Ironically, the very fact that the Republican Party invoked US national security in relation to climate change likely normalised the US military's involvement to some degree. It was, in some respects akin to the idea that only Nixon could 'go to China'. (This situation might have been vastly different if the Democratic Party had instigated US military response to climate change). In some respects, the two Bush initiatives identified by this research were also not dissimilar to the raft of (albeit non-security-related) climate change policy measures proposed by Howard throughout 2007 that—pressed by his political opponent Kevin Rudd—seemingly belied previously held (minimalist) views on the subject.

The third possible reason is more straightforward and suggests that the US military had (by early 2008) been sufficiently convinced by the science of climate

[4]Although this book did not examine wider US institutions, many featured prominently in US climate literature and debates (e.g., US State Department, EPA, NASA, NOAA).

[5]For Admiral Barrie contributions see the following sites at: https://www.youtube.com/watch?v=opw0OETYOo4 (14 July 2016), http://www.abc.net.au/news/2013-09-12/barrie—defence-force-and-climate-change/4953150 (14 July 2016), and https://cpd.org.au/2015/06/widespread-coverage-of-cpds-new-report-on-australias-climate-security-challenge-2/ (14 July 2016).

change—and the risks it posed to national security—that it successfully influenced the Bush Administration to take action.

Notwithstanding these efforts, the US military response (and the Australian military response for that matter) should not be overstated during the centre-right era of Bush and Howard. For, beyond these measures, there remained a distinct lack of military attention on climate change. Several factors were identified and, although this research did not focus on them per se, their importance should not be downplayed. Most prominent was that both the US and Australian militaries were heavily engaged in fighting two major wars (Iraq and Afghanistan) and were busy in a raft of other regional operations, training and various institutional reforms ('transformation' in the US military under Secretary Rumsfeld and initiatives such as 'Hardened Network Army' in the ADF). Iraq, in particular, occupied the majority of US strategic attention, especially during 2007, when US military deaths passed more than a hundred per month and the so-called surge of some 30,000 US troops took place. Leaders from both nations, military and political alike, often referred to the point that their countries were at war.

Not surprisingly, this dominated the security agenda of both nations during the conservative period and is modestly shown by this research through comparing the word count of terms such as *terrorism* to *climate change* in the speeches of the political and military elite. In more than 1,600 strategic documents examined in the US case study, 185 direct references to *climate change* were found against more than 4,300 for *terrorism*. As was noted by Secretary Gates in his memoirs as Secretary of Defence: 'My highest priority was to turn the situation around in Iraq' (Gates 2014: 25). The Australian situation was similarly summed up by the Minister for Defence, Brendan Nelson in 2007 when he remarked '[t]errorism … is the defining issue of my generation and that of my children' (Nelson 2007b).

Related to this observation was that climate change was also not yet really conceived of as being an issue requiring military involvement. This perspective remained amongst senior Australian military officials, who variously positioned the ADF as the 'agency of last resort' and identified climate change as a long-term issue which would not impact current force structures or capability (Alexander 2011: 30). Similar sentiments were to be found in the US, where Republican senators chided the (Democrat) Secretary of Navy for his energy measures as not focusing on core missions and capabilities and as an agenda to 'green' the military.

A final observation that might be made in relation to military climate response under centre-right government pertains to the interlinked ideas of energy security and climate change. The political focus on energy security witnessed a corresponding level of interest by the military. Arguably, however, they were for very different reasons. While the political sector focused on the generalities of energy security from a national perspective, the debates and discussion occurring within military circles were much more focused. Pointedly, although climate change and energy were conflated in the national-political debates, they were *not* a feature of the *energy* debates occurring within the militaries until at least 2008. Far from mitigating greenhouse gas emissions for the sake of contributing toward national emissions reductions, the military was rather focused on reducing energy costs, reducing the logistical overheads of fuel (re)supply lines, improving the endurance and sustainability of field deployable capabilities and, of particular focus, on saving lives of its soldiers, sailors and airmen through reducing logistic footprints and minimising requirement for convoy security. Dozens of high-level, wide-ranging reports from across many different parts of the US Department of Defense were undertaken with a view to 'solving the DoDs energy problem' (Defense Science Board Task Force 2008: 7). Climate change was not a feature of these discussions until at least 2008. During the same period, albeit on a much smaller scale, similar interest in energy matters (*absent of any discussion on climate change*) were reflected within the Australian military.

From 2007 onwards, however, climate change gradually emerged as a mainstream issue within the Australian and US militaries. This can be attributed to three main reasons. First—and as already outlined—was the election of centre-left federal governments in late 2007 (Australia Labor Party) and 2008 (US Democrat Party). Both Prime Minister Rudd and President Obama campaigned on progressive climate change platforms, vowing to take action, and framed climate change as a serious security threat of direct applicability to the military. The election of centre-left governments with a mandate for climate action therefore set the necessary political conditions for far more aggressive climate change policies to which the various departments of state (including the military) were expected to contribute. In the US, the Democrats were able to leverage off, and build upon, President Bush's climate security directives (particularly 2008 NDAA); consolidating climate change as a security issue of some importance for its military.

The second reason was the consolidation of climate change as an issue of international significance as a consequence of mounting scientific, social and economic consequences. These aspects were emphasised at the beginning of this book, mainly highlighted through globally prominent reports—notably IPCC AR4, *Stern Review* and, in Australia, the *Garnaut Review*—as well as its placement on the agenda at the highest political levels, including the G8, G20 and United Nations General Assembly and the Security Council.

The third reason was the growing public awareness and 'calls-to-act'. Influential in this sense were mainstream efforts resulting from movies such as Al Gore's *Inconvenient Truth*, as well as literature from academia, think-tanks, mainstream media, the intelligence community and congressional hearings; all establishing the case of climate change as a legitimate social, economic and security concern. Hurricane Katrina and other natural disasters also contributed to a sense that the climate was changing, justifying a requirement to act. Surveys and polling conducted around this period indicated elevated public concern for climate change.

Out of these three reasons, however, it was the political dimension that probably had most influence and impact on instigating a change in the military's response to climate change. More precisely, the election of centre-left governments created the necessary political conditions for each military to become more engaged in national climate discourses, whether through top-down legislation/regulation or via strategic guidance or from bottom-up initiatives that came from within the military bureaucracy itself. This assessment was more relevant for Australia and arguably less noticeable in the US where President Bush had already instigated some climate security initiatives. The following sections now turn to analysing military climate response, through the prism of securitisation theory, under centre-left governments.

7.4 Climate Securitisation in the ADF Under a Centre-Left Government

The election of a centre-left government(s) in Australia between late 2007 and 2013 resulted in a stark contrast of climate policy presented under John Howard. Whereas Howard opposed Kyoto, Kevin Rudd and Julia Gillard supported it. While Howard (at the last minute) was reluctantly manoeuvred into supporting an emissions trading scheme, Rudd initially embraced it, and a variation was subsequently legislated by Gillard. Where Howard did not view climate change as an urgent issue, Rudd and Gillard spoke frequently as though it was an immediate issue requiring swift, resolute action. To bolster their cause, Rudd and Gillard constantly described climate change as a social, moral, economic and security threat in which Australia was 'on the front line to suffer' (Rudd 2008f). Howard—except on the rarest occasions—never used such strong language. Whereas Howard was wary of 'individual reports', Rudd and Gillard regularly evoked 'the overwhelming global scientific evidence' (Rudd 2009c) amassed by the world's various scientific institutions and academies. Rudd and Gillard also cast themselves as champions of renewable energy, expanding on Howard policies and implementing a raft of new initiatives.

From an economic perspective, Rudd argued that the 'cost of inaction on climate change is far greater than the cost of action' and that 'failure to engage with the global community on climate change would exclude us from the chance to shape the global response in ways consistent with our national interest' (Rudd 2008a). This was vastly different from Howard, but it aligned with Rudd's broader strategic purpose of Australia as an activist Middle-Power in which he believed 'those that share the benefits of these systems must also share the responsibilities of supporting and enhancing them' (Rudd 2008g). Summarily, Rudd cemented climate change as a core *political* issue.

Having achieved this, Rudd became the first Australian political leader in government to present climate change as a major *security threat* to the Australian people. Again, the shift in rhetoric from Howard was dramatic. Rudd spoke of climate change in catastrophic and widespread terms that reached crescendo in late 2009, whereupon he was under significant pressure to help craft an international agreement at Copenhagen and ensure passage of his signature Carbon Pollution Reduction Scheme (CPRS) legislation through the Australian parliament. Inarguably, between his election in late 2007 and departure in mid-2010, Rudd used rhetoric and pursued actions that attempted to securitise climate change and render it—in the words of the Paris School—'governable'. While much commentary has been made of Rudd's political motives, what relationship and impact did all this have on the Australian military? Also, what does securitisation theory offer by way of an explanation?

Recalling the Copenhagen School framework, Rudd arguably sought to securitise climate change so that—in the words of securitisation theory—it became 'above' politics. By trying to raise climate change as an issue above politics, Rudd sought to neutralise sceptics, confound his political opposition and consolidate his centrepiece political agenda in the minds of the public (i.e., the audience) as being so crucial as to be a matter of national security. Undoubtedly, he also had a firm eye on widening the discussion (to his favour), such that he would attract political support for his climate change legislation. Enlisting the security institutions into climate security discourse for this cause might not have been beyond comprehension. In looking at the whole, securitisation was arguably born primarily out of his political considerations.

Thus, Rudd—as a legitimate securitising actor—became energetic in attempts to construct climate change as a security threat. From London to the Pacific to the United Nations, Rudd touted: 'Climate change is not just an environmental, economic and moral challenge. It is also a security challenge' (Rudd 2008b).

To the Pacific Island Forum, Rudd labelled it a 'matter of national survival' (2009f). To an Australian audience he argued that failure to act on climate change would have 'severe' consequences and 'resulting catastrophic events' (Rudd 2008g). The devastation portrayed was widespread; nine distinct categories of security were identified by this research in which Rudd framed climate change as impacting. Repeatedly, climate change was framed as a severe threat to the Australian economy, its society, and the environment and future generations. On occasion it was cast as a planetary-wide threat. His language intensified nearer the political culmination of his centrepiece climate legislation and the Copenhagen conference, 'the latest scientific research on climate change confirms our worst fears. Climate change is happening faster than we previously thought' (Rudd 2009d). In totality, his rhetoric approximated powerfully to the Copenhagen qualifier that to become securitised an issue must be presented as an existential threat.

Pursuing securitisation, Rudd called on climate change to become a legitimate issue within the country's key national security institutions and into 'policy and analysis process' (Rudd 2008g). In the language of the Paris School, Rudd now actively sought to render climate change governable as a security issue within the military-security bureaucracy. Thus, wherever he spoke of national security, climate change was prominent. Under Rudd and then Gillard, every major national security publication between 2008 and 2013 (2009 and 2013 Defence White Papers and the 2008 and 2013 National Security Strategies) described the risks posed by climate change. Linking climate change to expand Australia's HADR capabilities, it appeared to influence ADF capability acquisition through the purchase of naval ships and military training activities. As a minimum, senior Ministers in the Rudd government responsible for these initiatives at least spoke as though climate change was an influencing factor, even though the original purpose was really conceived in far broader terms by the ADF (e.g., the development of what the then CDF Angus Houston described as a 'balanced' force (Houston 2007: 60 and 61)). Nevertheless, the situation was strengthened by circumstance in which a succession of large scale domestic natural disasters (Victorian bushfires (2009), Cyclone Yasi (2011) and the Queensland floods (2010–2011)) witnessed unprecedented ADF responses.

In this context, the active mainstreaming of climate change within the national security institutions, as well as the intense focus by the Prime Minister, resulted in a situation where the ADF could no longer be exceptional to debates—and response measures—surrounding climate change. While the ADF had entered at the margins as early as 2006, by 2008 this was simply no longer tenable. Thus, from 2008 the Australian military (under successive governments of Rudd and then Gillard) initiated a broad response to climate change. By extension, it became partially engaged in the accompanying debates surrounding climate change and, for a time, had the appearance of a de facto securitising actor.

However, the ADF response to climate change was a marginal one. The senior leadership of the ADF barely spoke on climate change and never at any great length. This research analysed ninety-three speeches by the three Service Chiefs (Army, Navy and Air Force) between 2003 and 2013. Of the ninety-three speeches, just six contained any reference (brief as they were) to climate change. No organisational-wide mitigation or adaptation programme was implemented. Unlike in the UK and US, no senior military officer was appointed to oversee a climate response or act as a focal point. While mid-ranking military officers wrote on the various climate change threats, it was also acknowledged as being too intangible to have much policy relevance for the ADF (White 2008: 12). Energy security and its attendant issues once again became prominent, but even these arguably appeared piecemeal and lacking in strategic purpose. Contributions by the Australian military to establish emissions reductions targets (mitigation) were raised but then dropped. While one programme, *Combat Climate Change* claimed the ADF would do 'all it could' to reduce emissions (Defence 2008a); large and significant parts of the organisation (i.e., 'operational fuels' comprising about 50 of ADF emissions) were excluded (Commonwealth of Australia 2013a). Tellingly, ADF greenhouse gas emissions increased during the life of the programme.

Although the term *climate change* crept into doctrine, the penetration and extent appeared minimal. This was also the case with the ADF's major environmental and energy policies where—amongst literally dozens of initiatives—climate change was only rarely invoked as a pretext for action. Some evidence was also found of attempts to avoid using the term climate change, using instead 'climate variability'. Furthermore, although there emerged a deeper bureaucratisation ('mainstreaming') of climate change within and across the Australian military; the major statements (e.g., declaratory policy such as the Defence White Papers) emanated from the highest national strategic-levels (endorsed by the Minister of Defence and under the watchful sanction of the Prime Minister). Thus, while these high-level statements are part-military, they can be regarded as mostly political productions that reflected the agenda of the centre-left government more than necessarily that of the military.

Elsewhere, climate change was bureaucratised at the operational level in a largely piecemeal and uncoordinated fashion. There appeared no overarching directive; programme emerged across different areas but with no linking narrative or consistency of purpose. Where they did appear, they were ostensibly about mitigating long-term risks. Most prominent was a $2 million report on the risk of sea-level rise and inundation to Defence bases commissioned by Defence Support Group (DSG). However, because it was classified, the outcomes were never released to the public. No reference to the report (let alone climate change) was found in the 2011 *Australian Defence Force Posture Review*. DSG also oversaw *Combat Climate Change*, a reasonably prominent but narrowly conceived energy efficiency programme that literally targeted *individuals* on how to reduce their emissions footprint. An organisational-wide strategy was absent. The Vice Chief of Defence Group initiated another programme (*Global Change and Energy Sustainability Initiative*) examining the impact on ADF preparedness and readiness; this pointed towards organisational-level adaptation to the risks of climate change and broader levels of engagement with industry and academia. Defence Infrastructure also oversaw some medium-scale infrastructure initiatives, including a joint government-industry wave energy development at Fleet Base West. A suite of minor energy efficiency measures were also introduced that related more to cost saving than climate change.

More subtle, and not as visible as those just mentioned, the ADF also began to adapt its organisation to meet governance requirements of climate change. In 2011, Deputy Secretary Strategy Peter Jennings singled out discrete areas within the military bureaucracy to deal with what he saw as three key aspects impacting Defence: rising sea-levels, greenhouse gas emissions and ADF energy requirements. He particularly cited Strategic Policy Division as 'the lead' and that the strategic impacts of climate change would now be considered part of 'annual Defence planning' (Jennings quoted in Alexander 2011: 30 and 31). Some minor changes were also made to the Defence procurement guidelines that identified 'sustainable procurement' and 'energy efficiency' as procurement selection factors (Defence 2014: 1.2-2). Lastly, Defence also became a member of an Inter Departmental Committee on Climate Change, though its level of participation was cast in doubt when the final report released by the committee failed to mention any actual involvement by Defence. These initiatives represented the core ADF response to climate change; decidedly they were not policies that securitised climate change but rather piecemeal bureaucratic responses geared to reduce strategic risk, improve operating efficiency and save money. Arguably, while many initiatives appeared credible as stand-alone projects, the lack of an overall strategy tying them together gave the impression that such a policy was *deliberately* avoided. That is to say, for an organisation that prided itself on being strategy-led and one that had literally dozens of other strategies and policies of every kind, it is incredulous to think that climate change was somehow missed.

Furthermore, the ADF never exceeded political expectation that it would progress any major climate change initiatives. While one scholar (Zhang 2009) posed the idea that the Government might have deliberately talked-up climate change threats to create strategic-military opportunities, reality suggested otherwise. Just as climate change emerged as a national security threat of any significance, the Defence budget contracted to the lowest proportion of GDP since pre-World War Two (Thomson 2013: vi). Additionally, although climate change became established as an issue within discreet areas of the ADF, the broader picture suggested minor take-up. From a force-structure and capability perspective, when Rudd and his Ministers talked up the prospect of climate change increasing HADR and stability missions, the ADF readily added the requirement to create a 'balanced force … giving us the option to build a more potent force structure relatively quickly' (Houston 2010). Warfighting and its attendant capabilities that were geared to 'defeat the nation's enemies' remained priority (Morrison 2013: 6–8). Not surprisingly, climate change was never declared to be a direct driver of force-structure or capability. Neither, though this time perhaps surprisingly, was it included as a factor in the 2012 *Force Posture Review* (FPR) that (amongst other issues) examined future defence estate requirements. Given that the ADF already knew in 2011 from a first-pass study that three of its major bases (RAAF Base Townsville, HMAS Cairns and HMAS Stirling) might be subject to 'inundation by storm surges occurring within the context of rising sea levels from climate change', its absence from the FPR was, on the surface at least, questionable (Commonwealth of Australia 2012a: 261).

So, on the one hand, this reflected the Paris School framework that emphasises how security institutions create their own bureaucracies as a means to govern security issues (consisting of new policies, roadmaps, doctrine and so on). But on the other hand, the tentative ADF response—well behind the political sphere and industry response—stood in contrast to a central aspect of the Paris School, which portrays an almost Orwellian-like security apparatus exerting disproportionate influence over the political domain. If anything, on the evidence examined by this research, it was the political direction at the instigation of the centre-left Labor Government from 2007 onwards that prompted a military response. Equally, when it became clear that climate change was a partisan issue between the two major federal parties, the military began to limit its involvement and exposure to the issue of climate change.

A final, somewhat telling anecdote of the ADF's ambivalence to engage with the issue of climate change occurred between the period of Rudd's securitisation apogee in late 2009 and his removal as Prime Minister in mid-2010. Just as the pace of climate securitisation gathered, the ADF declared in its 2009 end-of-year annual review that it would set climate change emissions targets, promote an internal discussion paper on climate change impacts and develop the 'Defence Climate Change and Sustainable Development Strategy' for publication (Defence 2009a: 186). But just as the ADF made this announcement, political momentum changed. The Liberal Party of Australia, then in opposition, elected a leader (Tony Abbott) who fervently rejected Labor's climate agenda. When the Copenhagen conference failed to deliver a unified and comprehensive international climate agreement and Rudd's CPRS failed to gain passage through the Senate, climate change was consolidated as a particularly bitter and divisive political issue. While Rudd sought to minimise reference to it, his political opposition capitalised on his strategic back-down (Macintosh et al. 2010). When news broke on 27 April 2010 that Rudd had abandoned the scheme, a subsequent News poll found Labor had lost around one million supporters in a fortnight (Chubb 2014a). It was a contributing factor for Rudd being deposed as Prime Minister in June 2010 and replaced by Julia Gillard (Aulich 2010).

Hyper attentive to the partisan and polarised political debates, the ADF then reneged on its pledge to set emissions targets and publish a climate strategy. From what this research found, both remained unpublished and unspoken of by anyone within the ADF. Thereafter, and with the exception of the *Global Change and Energy Sustainability Initiative* and the piecemeal initiatives identified above, climate change remained a low priority for Defence (no other major climate change programme were initiated by Defence from 2010–11). Correspondingly, from 2011 onwards, this research found climate change was scarcely being framed as a security *challenge* or *threat* by Australia's political and military elite (see Fig. 5.4).

From the perspective of securitisation theory, and in relation to climate change, the Australian military thus appeared—at first glance—as a reluctant securitising actor. Arguably, this was atypical for the military where, in traditional security affairs, it was (and remains) a proactive if not central, securitising actor. And so, apart from being 'busy elsewhere' and not the lead agency to respond to climate change, what other factors were at play that fed ADF reluctance? Also, on the evidence available, was the ADF really a securitising actor at all, reluctant or otherwise?

To answer this question, a deeper assessment of Rudd's securitisation agenda is required. Arguably, although Rudd pursued securitisation of climate change, he never truly sought to pursue it to the lengths defined by Copenhagen's framework. Moreover, Rudd never sought 'emergency measures' that required actors to move 'beyond rules'. This was true of his political measures—he baulked at the prospect of a double-dissolution election to seek passage of his climate legislation—and it was also true of his securitisation measures. For instance, although Rudd spoke often on the magnitude of the security threats, he simultaneously reassured people that they would only manifest 'over the long run' (2008g). The 2009 *Defence White Paper* as a case in point referred to decades-long timescales. In the 2008 *National Security Statement* Rudd (2008h) doubled-down, '[o]ver the long-term, climate change represents a most fundamental national security challenge, over the long run'. Thus, a distinction became evident from this research between Rudd's politicisation time frame (urgent) and his securitisation time frame (long-term). The long time frame of securitisation further rendered the idea of taking emergency measures (in a securitisation sense) a nonsensical one and of marginal interest to the Australian military that dealt foremost with short and near-term time-horizons. As a result, climate change was partly framed by the military as an issue largely for capability development—an area of equally long duration.

Relevant to this point was the idea that climate change cannot be easily reconciled under Copenhagen securitisation theory. Apart from the decades-long time-scales, few genuinely plausible options existed as to what may have constituted emergency measures, particularly where military involvement was concerned. This equally applied to the idea of getting 'actors' to move 'beyond rules'. What actors, what rules? Were *green* (developed) nations to adopt punitive measures against *dirty* (undeveloped) nations? What role would the military play in this scenario? As absurd as it is to suggest, might the military be used to strike industrial greenhouse gas emissions sites of dirty nations? Might the military be used as an instrument for geo-engineering purposes (either unilaterally or multilaterally) such as releasing atmospheric particulates to reduce the amount of incoming solar radiation? Such a case was never truly made by Rudd. Indeed, it never could have been, since every nation was (and is) implicated. As remarked by former UK Foreign Secretary Margaret Beckett, 'we are all our own enemies' (Beckett 2007). Moreover, securitisation, as understood in traditional geopolitics, has typically witnessed military involvement in some form or another. But its role is far less certain in climate change; the most activist offerings stretch the realms of credibility.

Accordingly, while Rudd spoke in generalities of climate security threats, he never invoked the military as a solution to those threats. Rather, the Australian military was partly framed by some as a type of 'first responder' in the event of major climate disasters and the likelihood of regional stability operations. Both of these also fitted comfortably with Labor's enduring strategic narrative that has historically placed primacy on the Australian continent and the immediate region (Evans 2005). Nested within this strategic outlook was an increased emphasis by the ADF on HADR and stability operations. While these might be viewed as *involving* the security establishment, they were hardly *dominated* by the security establishment. Furthermore, despite efforts by some scholars (Hartmann 2010), it is difficult to argue that such measures were *not* in the interest of either Australians or those from across the region.[6]

Having now teased out Rudd's agenda, a more nuanced picture emerges. Summarily, securitisation was pursued by Rudd to support his political agenda. He did not securitise it to be 'above' politics, it *was* politics. Consequently, his securitisation agenda had intended limits; it was never invoked with the commitments described by Copenhagen securitisation theory. It was, in some respects, a securitisation feign where rhetoric exceeded any genuine willingness to act. How, then, can the Australian military response to climate change be summarised in the context of these theoretical positions over the period 2008–2013?

The Australian military, acute to the political sensitivities and cognisant of the inherent limits to Rudd's securitisation agenda (and the military's role within it), thus presented as asomething as an opportunity reluctant but nonetheless calculating actor in a scientifically informed but politically-dominated climate securitisation agenda. To be sure, it was being 'consistent' with Government policy, but it also had a firm grasp of the possibility that a change of government at either the 2010 and then 2013 elections might render any climate change investments or initiatives null and void or risky. Thus, from a Copenhagen conceptual perspective, the Australian military was not a securitising actor per se; rather, it sought to

[6]See, for example, the ADF's contribution to assisting Papua New Guinea during the 1997–98 El Niño drought and other regional HADR operations (Barrie et al. 2015: 39 and 60).

position itself as a benign participant—actively avoiding the political dimension—and focusing on a cautious, low-profile, low-cost, gradualist response to only the most practical and strikingly obvious of long-term risks posed by a changing climate. It was a strategy that minimised any investment (resource or reputational) lest a change of government rendered them invalid or it drew unwanted criticism. The non-securitised dimensions (the 'governance dimensions') of climate change subsequently became a focus for the operational and strategic levels of the ADF; bureaucratic programme were developed to oversee their governance but were cognisant not to over-step the mark. This outcome depicted aspects of the Paris School (particularly its ideas on 'bureaucratisation' and Olaf Corry's concept of 'riskification'), but it repudiated what this author understood as the *sine qua non* of this doctrine that conceived a framework in which the military exerted disproportionate influence over the national (political) security agenda. At the very heart of this analysis existed the challenge of an avowedly apolitical institution responding to what emerged in the Australian context as a politically partisan security issue.

Having reached this assessment, it might be asked: was (is) this situation acceptable? Moreover, if climate change was (and remains) an existential threat and if the current decade represents the so-called critical decade to avoid these threats, is the Australian military justified in adopting a minimalist climate change strategy? Should it play a more activist role? If so, what is this role? More broadly—and looking to the future—do the Australian military and associated federal intelligence and assessment institutions have a unique obligation to extricate themselves from the politics of climate change so that they might provide the public with an unfettered assessment of the security risks involved? Furthermore, what are the broader implications for Australian strategic policy and how might climate change affect the future relationship between the military, the Government and the people (particularly where the security threats are viewed in such a partisan manner)? As was put by one commentator: 'How does the military, hoping to develop intellectual grit, manoeuvre around the very institution that feeds it?' (Terreu 2014). Of more currency, however, is the requirement for government agencies to speak truth to power and adhere to what US political scientist Richard K. Betts identified what should be the 'irrevocable norm' for all military and intelligence analysis, in which: '[political] policy interests, preferences, or decisions must never determine intelligence judgements' (Betts 2005: 58). Similar arguments exist in relation to the security risks of climate change.

7.5 Climate Securitisation in the US Under a Centre-Left Government

The November 2008 election of President Obama signified a major shift in US political outlook regarding climate change. Vowing to 'mark a new chapter in America's leadership on climate change' (2008), Obama repudiated the Bush era where 'rigid ideology has overruled sound science' and accorded climate change one of his 'highest priorities' (Obama 2009j: 1). From an international relations perspective Obama even 'put climate at the *top* of our diplomatic agenda' (Obama 2009f; emphasis added). Considering that the US was still engaged in two wars, this was a striking comment.

In general, there existed many similarities between the climate policies of the Obama Administration and the Rudd-Gillard governments. Foremost in this sense was the introduction (and subsequent Senate defeat) of a cap and trade/emission trading scheme as well as the introduction of a number of policies designed to transform each economy towards a sustainable and clean-energy future. Following the GFC, both the Obama and Rudd administrations directed significant stimulus spending towards supporting this transformation. In this regard, both leaders used the GFC as an opportunity to make good on their campaign promise to address climate change. For Obama, it meant turning a 'moment of peril … into one of progress' (Obama 2009i: 3).

Central to this thesis were the strikingly similar rhetorical outlooks adopted by both Rudd and Obama in relation to climate securitisation. Like Rudd, Obama adopted a broad interpretation of climate change as a security threat. While this research only identified three separate climate securitised categories (to which he framed climate change on nine separate occasions), Obama's speeches on the matter arguably conveyed a much deeper understanding of climate security impacts than the coding inferred. To the Ghanaian Parliament, for instance, Obama argued that a 'warming planet will spread disease, shrink water resources, and deplete crops, creating conditions that produce more famine and more conflict' (Obama 2009k: 1090). To the UN General Assembly he reckoned climate change a 'danger' which 'must not be deferred' for want of 'wars over refugees and resources … drought and famine. Land that humans beings have lived on for millennia will disappear' (Obama 2009l: 1445). In Italy at the Major Economies Forum he declared '[e]very nation is at risk' (Obama 2009j: 1073). On many occasions he said the world was at a 'crossroad of history' and that 'America will not be held hostage' (Obama 2009i: 3). Overall, Obama surmised that the dangers posed by climate change rendered current efforts to end wars fruitless since they 'will be eclipsed by wars over refugees and resources' (Obama 2009l: 1445). Many of Obama's speeches thus echoed Rudd's vision of planetary scale devastation framed, as it were, in the language of security.

Like Rudd, Obama also sought to frame climate change as an issue requiring immediate (even 'urgent'), and sometimes extraordinary action. In one statement, Obama endorsed the idea that 'climate change poses a clear danger requiring an extraordinary global response' (2009j: 1075). In another he implored '[t]he threat from climate change is serious, it is urgent, and it is growing … if we fail to meet it boldly, swiftly, and together, we risk … irreversible catastrophe' (2009f). He further added 'the security and stability of each nations and all peoples … are in jeopardy … time we have to reverse this tide is running out' (Obama 2009f). Unlike the Australian case study, the research did not discern a difference between the urgency of acting on climate change for political reasons over security reasons. Put another way, *security* appeared part-and-parcel of the broader arguments being advanced by Obama to act on climate change; they were not held as separate or distinct issues operating on their own time-logics (as was characterised by Rudd and the Australian military where climate security were specifically called-out as long-term prospects). This difference may have been one reason why the US military (or parts thereof) acted in a much more emphatic manner than Australian military to climate change.

To what purpose, then, did climate securitisation fit into the broader US climate narrative under Obama? Like Rudd, Obama arguably pursued climate securitisation ostensibly to achieve his *political* climate change objectives. His overarching strategy was to ensure that he met his 2008 campaign-election promises to the American people by being seen to take action on climate change. Thus, and like Rudd, Obama's climate securitisation objective was not an end in itself, but rather a tactic nested within a much broader political strategy. At the broadest level, this was *not* securitisation as conceived by Copenhagen theory where emergency measures were sought to move beyond rules. Rather, Obama's strategy evolved to become a set of regulatory measures that sought a level of progress, control and governance over a manifestly complex public policy issue. Several examples are explored in the following passages that highlight the trajectory of Obama's climate strategy through the lens of securitisation.

Firstly, from a domestic perspective, Obama initially used climate securitisation as part of his broader strategy to act on climate change as a differentiator between himself and his political opposition.[7] Given that the Bush Administration (and the Republican Party more generally) were largely viewed as sceptical towards climate change *and* that the 2008 election campaign coincided with an IPCC honeymoon period in which climate change ranked highly as a US public policy concern, such differentiation undoubtedly built on existing sentiments and further played to Obama's political advantage. This research showed that whereas Bush *never* framed climate change as a security issue, Obama did so with regularity (if not explicitly, then implicitly).[8] Also, Obama was handed something of a political gift in the form of Bush's 2008 climate security directives, which played comfortably into his broader climate change agenda.

Following his election, Obama pursued climate securitisation with the aim of building public and political consensus around the subject with the specific aim of favourably influencing the national debate and impressing on Congress the requirement to pass his climate legislation. As occurred in Australia, by framing climate change as an issue of national security importance, Obama sought to 'raise the stakes' as though elevating it above (partisan) politics and as a necessary

[7]Although, to be fair, even John McCain sought to distance himself from President Bush's climate policies. See for example: http://www.nytimes.com/2008/05/13/us/politics/12cnd-mccain.html (10 January 2017).

[8]With the exception, of course, of Bush's Climate Security directives outlined earlier.

requirement to act in the national interest. In this context, the Administration's signature climate policy, the American Clean Energy and Security Act (the *Waxman-Markey Bill*) was passed by the House of Representatives on 26 June 2009. Reflecting Obama's outlook, the *Waxman-Markey Bill* noted '[g]lobal warming poses a significant threat to the national security, economy, public health and welfare and environment of the United States' and called for the legislation of an economy wide cap and trade system. Indeed, the very title of the Bill (as a *Security* Act) provided a telling example of the importance attached to ensuring climate change was framed as a security issue. In some respects, that climate change was presented as a security issue to the American people where *their very own actions* played a contributing role was probably unique in terms of security issues grappled by any US president. Furthermore, the role of securitisation as a means to influence the passage of domestic legislation might be a unique manner in which Copenhagen securitisation theory might be applied.

The timing of the ACES Bill through the House was also arguably about positioning the US to leverage climate commitments from other major emitting nations. Thus, the other audience of Obama's securitisation agenda was the international one. Within a fortnight that ACES was passed by the House, Obama attended a G8 summit that pledged to reduce their emissions by 80% by 2050 (Obama 2009j). Success of the bill through the House also delivered an important signal to developing nations (read China and India) that the US was serious about securing a far reaching and binding international agreement at COP 15 (held in November of 2009). To round out his message to an international audience, Obama spoke forcibly on the international stage—particularly across 2009—by framing climate change as a 'grave', 'serious', 'urgent' and 'growing' threat to international peace and security. This research empirically confirmed this assessment, revealing that of the thirty-three references to climate change as a security issue in the US strategic domain between 2003 and 2013, eleven occurred in 2009 (the most of any year) and most were uttered directly by President Obama himself.

In addition to the *Waxman-Markey Bill*, Obama used the American Recovery and Reinvestment Act of 2009 (*Recovery Act*) to present climate change as an especially urgent issue in the aftermath of the GFC. Arguably, the GFC transformed the nature of debate on climate change, not just in the US, but also around the world. Once again, the pragmatic issues of politics—levels of employment/ unemployment and the state of the economy overall—dominated the domestic and international agendas. Climate change, prominent from 2007 and through 2008, slipped somewhat from public and political consciousness and was reflected in polling (Pew Research found 'global warming' came in last in a survey of top policy priorities in January 2009 (PEW 2009)). Despite this, Obama used the *Recovery Act* to deliver on campaign promises to address climate change by directing tens of billions of dollars in investment toward renewable energy and other climate related projects. By circumstance, the GFC delivered Obama a window of bi-partisan political support that enabled some of his climate commitments to be delivered under the mantra of US economic recovery. In a domestic context, economic justification (following the GFC) became a dominant narrative and justification for action.[9]

Although the *Recovery Act* was presented as a 'win-win' outcome, mixed signals began to emerge. On the one hand, while Obama set about framing climate change as an existential security threat, on the other hand he placed much greater emphasis on framing it under the banner of job creation and economic growth. Thus, when Obama now spoke of emergency measures to solve the so called climate crisis, it was primarily in the context of reinvigorating the American economy. The shift in priority was evident: 'If the message is somehow we're going to ignore jobs and growth simply to address climate change, I don't think anybody is going to go for that. I won't go for that' (Obama 2012b). While it may seem obvious that he would steer such a course, a perception emerged that he 'grew timid' in his climate response (Lizza 2010). Also, from a lay-person perspective accustomed to security-deliverance by American hard-power, how was it possible that such a significant security threat had now been passed over for job creation? When, in 2010, the *Waxman-Markey Bill* failed to pass the Senate, Obama experienced a 'Rudd moment' whereupon rhetoric began to exceed substantive action. In a pattern of resemblance, Obama turned to 'financial reform' while Rudd turned to 'hospitals' (Lizza 2010; Chubb 2014b).

The mismatch between the *talk* of emergency measures and an inability (or willingness) to *act* on them was especially evident at the international level. The Kyoto Protocol was a good example. While Obama was inclined to say he favoured the principal of the Kyoto Protocol, he never set about actually taking executive action to ratify it. Similarly, he never pressed the issue on Congress. The failure of the Copenhagen conference, however, was arguably the most significant moment that exposed preference for rhetorical posturing over action. Until this point, the US (through the speech-acts of Obama himself) had helped to frame climate change as an existential international security threat that required immediate global action. However, that Obama, as the dominant presence at the Copenhagen conference, failed to secure a credible international agreement reflected not just a sense of US *powerlessness* but also a lack of US *willingness* to actually frame climate change as an issue of sufficient critical importance, let alone one of existential global importance. By any measure, when the crunch came, climate change was not presented by the US at this crucial juncture as being above politics nor one that signalled that nations must move beyond normal rules.

What form such action may have taken is open for speculation, but the US has previously demonstrated a propensity to adopt a range of diplomatic, economic or military measures against nations that do not accord to their world-views. None of these options were remotely evident at the Copenhagen conference, where the developing nations (primarily China and India) refused to endorse the more

[9]Notably, this was not an explicit finding found by the empirical phase of research, but it was evident on re-examining passages that were coded and was an identified shortcoming of the research.

substantial proposals favoured by the West (Lynas 2009). That the head of the Chinese delegation later argued that 'our sovereignty and our national interest' were at stake only served to reinforce the sensitivity surrounding the issue and demonstrate the true limits to which nations—including the US—might pursue action on climate change (Watts/Vidal 2009). All told, US efforts to present climate change as a major international security threat requiring emergency action once again collapsed in the face of *Realpolitik*.

All of this, however, contrasted somewhat with the actions of Prime Minister Rudd and pointed towards a small but perceptible difference in the relative political importance attached to climate change between the two leaders. On Kyoto, while Obama never attempted to ratify it, Rudd had literally signed it within minutes of taking office. In Copenhagen, while Obama used his proxies to do all the dealing until his arrival on the final day, Rudd had personally toiled in situ (and apparently without rest) for 'three days and three nights' (Rudd 2010b).

The personal toll exerted by Rudd at Copenhagen showed when he derided the Chinese delegation: '[t]hose Chinese fuckers are trying to rat fuck us' (Chubb 2014b: 90). His exertion on the issue of climate change had clearly taken its toll and he was reported to have suffered a 'form of breakdown' in the conference aftermath (Chubb 2014b: 90). Although this research never examined such personal aspects in detail, perhaps one small measure of his desperation recorded by this research was the number of occasions that Rudd framed climate change as a security imperative compared with Obama (Rudd framed it on 27 distinct occasions while Obama framed it on nine such occasions).

As Rudd became ever more dependent and politically desperate, Obama navigated a slightly more nuanced pathway that produced a series of smaller but *politically manageable* actions. Critical to Obama's approach was the need to placate powerful interest groups who represented the spectrum of climate action, both for and against. Ultimately—to break the gridlock—it led to a strategy that favoured executive regulatory action over the Congressional legislative route. Such a strategy had arguably been evident from 2009 (*Executive Order 13514 Federal Leadership in Environmental, Energy and Economic Performance* was a good example), but the doctrine was perhaps best encapsulated in his 2012 State of the Union Address. To those against climate action, Obama argued for an 'all-of-the-above' strategy which simultaneously oversaw an *expansion* of fossil fuel investments that had 'opened millions of new acres for oil and gas exploration' and that 'American oil production is the highest that it's been in eight years' (Obama 2012a). To those in favour of climate action he declared 'a comprehensive plan to fight climate change' which directed the EPA to cut emissions on future and existing US coal-fired power plants. As was noted by Obama: 'The differences in this Chamber may be too deep right now to pass a comprehensive plan to fight climate change … So far, you haven't acted. Well tonight I will' (Obama 2012a). In essence, however, such a strategy eschewed securitisation.

A concluding observation of US climate change debates was the central importance of *energy* (security, independence, transformation and so on). This research found that energy (framed within the context of climate change) formed the most dominant narrative in US strategic circles (from documents examined). This was particularly the case at the presidential level where roughly one-half of all securitised and non-securitised climate coded frames were done so in the context of energy. Moreover, energy matters (within the context of climate change) remained paramount to both President Bush *and* President Obama. This contrasted somewhat with the situation in Australia and reflected an important difference in emphasis between the progressive governments of each nation. Whereas Rudd framed climate change *less* than Howard had done in relation to energy, Obama actually framed it *more* than his predecessor. What did this mean in practical terms? Firstly, it indicated that Obama actively and deliberately framed climate change in terms of energy in order to establish common ground between those opposing climate action (generally Republicans) and those supporting climate action (generally Democrats). Energy, and particularly ideas of energy independence and energy security, has long represented a unifying issue that both sides of US politics could relate to and agree. Framing climate change in this manner afforded Obama a degree of (bi-partisan) political capital that he could leverage when pursuing his broader climate agenda. Moreover, linking the 'climate wagon' to the 'energy security horse' made an 'attractive route towards achieving [policy] success' that might otherwise not have been obtained by a 'retice[nt]' American Congress and public (Hayes/Knox-Hayes 2014: 89 and 90). Secondly, the centrality of *energy security* conferred an active role for the US Department of Defense and the US military more specifically. The US military was regularly cited by US politicians as the world's single largest consumer of energy, as a major incubator of cutting-edge energy-efficient technologies and as an institution blessed with infrastructure, land and capacity to act as a pre-commercial test-bed for new forms of energy. The US military thus became politically prominent in energy security and climate change debates in a manner not witnessed in Australia.

How then, can all this analysis be summarised in terms of the theory? Firstly, in terms of Copenhagen securitisation theory, climate change was never formed in the US as a truly securitised issue (in the sense it is described in the theory). While there were instances that could be considered securitising moves and use of strong language dramatising the urgency, there was no instance where emergency measures beyond rules were actually sought. In the US, this was evident at the domestic level as it was at the international, and it also correlated with Matt McDonald's observation of the Australian situation where there existed a gap between political rhetoric and political action (McDonald 2012: 580). And so, while President Obama was strong on the rhetoric of framing climate change as a security issue, he failed to match it with any commensurate or substantive actions that might genuinely have elevated the issue above politics and beyond normal rules. In many respects, merely framing climate change to the US public as a security threat posed many challenges—not least of all, communicating to the American public that they bore some responsibility for the very threats the president described. In this sense, framing climate change as a security threat revealed the limits of securitisation.

Obama's unwillingness (and inability) to pursue stronger actions at Copenhagen—even after the lengths he had gone towards actually framing it as an existential and planetary wide threat—was arguably a turning point. Once again, the US case study demonstrated the difficulty of securitising an issue like climate change to the definitions set by the Copenhagen securitisation framework. Limits of securitisation theory were further exposed by the argument that the failure to securitise should be viewed as a *triumph* of politics. A counter-argument to this would suggest that the failure to securitise climate change (and thereby elevating the necessity for action to reduce emissions) might rather be viewed as a *failure* of politics.

Compared to the Paris School framework, the US case study (at the strategic-political level) did display some congruence, particularly on the idea of deepening bureaucratic practices through regulatory action of federal government agencies. A direct (and ironic) consequence of a bureaucratic-led response in the US was to accelerate and expand involvement of the US military. In some respects, whereas ideas of climate securitisation never progressed military involvement, bureaucratisation had succeeded in consolidating the military as an important actor in US climate change (security) discourse. As has been shown, and discussed below, this was especially the case in relation to climate-energy debates and the Arctic. Given this political backdrop, how did the US military respond to climate change? Also, was there any evidence of securitisation (in spite of the lack of it at the political level)? Lastly, does the Paris School framework offer any insights into the US military climate change response, particularly in terms of bureaucratisation but also on the idea that the US military sought to use the issue for its own purposes and thereby dominate the political elite?

7.6 US Military Climate Response Under Centre-Left Political Leadership

The US military response—specifically the US Navy response—to climate change was larger, more consistent and more widespread than the response registered by the Australian military. This broad assessment, however, consisted of numerous sub-stratum of issues that, on occasion, highlighted both differentiation but also some similarities. The following sections examine the broad similarities and differences (mainly at the strategic level) between the Australian and US militaries. The final section examines the US Navy response to climate change in detail.

7.6.1 Similarities in Military Climate Response

The strategic levels of the Australian and US militaries exhibited a number of similarities in relation to their climate response. The first was the relative inattention to climate change by the most senior departmental defence officials (although this was *not* the case with US Navy leaders). This has been already covered from the Australian side, but the lack of articulate discussion on climate change (in the timeframe considered for this research) by both the US Secretary of Defense and by the Chairmen of the Joint Chiefs of Staff was also evident. For instance, climate change (in the speeches analysed) was never mentioned by Donald Rumsfeld (2001–2006) and was only mentioned in passing by Secretary Gates (2006–2011) and Panetta (2011–2013). Gates, in particular, provided an interesting case study. As the US Defense Secretary, who worked under both Bush and Obama across the peak period of interest in climate change, it is remarkable that he mentioned climate change on just *four* occasions in five years (from 365 speeches examined); this provides an insight into where it ranked in terms of his priorities. Most commentary on the subject found by this research at the most strategic-military level was made by Under Secretary for Defense, William Lynn, whose speeches were only included for analysis on the occasions he formally stood-in for Secretary Gates. In summary, in 807 speeches analysed by this research between 2003 and 2013 for US Defense Secretaries only *six* direct references to climate change were made (this excluded the eleven references made by Under Secretary Lynn). This contrasted strongly with the political sphere, particularly during the Obama presidency, but it was consistent with evidence of the most senior Australian military officials (such as the CDF, Secretary and Service chiefs). Notably, this did change somewhat under Secretary Hagel, but was not apparent until at least some way into his tenure. For instance, of the 22 speeches examined here and delivered by Hagel until April 2013, climate change was not mentioned. However, by 2014 Hagel had delivered several important speeches outlining climate change as a security threat and the US DoD response.

As for the Secretary, so for the Chairman of the Joint Chiefs of Staff. In the 609 speeches analysed (all delivered by Admiral Mullen and General Dempsey) between 2007 and 2013, just twenty-one direct references were found. Mullen was the most prominent of senior US Defense (military) officials examined to describe the security impacts of climate change in any detail. His replacement, General Dempsey, appeared to downplay notions of climate change by casting it alongside 'the Mayan prophecy' and 'life on other planets' (Dempsey 2011). For Dempsey, climate change was simply *not* an urgent enough issue that warranted near or mid-term policy prioritisation.

The lack of prominence of climate change in the speech-acts of the most senior US military officials was also reflected in some *military*-strategic publications. The most prominent was the 2011 *National Military Strategy*, which only made passing reference to climate change and, in any case, as having only 'uncertain impact' (US DoD 2011: 2). Once again, this contrasted with the urgency, certainty and extent to which climate change was framed in the *political*-strategic publications, notably the 2010 *Quadrennial Defence Review* and the 2010 *National Security Strategy* (i.e. politically oriented strategic documents), the latter boldly asserting: 'The danger from climate change is real, urgent and severe. The change wrought by a warming planet will lead to new conflicts' (White House 2010: 47). Arguably, these publications reflected a difference of priorities. The US political sphere framed climate change as having significant strategic (and military) impacts in near term time frames that justified an urgent requirement to act (though these were arguably of a political nature). In contrast, the 2010 NMS—primarily the domain of the Chairman of Joint Chiefs—cast climate change as less relevant and as a longer-term issue only. This aspect pointed towards some discrepancy in US military climate policy, with some areas pressing ahead (US Navy) and others acting as though there was nothing happening (CJCS). Such organisational unevenness was comparable to the Australian military situation, and was evidently not a unique circumstance as far as bureaucratic and government institutions were concerned (Hulme 2009). Moreover, such differences may also be accounted for by the fact that different parts of the military have different 'horizons'; an active service arm like the US Navy having a shorter operational horizon against the more long-term strategic outlook of the Joint Chiefs.

On a wider level, a general trend emerged that while the political sphere in the US and Australia under centre-left governments tended to frame climate change as a matter of policy urgency, the militaries of both nations tended to refrain from casting climate change in such terms. This was even evident in the US Navy where, more than any other agency examined, climate action was progressed with high vigour. Rear Admiral David Titley, arguably the most authoritative voice on climate change in the US military noted, 'I am not sure I would call it a crisis. It is a strategic challenge' (Titley 2010a).

Other policy similarities existed. Both the US military and the Australian military were partly framed as 'first responders' in the event of major humanitarian aid and disaster relief (HADR). This aspect struck a particular chord following the deployment of either military in response to domestic natural disaster events that came to be increasingly viewed in the context of climate change. That said, HADR arguably had a greater presence and influence in Australian-military climate debates, where it became a *leitmotif* of the Australian Labor Party under Kevin Rudd. This contrasted somewhat with the US Navy experience where HADR was but one of many factors. As was noted by the Head of US Navy Oceanography, RADM David Titley, climate change was 'more than humanitarian disaster relief' (Titley 2010c: 27).

Perhaps more broadly than HADR, however, both militaries tended to frame climate change as a matter of national security over any other form of security (for example human security or environmental security). This aspect underpinned the broader Australian and US diplomatic efforts to frame climate change as a security issue in the UNSC, whereupon it was strongly contested by other countries that tended to view climate change from sustainability, economic, social and other such perspectives.

Another broad similarity between the US and Australian military was that climate change became prominent as a security issue from 2008 onwards but tended to decline in (military) prominence from 2011 onwards. That is to say, the issue in a military sense seemed to emerge (2008), to peak (2009) and then decline (2011). In the Australian case study, political partisanship was argued as a major factor for both the rise and decline in the prominence of climate change by that military. In the US case study, additional factors emerged as having significance (to varying degrees, these were also present in the Australian instance, but were arguably less visible). First was the 'too busy elsewhere' paradigm in which major revisions of US national strategic and military priorities witnessed operational drawdown from the Middle East, subsequent rebalance toward Asia-Pacific as well as historic reductions in US Defense spending through planned austerity measures but also additional funding cuts, known as sequestration. The second was a declared reframing paradigm by senior US military and civilian leaders moving away from climate change towards concentrating on concepts of energy security. This was detected as occurring in both the Australian and US militaries, but the turn was particularly strident in the US Navy, particularly after about 2011, where Secretary Ray Mabus publicly argued moving away from climate change as justification for his energy transformation plans:

> I think our chances are less certain if we base it on what to most people are a little more nebulous ideas like climate change – harder to see, harder to understand and harder to qualify (Mabus 2010a: 3).

Given the shift in Obama's outlook on climate change following the failure of the *Waxman-Markey Bill* to pass the US Senate in 2010, Mabus' strategic shift might be interpreted as a response to ensuring he was aligned with the President's broader strategic aims on climate policy. Thirdly, the majority of studies on the impacts of climate change to the US military were—by 2011—already under way, were scheduled to occur sometime in the near future or had already been commissioned or been completed elsewhere (for example from military affiliated institutes, academia, think-tanks or NGOs).

Nevertheless, the downturn (if it can be called that) in US military climate security detected by this study did not last. From 2013 onwards—beyond the timeframe considered by this case study—the US military continued to advance the climate security agenda through a series of major climate change policies that impacted the breadth and depth of the organisation (as reviewed in Chap. 3). This marked a major divergence with the Australian military, and it is to these differences that the final section turns.

7.6.2 Differences in US and Australian Military Climate Response

The US military response to climate change differed to that of the Australian military in several important ways. In the broadest sense, the US military response to climate change was more prominent, more sustained and had far greater organisational penetration in terms of its reach downwards into the organisation and across the various sub-agencies within it. The response was also multifaceted, with a mix of adaptation and mitigation measures evident. On occasion, however, it was difficult to distinguish between US priorities; although climate change was invoked as a pretext for action, deeper examination revealed a combination of justifications for the size and dispersion of action.

The 2010 *National Security Strategy* and the 2010 *Quadrennial Defense Review* stood as milestone documents that heralded a shift in US military thinking on climate change. Until their publication, US military policy had considered climate change largely in economic terms or as part of broader global change issues. Both the 2010 NSS and QDR now identified climate change as a major strategic driver for the twenty-first century (in its own right) with potential to create new sources of instability and exacerbate existing ones (i.e., as a threat multiplier) and/or by directly degrading US military homeland and global estate, facilities, infrastructure, training and testing grounds (i.e., as a burden multiplier). The consideration that climate change might diminish US military global capacity also underlined a fundamental difference between the Australian and US militaries. The Australian military, of limited strategic reach and far smaller geographical footprint, had much less to lose in relation to climate threats than the US military through its vast global system of bases, alliances and interests. As a consequence, this research found that the US strategic and military spheres framed climate change in terms of the international context far more than their Australian counterparts. Overall, the 2010 NSS and QDR might be viewed as a defining moment—one in which the US military signalled its strategic commitment to respond to climate change. The audiences for this signal were twofold: domestic audiences (departmental but also the political class and American public) and international audiences (other militaries for example).

Preceding the release of these capstone policy documents, however, was Executive Order 13514. This directed the US military (amongst other things) to publish an annual strategic sustainability performance plan that placed targets on non-operational US military greenhouse gas emissions and a climate change adaptation roadmap (first published by the US military in 2010). Although both of these aspects have received very little media or scholarly attention, they reflected an emerging strategic difference between the response to climate change by both the US and Australian militaries. Stand-out examples were the goals of the US military to reduce Scope 1 and Scope 2 greenhouse gas emissions by 34% by FY 2020 relative to 2008 and Scope 3 emissions by 13.5% in equivalent time frames. This was in stark contrast to the Australian military's aborted promise to place targets on its emissions. Another stark difference was how climate change was promoted within US Defense departmental processes and procedures such that they would 'fully integrate climate change considerations into its extant policies, planning, practices, and program' (US DoD 2013a: D-6).

The broad impact of these capstone documents was to lay a strategic framework across the US military that empowered lower-level departments and agencies (such as the service arms) to proceed with developing and implementing their own climate policies. In this context, climate change became (from at least 2009) a legitimate justification for the instigation of a range of new US military energy programme, sustainability initiatives and environmental policies. Furthermore, climate change emerged as a rallying point for elevating the importance of a range of existing and disparate energy/environmental/sustainability programme that might otherwise have remained buried in discreet pockets of the US military bureaucracy. To some extent, climate change revived these programme by providing renewed emphasis and momentum that they might not otherwise have received. (The 2010 QDR provided several examples, whereupon it raised the Strategic Environmental Research and Development Program, Defence Environmental International Cooperation Program, Environmental Security and Technology Certification Program and the Energy Conservation and Investment Program as all contributing to the US military's strategic climate change response. Other examples included US military greenhouse gas mitigation efforts and the use of US military land for green energy initiatives.) This was not the case in the Australian military. Whereas the ADF had developed a number of impressive environmental and sustainability initiatives, few invoked climate change as a pretext for taking such action.

Moreover, having laid the broad framework, lower levels within the US military underwent a rapid expansion of climate and energy policies (from 2009) that once again contrasted sharply with the Australian experience. Climate change was often a key aspect used to justify the policy action (refer to footnote below for some examples).[10]

This research specifically examined the US Navy climate policy documents in detail, revealing a deep (vertical) penetration and wide (horizontal) spread of climate and energy initiatives within the US Navy. Examples of the vertical penetration of climate policy was evident in *Task Force Climate Change* and *Climate Roadmap*, which gave detailed tasking for mid-ranking Navy officers to complete a range of climate initiatives to be funded from 2014 onwards. The horizontal spread was likewise evident, whereupon climate action was tasked across US Navy operational, logistical, capability, communications and intelligence portfolios. On the whole, these actions pointed more towards the bureaucratisation of Navy Arctic climate change policy rather than one of securitisation. Arguably, these initiatives were only permissible in the political environment established through the election of Obama and built upon by the climate security directives enacted by Bush late in his final term. The vertical penetration and horizontal spread of climate policy within the US Navy once again contrasted sharply with the Australian military where it scarcely impacted.

---

[10]Examples included the US *Army Energy Initiatives Task* Force (2009), US DoD *Installation Energy Management* (DoDI 4170.11 of 2009), USMC *Facilities Energy and Water Management Campaign Plan* (2009), US Navy *Task Force Climate Change* (2009), US *Navy Arctic Roadmap* (2009), US *Navy Climate Change Roadmap* (2010), US *Navy Energy Vision* (2010), US DoD *Operational Energy Strategy* (2011), US Army *Net Zero Initiative* (2011), US DoD *Operational Energy Strategy Implementation Plan* (2012), US *Army Renewable Energy Development Guide* (2012), USAF *Sustainable Design and Development Implementing Guidance* (2012), US *DoD Climate Change Adaptation Working Group* (2012) and US DoD *Climate Change Adaptation Roadmap* (published as an annex to the US DoD *Strategic Sustainability Performance Plan* from 2012), US Army Corps of Engineers *Climate Adaptation Statement* (2011), revised and updated as the *Climate Preparedness and Resilience Policy Statement* (2014).

A final indication of the significance with which climate change had come to be viewed in US military circles occurred in relation to austerity. When austerity measures were imposed on the US military from 2011, there would have been a reasonable expectation that climate-related programme would likely be the first to face cancellation. When asked if his energy and climate goals were attainable in view of 'sequestration', Mabus defiantly responded that 'we cannot afford *not* to do this now' (Mabus quoted in US House of Representatives 2013; emphasis added). Moreover, even in the face of unprecedented budget cuts (and some years after the GFC stimulus), that climate change programme still proceeded gave it a level of priority previously unseen.[11] Once again, this contrasted sharply with the Australian military experience.

Despite this constructive outlook, this research also detected that US military climate policy was not an entirely altruistic enterprise. In short—and of no revelation—beneath the veneer of US military climate policy, realpolitik and self-interest were ever-present. The vigorous manner in which many senior US military and US Defense civilian leaders were forced to defend their climate policies on the grounds that it was *not* an environmental driven agenda but was rather done in the national self-interest differed from the Australian situation. Possibly, it differed precisely because the US military had implemented climate policies beyond what might have been expected of it, particularly from hostile conservative politicians. This was not evident in the Australian military, where its minimalist policy failed to invite sufficient criticism to warrant any such defence.

Two aspects, in particular, provide a window into the true motives lying behind US military climate policy. The first was ensuring primacy of US (military) power in the context of global change. Climate change formed a part of this narrative, but so too did a range of global change issues that, since the end of the Cold War, had expanded to include population growth, rapid urbanisation, pollution, cyber, resource depletion and a range of other interconnected, globalised and common security challenges. The Arctic also formed a core part of the broader global change dimension and was particularly associated with the onset of climate change. In this regard, the US Navy began shifting strategic weight toward developing policies and capabilities to ensure Arctic primacy. The Chief of (US) Naval Operations, Admiral Gary Roughead, was fervent in his appreciation that the opening of the 'fifth ocean' meant 'trillions in economic potential' which pointed 'undeniably towards a new venue of operations and responsibility' (Roughead 2011b). Undoubtedly, the possibility of its long-term strategic rival (Russia) or even newly emergent competitor nations (China), moving to fill an Arctic vacuum would have provided another incentive. Although the Australian military established a small section to

[11]An additional factor not considered in detail was the strategic certainty of having a single President who was very supportive of military climate change programmes (as distinct from the Australian situation in which Rudd was replaced by Gillard, who was ousted by Tony Abbott in September 2013).

examine global change issues, it never had an 'Arctic' that brought climate change into sharp strategic focus in the way that it had in the US military. Though some have tried to present Antarctica as 'Australia's Arctic', the two are not analogous for a range of reasons, not least of which is that Australia has no military presence, the location is not (yet) a site of geopolitical contest surrounded by the great powers (US, China, Russia and Europe), Antarctica holds no commercial shipping advantages offered by an ice-free channel, and prospects for resource exploitation remain negligible, given the long-standing global treaty and access difficulties.

The second motive behind US military climate policy was its inseparable coupling with concepts of energy security. Although climate change was presented as a factor, the main debates centred on maintaining US (military) energy security in the context of fears of reduced global supply, increasing energy costs and efforts to find efficiencies (savings), particularly after austerity took hold. In some policy areas and in some speeches, military mitigation efforts often appeared as an afterthought. Thus, a contradiction emerged between the seeming altruistic efforts to reduce emissions (for the greater good) against the reality of improving its energy use to cut cost while maintaining global military supremacy. This was particularly the case after roughly 2010, when it became evident that the broader political strategy on climate change required a makeover.

And so, embedded throughout the various speech-acts on US military climate initiatives came another clear message: the US military was not responding to climate change purely for environmental reasons, it was responding to protect US national interests and preserve global military dominance. Thus, for Obama, 'the Pentagon isn't seeking these alternate fuels just to protect the environment; they're pursuing these home-grown energy sources to protect national security … *clean energy is about our security*' (Obama 2010; emphasis added). For Secretary of Navy Ray Mabus: 'The Great Green Fleet is not about some environmental agenda. *It is about maintaining America's military and economic leadership across the globe in the 21st century*' (Mabus 2012c; emphasis added). These factors pointed to a deeper agenda—a 'hidden truth' (Bigo/Tsoukala 2008: 13)—that only emerged in the speech-acts of senior US leaders on rare occasions. It hinted that US military climate mitigation efforts for environmental reasons only were a ruse, that the maintenance of American power was paramount. Though not to be overplayed, these ideas connect with Paris School notions suggesting that, concerning matters of security, politicians and the military alike resort to type by wielding their institutional power to influence national debates to their advantage.[12]

[12]For a further discussion of this as it relates to the contemporary neo-liberal period, see Dalby et al. (2009).

7.7 Chapter Conclusion

The Australian and US militaries exhibited both similarities and differences in their response to climate change. Political influence was a dominant theme where the election of centre-left governments in 2007 (Australia) and 2008 (US) both enabled and prompted (sometimes through regulation but other times through strategic guidance via formal speeches and directives) the respective militaries to take stronger action. In the ADF, however, a strategic response to climate change became stillborn as the Rudd Labor government's broader climate policy platforms unravelled. As bi-partisan political support collapsed in 2010, any form of climate policy in the Australian military remained on the margins. In the US, President Obama used climate securitisation as part of his broader climate platform and he built upon the climate security directives enacted by President Bush in late 2008.

Given some political bi-partisanship concerning climate security matters, the US military response became much more emphatic than the ADF. Apart from the political dimensions, this book identified three main reasons. First, climate change policy response was tightly coupled to issues surrounding energy security. Thus, when climate action in the US weakened at the political level, mitigation could still be tied to current operational and broader national security interests by ensuring improved war-fighting, greater efficiency and reduced operating costs. The long-standing nature of energy security as a prominent strategic issue in the US (particularly regarding concepts of energy independence) also resonated with both sides of politics (thereby providing a degree of bi-partisanship). Secondly, climate change was largely accepted as the key reason for driving significant change across the Arctic. Australia never had an equivalent geo-strategic location generating such attention; particularly one which borders with a long-standing competitor-adversary in the form of Russia. The US Navy used the Arctic as a centrepiece issue to drive a range of climate programme that impacted operations, capability development and other functional areas. In the context of enhancing energy security, reducing emissions, reducing costs and increasing its presence in the Arctic, the US Navy response was therefore framed as something as an opportunity—but there were also indications its motives were not entirely altruistic. The last of these possibilities contemplated an alternate US military climate response driven by maintaining military dominance and protecting national self-interest in an era of global change. On occasion, such bluntness was revealed in the speeches from the President and was traceable down through the chain of command. The final reason for motivating a larger US military response was the global nature of its operations, bases, infrastructure, alliances and interests. In short, the US had—and has—much more to lose and far more at risk if climate change unfolds on a Business-As-Usual trajectory. Relatedly, the US military also advanced climate change as an opportunity to enhance bi-lateral and multilateral relationships. As a far smaller military, both in physical footprint, energy consumption, strategic reach and strategic necessity, the Australian military never had as much at stake.

References

Alexander, Natalie. 2011. "Climate Change on radar." In *Defence Magazine*, edited by Michael Weaver, vol. 3, pp. 30–31. Canberra: Department of Defence. Accessed on 12 April 2013, http://www.defence.gov.au/defencemagazine/editions/2011_03/mag.pdf.

Aulich, Chris. 2010. "It was the best of times; it was the worst of times." In *The Rudd Government: Australian Commonwealth Administration 2007–2010*, edited by Chris Aulich and Mark Evans, pp. 3–15. Canberra: ANU E Press. Accessed on 15 June 2013, http://press.anu.edu.au/apps/bookworm/view/The+Rudd+Government%3A+Australian+Commonwealth+Administration+%092007+-+2010/5091/ch01.xhtml.

Barrie, Chris, Will Steffen, Alix Pearce, and Michael Thomas. 2015. *Be Prepared: Climate Change, Security and Australia's Defence Force*. Sydney: Climate Council. Accessed on 20 September 2015, https://www.climatecouncil.org.au/uploads/fa8b3c7d4c6477720434d6d10897af18.pdf.

Beckett, Margaret. 2007. "The Urgent Case for Climate Security." London: Royal United Services Institute. Published online 10 May 2007. Accessed 15 May 2017, https://rusi.org/publication/urgent-case-climate-security.

Betts, Richard K. 2005. "Politicization of Intelligence Costs and Benefits." In *Paradoxes of Strategic Intelligence: Essays in Honor of Michael I. Handel*, edited by Richard K. Betts and Thomas G. Mehnken, pp. 57–76. Portland: Frank Cass Publishers.

Bigo, Didier, and Anastassia Tsoukala. 2008. "Understanding (in)security." In *Terror, Insecurity and Liberty: Illiberal practices of liberal regimes after 9/11*, edited by Didier Bigo, Elspeth Guild and R. B. J. Walker, pp. 1–9. Abingdon: Routledge.

Bigo, Didier, Anastassia Tsoukala, Laurent Bonelli, Emmanuel-Pierra Guittet, and Christian Olsson. 2008. *Terror, Insecurity and Liberty: Illiberal practices of liberal regimes after 9/11*. Edited by Didier Bigo, Elspeth Guild and R. B. J. Walker, *Routledge Studies in Liberty and Security*. Abingdon: Routledge.

Brauch, Hans Günter. 2011b. "Security Threats, Challenges, Vulnerabilities and Risks in US National Security Documents (1990–2010)." In *Coping with Global Environmental Change, Disasters and Security: Threats, Challenges, Vulnerabilities and Risks*, edited by Hans Günter Brauch, Ursula Oswald Spring, Czeslaw Mesjasz, John Grin, Patricia Kameri-Mbote, Bechir Chourou, Pal Dunay and Jorn Birkmann, vol. 5, pp. 249–274. Berlin: Springer.

Buzan, Barry, Ole Wæver, and Jaap de-Wilde. 1998. *Security: A New Framework for Analysis*. Boulder: Lynne Rienner Publishers.

Chubb, Philip. 2014a. "The day the Rudd Government lost its way on climate change." *The Age*, Fairfax Media, 9 May 2014. Accessed on 26 May 2014, http://www.theage.com.au/insight/the-day-the-rudd-government-lost-its-way-on-climate-change-20140509-zr7fm.html.

Chubb, Philip. 2014b. *Power Failure: The Inside Story of Climate Politics Under Rudd and Gillard*. Melbourne: Black Inc. Agenda.

A Bill to address security risks posed by global climate change, and for other purposes. Global Climate Change Security Oversight Act (110th Congress, First session, H.R. 1961, (Bill was not enacted). Accessed on 3 March 2013, http://www.gpo.gov/fdsys/pkg/BILLS-110hr1961ih/pdf/BILLS-110hr1961ih.pdf.

Commonwealth of Australia. 2006. "Supplementary Budget Estimates." In *Standing Committee on Foreign Affairs, Defence and Trade Official Committee Hansard*, 1 November 2006. Canberra: Commonwealth of Australia. Accessed on 19 March 2012, http://www.aph.gov.au/Parliamentary_Business/Senate_Estimates/fadtctte/estimates/sup0607/index.

Commonwealth of Australia. 2007a. "Budget Estimates." In *Standing Committee on Foreign Affairs, Defence and Trade Hansard*, 31 May 2007. Canberra: Commonwealth of Australia. Accessed on 13 February 2011, http://www.aph.gov.au/~/media/Estimates/Live/commttee/S10269.ashx.

Commonwealth of Australia. 2012a. "Senate Budget Estimates Questions On Notice—Committees." In *Senate Standing Committee on Foreign Affairs, Defence and Trade Hansard*, 28 & 29 May 2012. Canberra: Commonwealth of Australia. Accessed on 17 July 2013, http://www.aph.gov.au/Parliamentary_Business/Senate_Estimates/fadtctte/estimates/bud1213/def/Department-of-Defence/defenceqonsindex.

Commonwealth of Australia. 2013a. *Energy Use in the Australian Government's Operations 2010–11*. Department of the Environment and Energy. Canberra: Commonwealth of Australia. (Available at: http://content.webarchive.nla.gov.au/gov/wayback/20160615122731/http://www.industry.gov.au/Energy/EnergyEfficiency/Non-residentialBuildings/GovernmentBuildings/Documents/report-2011.pdf, accessed 13 June 2017).

Council on Foreign Relations. 2007. "Political Interference with Climate Change Science Under the Bush Administration, December 2007." Online: Council on Foreign Relations. Accessed on 25 January 2017, http://www.cfr.org/climate-change/political-interference-climate-change-science-under-bush-administration-december-2007/p15079.

Dalby, Simon, Hans Günter Brauch, and Úrsula Oswald Spring. 2009. "Environmental Security Concepts Revisited During the First Three Phases (1983–2006)." In *Facing Global Environmental Change Environmental, Human, Energy, Food, Health and Water Security Concepts*, edited by Hans Günter Brauch, Úrsula Oswald Spring, John Grin, Czeslaw Mesjasz, Patricia Kameri-Mbote, Navnita Chadha Behera, Béchir Chourou and Heinz Krummenacher, vol. 4, pp. 781–790. Berlin: Springer.

Defence, Department of. 2008a. *Combat Climate Change*. Canberra: Department of Defence.

Defence, Department of. 2009a. *Defence Annual Report 2008–09*. Department of Defence. Canberra: Department of Defence. Accessed on 10 July 2012, http://www.defence.gov.au/AnnualReports/08-09/.

Defence, Department of. 2014. *Defence Procurement Policy Manual*. Canberra: Department of Defence.

Defense Science Board Task Force. 2008. *Report of the Defense Science Board Task Force on DoD Energy Strategy: More Fight—Less Fuel*. Washington, DC: Department of Defense.

Dempsey, Martin. 2011. "General Dempsey at the Military Reporters and Editors Conference." In *Joint Chiefs of Staff Speeches*, 18 November 2008, vol. 2013. Arlington: Department of Defense. Accessed on 12 July 2012, http://www.jcs.mil/Media/Speeches/tabid/3890/Article/571837/gen-dempseys-remarks-and-qa-at-at-the-military-reporters-and-editors-conference.aspx.

Evans, Michael. 2005. *The Tyranny of Dissonance: Australia's Strategic Culture and Way of War 1901–2005*. Land Warfare Studies Centre Study Paper Number 306: Canberra: Land Warfare Studies Centre.

Floyd, Rita. 2010. *Security and the Environment: Securitisation Theory and US Environmental Security Policy*. Cambridge: Cambridge University Press.

Gates, Robert M. 2014. *Duty: Memoirs of a Secretary at War*. New York: Alfred A. Knopf.

Goldenberg, Suzanne. 2009. "The worst of times: Bush's environmental legacy examined." *The Guardian*, Guardian News and Media, 17 January 2009. Accessed on 25 January 2017, https://www.theguardian.com/politics/2009/jan/16/greenpolitics-georgebush.

Hartmann, Betsy. 2010. "Rethinking Climate Refugees and Climate Conflict: Rhetoric, Reality and the Politics of Policy Discourse." *Journal of International Development*, vol. 22, no. 2:233–246. doi:10.1002/jid.1676.

Hayes, Jarrod, and Janelle Knox-Hayes. 2014. "Security in Climate Change Discourse: Analysing the Divergence between US and EU Approaches to Policy." *Global Environmental Politics*, vol. 14, no. 2:82–101. doi:10.1162/GLEP_a_00230.

Houston, Angus. 2007. "The ADF of the Future." *Australian Defence Force Journal: Journal of the Profession of Arms*, no. 173.

Houston, Angus. 2010. "Future security prospects, future capability plans and how the ADF makes a contribution to a 'whole of government' approach to national security." In *Address to National Security Lecture Series*, 29 October 2010. Canberra: University of Canberra.

Hulme, Mike. 2009. *Why we disagree about climate change: Understanding controversy, inaction and opportunity*. Cambridge: Cambridge University Press.
Lizza, Ryan. 2010. "As The World Burns." *The New Yorker*, Condé Nast, 11 October 2010. Accessed on 10 November 2014, http://www.newyorker.com/magazine/2010/10/11/as-the-world-burns.
Lynas, Mark. 2009. "How do I know China wrecked the Copenhagen deal? I was in the room." *The Guardian*, The Guardian, 23 December 2009. Accessed on 10 May 2013, http://www.theguardian.com/environment/2009/dec/22/copenhagen-climate-change-mark-lynas.
Mabus, Ray. 2010a. "Remarks by the Honourable Ray Mabus Secretary of the Navy, Center for Naval Analysis Earth Day Luncheon, Alexandria, VA (29 April)." In *Transcripts of the Secretary of Navy*, 29 April 2010. Washington, DC: United States Navy. Accessed on 8 September 2013, http://www.navy.mil/navydata/people/secnav/Mabus/Speech/Center%20for%20Naval%20Analysis%20Earth%20Day%20Luncheon%2029%20Apr%2010.pdf.
Mabus, Ray. 2012c. "Remarks by the Honourable Ray Mabus Secretary of the Navy, Senate Committee on Energy and Natural Resources, Norfolk, VA." In *Transcripts of the Secretary of Navy*, 12 March 2012. Washington, DC: United States Navy. Accessed on 4 October 2013, http://www.navy.mil/navydata/people/secnav/Mabus/Speech/Senate%20Committee%20on%20Energy%20and%20Natural%20Resources%20oral%20testimonyAsPrepared.pdf.
Macintosh, Andrew, Deb Wilkinson, and Richard Denniss. 2010. "Climate Change." In *The Rudd Government: Australian Commonwealth Administration 2007–2010*, edited by Chris Aulich and Mark Evans, pp. 596–652. Canberra: The Australian National University E Press. Accessed on 3 March 2014, http://press.anu.edu.au/apps/bookworm/view/The+Rudd+Government%3A+Australian+Commonwealth+Administration+%092007+-+2010/5091/ch11.xhtml#toc-anchor.
Major Economies Forum Declaration. 2009. "Declaration by the Leaders of the Major Economies Forum on Energy and Climate." Washington, DC: Office of the Federal Register, National Archives and Records Administration. Accessed on 11 October 2013, http://www.gpo.gov/fdsys/granule/PPP-2009-book2/PPP-2009-book2-Doc-pg1072/content-detail.html.
McDonald, Matt. 2012. "The Failed Securitization of Climate Change in Australia." *Australian Journal of Political Science*, vol. 47, no. 4:579–592. doi:10.1080/10361146.2012.731487.
Morrison, David. 2013. "Speech to Royal United Services Institute, Sydney." In *Chief of Army Speeches and Transcripts*, 26 March 2013. Canberra: Department of Defence.
Nelson, Brendan. 2007b. "Opening Address to D+I Conference Adelaide." In *Transcripts of the Minister for Defence* 22 August 2007. Canberra: Department of Defence.
Obama, Barack H. 2008. "Videotaped Remarks to the Bi-Partisan Governors Global Climate Summit November 18, 2008" In *Obama-Biden Transition Project Press Releases*. Washington, DC: The Office of the President Elect. Accessed 20 May 2017, http://www.presidency.ucsb.edu/ws/index.php?pid=84875.
Obama, Barack H. 2009f. "Remarks at the United Nations Climate Change Summit in New York City." In *The American Presidency Project*, edited by Gerhard Peters and John T. Woolley, 22 September 2009. Published online: The American Presidency Project. Accessed on 3 March 2014, http://www.presidency.ucsb.edu/ws/?pid=107665.
Obama, Barack H. 2009h. "Remarks by the President to the United Nations General Assembly (United Nations Headquarters, New York)." In *President Obama, Speeches and Remarks*, 23 September 2009. Washington, DC: White House. Accessed on 8 November 2013.
Obama, Barack H. 2009i. "Remarks on Energy." In *Compilation of Presidential Documents*, 26 January 2009. Washington, DC: US Government Printing Office. Accessed on 3 September 2013, http://www.gpo.gov/fdsys/pkg/DCPD-200900019/content-detail.html.
Obama, Barack H. 2009j. "Remarks on the Declaration by the Major Economies Forum on Energy and Climate Change in L'Aquila, Italy." In *Public Papers of the Presidents of the United States: Barack Obama* 9 July 2009, vol. (Book) II, pp. 1072–1075. Washington, DC: US Government Printing Office. Accessed on 11 October 2013, http://www.gpo.gov/fdsys/granule/PPP-2009-book2/PPP-2009-book2-Doc-pg1072/content-detail.html.

Obama, Barack H. 2009k. "Remarks to the Ghanian Parliament in Accra." In *Public Papers of the Presidents of the United States: Barack Obama*, 11 July 2009, vol. (Book) II, pp. 1087–1092. Washington, DC: US Government Printing Office. Accessed on 8 November 2013, http://www.gpo.gov/fdsys/granule/PPP-2009-book2/PPP-2009-book2-Doc-pg1087/content-detail.html.

Obama, Barack H. 2009l. "Remarks to the United Nations General Assembly in New York City." In *Public Papers of the Presidents of the United States: Barack Obama* 23 September 2009, vol. (Book) II, pp. 1440–1447. Washington, DC: US Government Printing Office. Accessed on 4 May 2013, http://www.gpo.gov/fdsys/granule/PPP-2009-book2/PPP-2009-book2-Doc-pg1440/content-detail.html.

Obama, Barack H. 2010. "Remarks by The President on Energy Security at Andrews Air Force Base." In *President Barack Obama, Speeches and Remarks*, 31 March 2010. Washington, DC: White House. Accessed on 23 October 2014, http://www.presidency.ucsb.edu/ws/?pid=107665.

Obama, Barack H. 2012a. "Address before a Joint Session of Congress on the State of the Union." In *The American Presidency Project*, edited by Gerhard Peters and John T. Woolley. Published online: The American Presidency Project. Accessed on 18 August 2013, http://www.presidency.ucsb.edu/ws/?pid=99000.

Obama, Barack H. 2012b. "Remarks by the President in a News Conference." In *President Barack Obama, Speeches and Remarks*, 14 November 2012. Washington, DC: White House. Accessed on 7 October 2014, http://www.whitehouse.gov//the-press-office/2012/11/14/remarks-president-news-conference.

PEW. 2009. "Economy, Jobs Trump All Other Policy Priorities." In *2009: Environmental, Immigration, Health Care Slip Down the List*, edited by Andrew Kohut. Washington, DC: The PEW Research Cente.

Reich, Jenny, John Hearps, Andrew Cohn, Jeromy Temple, and Peter McDonald. 2006. *Defence Personnel Environment Scan 2025*. Canberra: Director Strategic Personnel Planning and Research for Department of Defence. Accessed on 13 February 2012, http://trove.nla.gov.au/work/26715104?selectedversion=NBD41326420.

Revkin, Andrew C. 2005. "Bush Aide Softened Greenhouse Gas Links to Global Warming." *The New York Times*, Arthur Ochs Sulzberger, 8 June 2005. Accessed on 10 January 2017, http://www.nytimes.com/2005/06/08/politics/bush-aide-softened-greenhouse-gas-links-to-global-warming.html.

Roughead, Gary. 2011b. "Chief of Naval Operations Adm. Gary Roughead delivers remarks at the 4th Symposium on the Impacts of an Ice-Diminishing Arctic on Naval & Maritime Operations." In *2011 Speeches: ADM Gary Roughead*, 20 June 2011. Washington, DC: US Navy. Accessed on 7 September 2013, http://www.navy.mil/navydata/people/cno/Roughead/Speech/110620%20Arctic%20Symposium.pdf.

Rudd, Kevin M. 2008a. "Address to the East Asia Forum in conjunction with the Australian National University, Advancing Australia's Global and Regional Economic Interests." In *PM Transcripts: Transcripts by the Prime Ministers of Australia*, 26 March 2008. Canberra: Department of Prime Minister and Cabinet. Accessed on 11 February 2014, http://pmtranscripts.pmc.gov.au/release/transcript-15823.

Rudd, Kevin M. 2008b. "Address to the London School of Economics: Australia and the UK—Global Partners in shaping the future Global Order, London." In *PM Transcripts: Transcripts from the Prime Ministers of Australia*, 7 April 2008. Canberra: Department of Prime Minister and Cabinet. Accessed on 18 February 2014, http://pmtranscripts.pmc.gov.au/release/transcript-15847.

Rudd, Kevin M. 2008f. "Climate Change and Water." Canberra: National Library of Australia. Accessed on 10 February 2014, http://pandora.nla.gov.au/pan/79983/20080812-0001/www.pm.gov.au/topics/climate.html.

Rudd, Kevin M. 2008g. "The First National Security Statement to the Parliament Address by the Prime Minister of Australia The Hon. Kevin Rudd MP." In *PM Transcripts: Transcripts of the Prime Ministers of Australia*, 4 December 2008. Canberra: Department of Prime Minister and Cabinet. Accessed on 5 March 2013, http://pmtranscripts.pmc.gov.au/release/transcript-16289.

Rudd, Kevin M. 2009c. "Address to the Lowy Institute." In *PM Transcripts: Transcripts from the Prime Ministers of Australia*, 06 November 2009. Canberra: Department of Prime Minister and Cabinet. Accessed on 12 March 2014, http://pmtranscripts.pmc.gov.au/release/transcript-16904.

Rudd, Kevin M. 2009d. "Focus on Climate Change—Prime Minister Blog." Canberra: National Library of Australia. Accessed on 11 February 2014, http://pandora.nla.gov.au/pan/79983/20091030-1529/www.pm.gov.au/PM_Connect/PMs_Blog/Climate_Change_Blog.html.

Rudd, Kevin M. 2009f. "Transcript of address at the close of the Pacific Islands Forum Cairns." In *PM Transcripts: Transcripts from the Prime Ministers of Australia*, 7 August 2009. Canberra: Department of Prime Minister and Cabinet. Accessed on 4 March 2013, https://pmtranscripts.pmc.gov.au/release/transcript-16737.

Rudd, Kevin M. 2010b. "Prime Minister Transcript of interview with Kerry O'Brien The 7.30 Report 12 May 2010." In *PM Transcripts: Transcripts from the Prime Ministers of Australia*, 12 May 2010. Canberra: Department of Prime Minister and Cabinet. Accessed on 4 March 2014, https://pmtranscripts.pmc.gov.au/release/transcript-17292.

Tauberer, Josh. 2015. "H.R. 1961 (100th): Global Climate Change Security Oversight Act." Ben Hammer. Published online: Civic Impulse, LLC. Accessed on 10 November 2015, https://www.govtrack.us/congress/bills/110/hr1961.

Terreu, Julia. 2014. "Can the Australian Military Think Out Loud?." In *The Diplomat*, edited by James Pach, 1 April 2014. Toyko: Trans-Asia Inc. Accessed on 12 March 2015, http://thediplomat.com/2014/04/can-the-australian-military-think-out-loud/.

Thomson, Mark. 2013. *The Cost of Defence: ASPI Defence Budget Brief 2013–14*. Canberra: Australian Strategic Policy Institute. Accessed on 15 May 2014, https://www.aspi.org.au/publications/the-cost-of-defence-aspi-defence-budget-brief-2013-2014/ASPI-CostDefence2013.pdf.

Titley, David W. 2010a. "Climate Change Finance: Providing assistance for vulnerable countries." In *Hearing Before the Subcommittee on Asia, the Pacific and the Global Environment of the Committee on Foreign Affairs House of Representatives One Hundred Eleventh Congress Second Session*, 111th Congress, Serial No. 111–120, 27 July 2010. Washington, DC: US Government Printing Office. http://www.gpo.gov/fdsys/pkg/CHRG-111hhrg57687/html/CHRG-111hhrg57687.htm.

Titley, David W. 2010c. "Global Climate Change." Climate and Energy Symposium 2010: Proceedings on Climate and Energy, Imperatives for Future Naval Forces, Baltimore. Accessed on 18 October 2012, http://www.jhuapl.edu/ClimateAndEnergy/Book/2010CEBook.pdf.

US DoD. 2011. *The National Military Strategy of the United States of America 2011: Redefining America's Military Leadership*. Washington, DC: Department of Defense. Accessed on 5 March 2013, http://www.defense.gov/Portals/1/Documents/pubs/2011-National-Military-Strategy.pdf.

US DoD. 2013a. *Department of Defense Strategic Sustainability Performance Plan FY-2013*. Washington, DC: Department of Defense. Accessed on 5 March 2014, http://www.denix.osd.mil/sustainability/loader.cfm?csModule=security/getfile&pageid=35931.

US House of Representatives. 2013. "President Obama's Fiscal 2014 Budget Proposal for the U.S. Navy and Marine Corps." In *House Appropriations Subcommittee on Defense Hearing*, 7 May 2013. Washington, DC: US Government Printing Office. Accessed on 6 March 2014, http://www.navy.mil/navydata/people/cno/Greenert/Testimony/130507%20HACD%20Hearing%20on%202014%20Navy%20and%20Marine%20Corps%20Budget.pdf.

Watts, Jonathon, and John Vidal. 2009. "China blamed as anger mounts over climate deal." *The Guardian*, The Guardian, 20 December 2009. Accessed on 5 June 2014, http://www.theguardian.com/environment/2009/dec/20/china-blamed-copenhagen-climate-failure?guni=Article:in%20body%20link.

White House. 2010. *The National Security Strategy of the United States of America*. White House. Washington, DC: Government of the United States of America. Accessed 20 May 2017, http://nssarchive.us/NSSR/2010.pdf.

White, Martin. 2008. "The Compelling Requirement to Energy-Proof the Australian Defence Force." *Journal of the Australian Profession of Arms*, no. 175:12–21.

Zhang, Chun. 2009. "Regional Reactions to the Defence White Paper 2009: Rebuilding middle power leadership for Australia." *Australian Defence Force Journal: Journal of the Australian Profession of Arms*, no. 180:7–9.

Part IV
Conclusion

Chapter 8
Conclusion and Outlook

This book set out to examine the securitisation of climate change within the political and military sectors of Australia, with a comparison to the United States. Using two well-known analytical frameworks—the Copenhagen School and Paris School—the book adopted a hybridised qualitative and quantitative method that analysed a large sample of documents from across the political and military levels. Research results indicated both differences and similarities in climate securitisation. In Australia, climate change became a security issue with the election of a centre-left government from late 2007 (Australian Labor Party led by Prime Minister Kevin Rudd and then Julia Gillard). Previous to that, the centre-right government, led by Prime Minister John Howard, did not view climate change as a security issue. The pattern in the US was slightly different. While President Bush never framed climate change as a security issue (like Howard) and his administration was generally against strong climate action, he did authorise the 2008 NDAA which served as the basis for a considered US DoD climate response. This segued neatly with the incoming Obama Administration, whereupon the president frequently framed climate change as a significant security threat. Contributing factors—other than ideological framing—included the point that climate change gradually rose to greater prominence in US security circles from 2003 but took-off notably from 2007 upon the release of a number of highly publicised reports, congressional hearings on the security implications of climate change and an increasing number of scholarly works, think-tank publications and media articles. Securitisation was also evident around this time in major international and regional forums, including divisive discussion at the UN Security Council.

The ascendancy of centre-left governments in Australia and the US favourable to climate action witnessed a significant uptake in the respective military climate responses. In the US military this was far more pronounced where strategic guidance in high-level capstone national security documents signalled approval for large-scale uptake by the lower levels within the military services and military-bureaucracy more broadly. Arguably, the US military, with a long-standing and pervasive interest in energy security, came to associate climate change both as a

M.D. Thomas, *The Securitization of Climate Change: Australian and United States' Military Responses (2003–2013)*, The Anthropocene: Politik—Economics—Society—Science 10, DOI 10.1007/978-3-319-49658-0_8

threat to its global strategic interests (in terms of sea-level rise, extreme weather, resource competition and so on) but also as an *opportunity* to absolve itself from the tether of fossil fuels and shift its energy basis toward a larger proportion of renewables (including nuclear, solar, geothermal, wind biofuels) for both its mobile platforms and its static bases. A sense that climate change was opening the Arctic was also a major driver in US military efforts on climate change. Underpinning this outlook, however, was the maintenance of US global supremacy in a time of rapid global change and the requirement to find cost savings in a time of fiscal austerity.

Overall, the US military did not seek to securitise climate change as a means to pressure politicians into adopting extraordinary measures. Accordingly, it never convincingly framed climate change as an existential threat but rather as a broader socio-political and national security challenge requiring a strategic, consistent and long-term approach. In this regard, the US military adopted a range of measures designed to enhance its resilience to climate change but that simultaneously increased its competitive edge over other militaries. Its focus on the Arctic stood out as the prime example. In this context only, the US military arguably *militarised* climate change in the context of national security discourse. It was, to some extent, a 'climatization of the security field' in which the US military was 'instrumentalized for adjusting the national security apparatus to new tasks by creating new military missions' (Oswald Spring/Brauch 2011: 1493).

Notwithstanding, while some measures might be construed as encroaching into areas over which the military had no traditional role, there were other aspects that may ultimately yield important outcomes. Increased research and development investments and large-scale roll-out of renewable energy programme and infrastructure was one such example. Another was the implementation of programme and policies aimed to reduce US military greenhouse gas emissions. Indeed, the role of the US military as an incubator for transformational technology has historical precedent and its ability to deliver transformational renewable technology, or as a minimum, a secure market for commercialisation, cannot be dismissed lightly (DiPeso 2010). The political influence that the US military can exert—by the very act of even addressing climate change in some form or another—also cannot be downplayed (particularly where linked to investments and jobs). As a minimum, the very act of the US military raising the issue of climate change can have a normalising affect, contributing to help set the possibility of the necessary political conditions in which Republicans can legitimately act on emissions reductions. Lastly, the notion of militaries expanding their HADR capabilities and commitments might also be viewed as a positive development; if increases in extreme weather unfold, then there are few institutions capable of offering the kind of large-scale rapid response required during disaster relief. If this so happens to expand military-to-military engagement, this may likewise be regarded in a positive manner and as a means to reducing tensions in other intractable traditional security areas. Such a perspective is now evident in the 2014 US *Quadrennial Defence Review* that has shifted from a narrow perspective of how climate change might impact US military infrastructure to understanding how the US military might use it to contribute to capacity building in vulnerable countries. US Pacific Command

(PACOM) partnering with small island developing states (recent examples include Kiribati and Maldives) to enhance engineering support against climate-change-induced degradation is one example (Olson 2014). More significantly for global security, this may be a centre ground for the US and China to act as a force for good throughout the Asia-Pacific.

The overarching trend here was the importance of the US military in establishing climate change as a mainstream national security issue that contributed to the development of new mechanisms and established norms. The creation of new normative standards—whereby climate change has been registered as a mainstream national security issue requiring resolute action—may be extended to emphasise other forms of security, including human and environmental.

Despite this assessment, the recent election of Donald Trump will bring greater clarity of the longer-term influence and commitment of the US military in climate change discourses. With the rise of the political right, manifestly against climate action, it might be anticipated that the US military 'role' in climate change will be scaled back. If this were to occur, then the Australian example might prove a timely case study. In this instance, while the Australian military expanded its climate change response in an effort to meet the government agenda, it then retreated from having larger involvement as the political situation became increasingly partisan and the new government changed tact. This situation gave rise to the difficulty of a *non-partisan* institution responding to what became a *politicised* security issue. Thus, climate securitisation in Australia lost all momentum following the failure of the Copenhagen conference and a domestic political stand-off to deliver Prime Minister Kevin Rudd the legitimacy to force his centrepiece Carbon Pollution Reduction Scheme through the upper house of the Australian Parliament. Perforce, while it continued to be framed as a national security issue in the major national security publications (2013 and 2016 Defence White Papers and the 2013 National Security Strategy), the Australian military was largely absent from the national discourse occurring on climate change. This situation was reinforced with the election of a centre-right government in late 2013 which set about repealing much of the broader climate change agenda implemented under the previous centre-left government. This has not just been limited to social and economic policy sphere, but also in the security sphere.

Having made this assessment, it can now be asked—is this situation desirable? Moreover, *if* climate change was (and remains) an existential threat and *if* the current decade represents the critical decade to avoid these threats, then it may become increasingly difficult for the Australian military, the security institutions and security professionals more broadly to continue to justify a minimalist climate change outlook. Should they play a more proactive role? If so, what is this role and do notions of climate-securitisation have any relevance in elevating it as a policy priority? More broadly—and looking to the future—do the Australian military and associated federal intelligence and assessment institutions have a unique obligation to extricate themselves from the *politics* of climate change, but not beyond its democratic processes, so that they might conduct the necessary strategic planning and provide the public an unfettered appreciation and clearer picture of the security risks involved? This assessment directly resonates with what US political scientist Richard Betts characterised as the susceptibility of government agencies to 'illegitimate politicization' that 'fabricates or distorts information to serve policy preferences or vested interests' rather than the greater good (Betts 2005: 57). It suffocates the universal norm that 'intelligence judgements be more objective, non-partisan, and scientific than other judgements' (ibid). Thus held, what are the broader implications for Australian strategic policy and how might climate change affect the future relationship between the military, the government and the people (particularly where the security threats continue to be viewed in such a partisan manner)? Navigating these strategic challenges will require political *nous*, since: 'How does the military, hoping to develop intellectual grit, manoeuvre around the very institution that feeds it?' (Terreu 2014).

The original authors of the Copenhagen securitisation theory stated that understanding securitisation is more difficult when it is 'moved out of the military' (Buzan et al. 1998: 1). This may or may not be so, but as the role, influence and capacity of military institutions in responding to climate change unfolds, there should be recognition that this remains a nascent discourse that presents both opportunity and risk. The irony of this situation should not be lost. So far as climate change is concerned, militaries that were once ostensibly designed to protect *us* from *them*, are incapable of protecting *us* from *ourselves*. Military institutions across the world should reflect deeply on this, and question what their contribution will be to addressing this unprecedented strategic challenge.

References

Betts, Richard K. 2005. "Politicization of Intelligence Costs and Benefits." In *Paradoxes of Strategic Intelligence: Essays in Honor of Michael I. Handel*, edited by Richard K. Betts and Thomas G. Mehnken, pp. 57–76. Portland: Frank Cass Publishers.

Buzan, Barry, Ole Wæver, and Jaap de-Wilde. 1998. *Security: A New Framework for Analysis*. Boulder: Lynne Rienner Publishers.

DiPeso, Jim. 2010. "Can DOD lead the way to a better energy future?" *Environmental Quality Management*, vol. 20, no. 2:95–100. doi:10.1002/tqem.20283.

Olson, Wyatt. 2014. "PACOM not waiting on politics to plan for climate change challenges." Hawaii: Stars and Stripes. Accessed on 14 December 2014, http://www.stripes.com/news/pacom-not-waiting-on-politics-to-plan-for-climate-change-challenges-1.297433.

Oswald Spring, Úrsula, and Hans Günter Brauch. 2011. "Coping with Global Environmental Change—Sustainability Revolution and Sustainable Peace." In *Coping with Global Environmental Change, Disasters and Security: Threats, Challenges, Vulnerabilities and Risks*, edited by Hans Gunter Brauch, Ursula Oswald Spring, Czeslaw Mesjasz, John Grin, Patricia Kameri-Mbote, Bechir Chourou, Pal Dunay and Jorn Birkmann, vol. 5, pp. 1487–1503. Berlin: Springer.

Terreu, Julia. 2014. "Can the Australian Military Think Out Loud?" In *The Diplomat*, edited by James Pach, 1 April 2014. Toyko: Trans-Asia Inc. Accessed on 12 March 2015, http://thediplomat.com/2014/04/can-the-australian-military-think-out-loud/.

Bibliography

Alexander, Natalie. 2011. "Climate Change on radar." In *Defence Magazine*, edited by Michael Weaver, vol. 3, pp. 30–31. Canberra: Department of Defence. Accessed on 12 April 2013, http://www.defence.gov.au/defencemagazine/editions/2011_03/mag.pdf.

ANAO. 2014. *The Auditor-General, Audit Report No. 24 2013–14 Performance Audit: Emergency Defence Assistance to the Civil Community*. The Auditor-General's Department. Canberra: Commonwealth of Australia. Accessed on 20 May 2017, https://www.anao.gov.au/sites/g/files/net616/f/AuditReport_2013-2014_24.pdf.

Andrews, Anthony. 2009. *Department of Defence Facilities Energy Conservation Policies and Spending*. Congressional Research Service. Washington, DC: US Government Printing Office. Accessed on 11 November 2013, http://www.fas.org/sgp/crs/natsec/R40111.pdf.

An act making supplemental appropriations for job preservation and creation, infrastructure investment, energy efficiency and science, assistance to the unemployed, and State and local fiscal stabilization, for the fiscal year ending September 30, 2009, and for other purposes. American Recovery and Reinvestment Act (ARRA). 111th US Congress, H.R. 1, Public Law 111-5, 17 February 2009. Accessed on 3 March 2013, http://www.gpo.gov/fdsys/pkg/PLAW-111publ5/html/PLAW-111publ5.htm.

Arrhenius, Svante. 1896. "On the Influence of Carbonic Acid in the Air upon the Temperature on the Ground." *Philosophical Magazine and Journal of Science*, vol. 41, no. 5:237–276.

Asian Development Bank. 2010. *Responding to Climate Change in the Pacific: Moving from Strategy to Action*. Pacific Studies Series: Manila: Asian Development Bank. Accessed on 18 July 2012, http://www.adb.org/sites/default/files/publication/28008/responding-climate-change.pdf.

ASP. 2016. "Global Security Defence Index on Climate Change." Washington DC: American Security Project. Accessed on 22 November 2016, https://www.americansecurityproject.org/wp-content/uploads/2014/09/Ref-0168-GSDI-Introduction-2014.pdf.

Aulich, Chris. 2010. "It was the best of times; it was the worst of times." In *The Rudd Government: Australian Commonwealth Administration 2007–2010*, edited by Chris Aulich and Mark Evans, pp. 3–15. Canberra: ANU E Press. Accessed on 15 June 2013, http://press.anu.edu.au/apps/bookworm/view/The+Rudd+Government%3A+Australian+Commonwealth+Administration+%092007+-+2010/5091/ch01.xhtml.

Australian Army. 2009. *Adaptive Campaigning: Army's Future Land Operating Concept*. Department of Defence. Canberra: Directorate of Future Land Warfare. Accessed on 20 May 2017, https://www.army.gov.au/sites/g/files/net1846/f/acfloc_2012_main.pdf.

Australian Army. 2014. *Future Land Warfare*. Department of Defence. Canberra: Directorate of Future Land Warfare. Accessed on 5 May 2017, https://www.army.gov.au/sites/g/files/net1846/f/flwr_web_b5_final.pdf.

Australian Bureau of Statistics. 2008. *Year Book Australia: A comprehensive source of information about Australia*. Vol. 90, *Australian Year Book*. Canberra: Commonwealth of Australia.

M.D. Thomas, *The Securitization of Climate Change: Australian and United States' Military Responses (2003–2013)*, The Anthropocene: Politik—Economics—Society—Science 10, DOI 10.1007/978-3-319-49658-0

Bächler, Günther, David J. Eaton, Joseph W. Eaton, Malin Falkenmark, Partha S. Ghosh, Jad Isaac, Nurit Kliot, John Markakis, Robert Mayor, Katrina S. Rogers, Vaclav Smil, Kurt R. Spillmann, Ashok Swain, Peter Wallensteen, and Arthur H. Westing. 1995. *Environmental Crisis: Regional Conflicts and Ways of Cooperation*. CSS Environment and Conflicts Project: 14. Zurich.

Baer, Paul, Tom Athanasiou, Sivan Kartha, and Eric Kemp-Benedict. 2010. "Greenhouse Development Rights: A Framework for Climate Protection That Is "More Fair" Than Equal Per Capita Emissions Rights." In *Climate Ethics: Essential Readings*, edited by Stephen M. Gardiner, Simon Caney, Dale Jamieson and Henry Shue. Oxford: Oxford University Press.

Baker & McKenzie. 2011. *Local Council Risk of Liability in the Face of Climate Change—Resolving Uncertainties A Report for the Australian Local Government Association*. Sydney: Australian Local Government Association. Accessed on 10 October 2015, http://www.environment.gov.au/system/files/resources/d9b2f9cf-d7ab-4fa0-ab0e-483036079dc7/files/alga-report.pdf.

Barker, T., I. Bashmakov, L. Bernstein, J. E. Bogner, P. R. Bosch, R. Dave, O. R. Davidson, B. S. Fisher, S. Gupta, K. Halsnæs, G.J. Heij, S. Kahn Ribeiro, S. Kobayashi, M. D. Levine, D. L. Martino, O. Masera, B. Metz, L. A. Meyer, G. J. Nabuurs, A. Najam, N. Nakicenovic, H. H. Rogner, J. Roy, J. Sathaye, R. Schock, P. Shukla, R. E. H. Sims, P. Smith, D. A. Tirpak, D. Urge-Vorsatz, and D. Zhou. 2007. "Technical Summary." In *Contribution of Working Group III to the Fourth Assessment Report of the Intergovernmental Panel on Climate Change*, edited by B. Metz, O. R. Davidson, P. R. Bosch, R. Dave and L. A. Meyer. Cambridge, United Kingdom: Cambridge University Press.

Barnett, Jon. 2003. "Security and climate change." *Global Environmental Change*, vol. 13, no. 1:7–17. doi:10.1016/S0959-3780(02)00080-8.

Barnett, Jon. 2009. "Environmental Security in the Asia-Pacific Region: Contrasting Problems, Places, and Prospects." In *Facing Global Environmental Change Environmental, Human, Energy, Food, Health and Water Security Concepts*, edited by Hans Günter Brauch, Úrsula Oswald Spring, John Grin, Czeslaw Mesjasz, Patricia Kameri-Mbote, Navnita Chadha Behera, Béchir Chourou and Heinz Krummenacher, vol. 4, pp. 939–950. Berlin: Springer.

Barnett, Jon, Richard A. Matthew, and Karen L. O'Brien. 2010. "Global Environmental Change and Human Security: An Introduction." In *Global Environmental Change and Human Security*, edited by Richard A. Matthew, Jon Barnett, Bryan McDonald and Karen L. O'Brien, pp. 3–32. Massachusetts: The MIT Press.

Barrie, Chris, and Will Steffen. 2015. "Statement from the co-chairs of the Australian Climate Security Summit." Sydney: Climate Council. Accessed on 8 January 2017, https://www.climatecouncil.org.au/statement-from-the-co-chairs-of-the-australian-climate-security-summit.

Barrie, Chris, Will Steffen, Alix Pearce, and Michael Thomas. 2015. *Be Prepared: Climate Change, Security and Australia's Defence Force*. Sydney: Climate Council. Accessed on 20 September 2015, https://www.climatecouncil.org.au/uploads/fa8b3c7d4c6477720434d6d10897af18.pdf.

Bartin, James T., and Larence Van Bibber. 2011. *Alternative Fuels for Military Applications*. Santa Monica: RAND Corporation. Accessed on 17 March 2014, http://www.rand.org/content/dam/rand/pubs/monographs/2011/RAND_MG969.pdf.

Beckett, Margaret. 2007. "The Urgent Case for Climate Security." London: Royal United Services Institute. Published online 10 May 2007. Accessed 15 May 2017, https://rusi.org/publication/urgent-case-climate-security.

Bergin, Anthony. 2016. "Defence must regard climate change as a serious issue." *The Australian*, News Limited, 2 December 2016. Accessed on 20 May 2017, https://www.aspi.org.au/opinion/defence-must-regard-climate-change-as-a-serious-security-issue.

Bergin, Anthony, and Jacob Townsend. 2007. *A change in climate for the Australian Defence Force*. Special Report: 7. Canberra: Australian Strategic Policy Institute. Accessed on 12 February 2013, https://www.aspi.org.au/publications/special-report-issue-7-a-change-in-climate-for-the-australian-defence-force/SR7_climagechange_ADF.pdf.

Berman, Alan. 2016. "A Vote For Turnbull Could Be A Vote For Abbott." *The Huffington Post*, AOL, 5 July 2016. Accessed on 9 December 2016, http://www.huffingtonpost.com.au/alan-berman/a-vote-for-turnbull-could-be-a-vote-for-abbott/.

Betts, Richard K. 2005. "Politicization of Intelligence Costs and Benefits." In *Paradoxes of Strategic Intelligence: Essays in Honor of Michael I. Handel*, edited by Richard K. Betts and Thomas G. Mehnken, pp. 57–76. Portland: Frank Cass Publishers.

Bigo, Didier. 2006. "Security, Exception, Ban and Surveillance." In *Theorizing Surveillance: The panopticon and beyond*, edited by D. Lyon, pp. 46–68. Davon: Willan Publishing.

Bigo, Didier. 2008. "Globalized (in)security: The field and the ban-opticon." In *Terror, Insecurity and Liberty: Illiberal practices of liberal regimes after 9/11*, edited by Didier Bigo, Elspeth Guild and R. B. J. Walker, pp. 10–48. Abingdon: Routledge.

Bigo, Didier, and Anastassia Tsoukala. 2008. "Understanding (in)security." In *Terror, Insecurity and Liberty: Illiberal practices of liberal regimes after 9/11*, edited by Didier Bigo, Elspeth Guild and R. B. J. Walker, pp. 1–9. Abingdon: Routledge.

Bigo, Didier, Anastassia Tsoukala, Laurent Bonelli, Emmanuel-Pierra Guittet, and Christian Olsson. 2008. *Terror, Insecurity and Liberty: Illiberal practices of liberal regimes after 9/11*. Edited by Didier Bigo, Elspeth Guild and R. B. J. Walker, *Routledge Studies in Liberty and Security*. Abingdon: Routledge.

Blasing, T. J. 2012. "Recent Greenhouse Gas Concentrations." Carbon Dioxide Information Analysis Centre (CDIAC). Accessed on 28 January 2013, http://cdiac.ornl.gov/pns/current_ghg.html.

Bolt, Andrew. 2014. "Tony Abbott has arrived as a true leader." *Herald Sun*, News Limited, 3 September 2014. Accessed on 9 December 2016, http://www.heraldsun.com.au/news/opinion/andrew-bolt/tony-abbott-has-arrived-as-a-true-leader/news-story/88238eab8e7da2705b6de64aec04853c.

Bourdieu, Pierre. 1969. "Intellectual field and creative project." *Social Science Information*, vol. 8, no. 2:89–119. doi:10.1177/053901846900800205.

Brauch, Hans Günter. 2009. "Securitizing Global Environmental Change." In *Facing Global Environmental Change: Environmental, Human, Energy, Food, Health and Water Security Concepts*, edited by Hans Günter Brauch, Úrsula Oswald Spring, John Grin, Czeslaw Mesjasz, Patricia Kameri-Mbote, Navnita Chadha Behera, Bechir Chourou and Heinz Krummenacher, vol. 4, pp. 65–102. Berlin: Springer.

Brauch, Hans Günter. 2011a. "Concepts of Security Threats, Challenges, Vulnerabilities and Risks." In *Coping with Global Environmental Change, Disasters and Security: Threats, Challenges, Vulnerabilities and Risks*, edited by Hans Gunter Brauch, Úrsula Oswald Spring, Czeslaw Mesjasz, John Grin, Patricia Kameri-Mbote, Bechir Chourou, Pal Dunay and Jorn Birkmann, vol. 5, pp. 61–106. Berlin: Springer.

Brauch, Hans Günter. 2011b. "Security Threats, Challenges, Vulnerabilities and Risks in US National Security Documents (1990–2010)." In *Coping with Global Environmental Change, Disasters and Security: Threats, Challenges, Vulnerabilities and Risks*, edited by Hans Günter Brauch, Ursula Oswald Spring, Czeslaw Mesjasz, John Grin, Patricia Kameri-Mbote, Bechir Chourou, Pal Dunay and Jorn Birkmann, vol. 5, pp. 249–274. Berlin: Springer.

Brown, Lester R. 1977. *Redefining National Security*. Washington, DC: Worldwatch Institute.

Brown, Lester R. 1986. "Redefining National Security." *Challenge*, vol. 29, no. 3:25–32.

Brull, Michael. 2016. "Into The Abyss: Right On Cue, Desperate Malcolm Turnbull Turns On Minorities." *New Matilda*, At Large Media, 31 October 2016. Accessed on 9 December 2016, https://newmatilda.com/2016/10/31/into-the-abyss-right-on-cue-desperate-malcolm-turnbull-turns-on-minorities/.

Brzoska, Michael. 2012a. "Climate change and the military in China, Russia, and the United Kingdom, and the United States." *Bulletin of the Atomic Scientists*, vol. 68, no. 2:43–54. doi:10.1177/0096340212438384.

Brzoska, Michael. 2012b. "Climate Change as a Driver of Security Policy." In *Climate Change, Human Security and Violent Conflict: Challenges for Societal Stability*, edited by Jurgen Scheffran, Michael Brzoska, Hans Gunter Brauch, Peter Michael Link and Janpeter Schilling. Berlin: Springer.

Bulkeley, Harriet, and Peter Newell. 2010. *Governing Climate Change*. Edited by Thomas G. Weiss, *Global Institutions Series*. London: Routledge.

Busby, Joshua W. 2007. *Climate Change and National Security: An Agenda for Action*. Council on Foreign Relations Special Report: 32. New York: Council on Foreign Relations Press.

Bush, George W. 2001. "Address before a Joint Session of the Congress on Administration Goals." In *Compilation of Presidential Documents*, 5 March 2001, vol. 37, no. 9, pp. 351–357. Washington, DC: US Government Printing Office. Accessed on 12 June 2013, http://www.gpo.gov/fdsys/granule/WCPD-2001-03-05/WCPD-2001-03-05-Pg351-2/content-detail.html.

Bush, George W. 2002a. "Address before a Joint Session of the Congress on the State of the Union." In *Compilation of Presidential Documents*, 4 February 2002, vol. 38, no. 5 pp. 133–139. Washington, DC: US Government Printing Office. Accessed on 14 July 2013, http://www.gpo.gov/fdsys/granule/WCPD-2002-02-04/WCPD-2002-02-04-Pg133-3/content-detail.html.

Bush, George W. 2002b. "Remarks announcing the Clear Skies and Global Climate Change Initiatives in Silver Spring, Maryland." In *Compilation of Presidential Documents*, 18 February 2002, vol. 38, no. 7, pp. 232–236. Washington, DC: US Government Printing Office. Accessed on 3 September 2013, http://www.gpo.gov/fdsys/granule/WCPD-2002-02-18/WCPD-2002-02-18-Pg232/content-detail.html.

Bush, George W. 2003. "Remarks On Energy Independence." In *Compilation of Presidential Documents*, 10 February 2003, vol. 39, no. 6, pp. 161–164. Washington, DC: US Government Printing Office. Accessed on 10 August 2013, http://www.gpo.gov/fdsys/granule/WCPD-2003-02-10/WCPD-2003-02-10-Pg161/content-detail.html.

Bush, George W. 2005. "Address before a Joint Session of the Congress on the State of the Union." In *Public Papers of the Presidents of the United States: George W. Bush* 2 February 2005, vol. (Book) I, pp. 113–121. Washington, DC: US Government Printing Office. Accessed on 4 August 2013, http://www.gpo.gov/fdsys/granule/PPP-2005-book1/PPP-2005-book1-doc-pg113/content-detail.html.

Bush, George W. 2006. "Address before a Joint Session of the Congress on the State of the Union." In *Public Papers of the Presidents of the United States: George W. Bush* 31 January 2006, vol. (Book) I, pp. 146–154. Washington, DC: US Government Printing Office. Accessed on 17 August 2013, http://www.gpo.gov/fdsys/granule/PPP-2006-book1/PPP-2006-book1-doc-pg146-2/content-detail.html.

Bush, George W. 2007a. "Address before a Joint Session of the Congress on the State of the Union." In *Compilation of Presidential Documents*, 29 January 2007, vol. 43, no. 4, pp. 57–64. Washington, DC: US Government Printing Office. Accessed on 12 September 2013, http://www.gpo.gov/fdsys/granule/WCPD-2007-01-29/WCPD-2007-01-29-Pg57/content-detail.html.

Bush, George W. 2007b. "The President's News Conference." In *Compilation of Presidential Documents*, 9 April 2007, vol. 43, no. 14, pp. 409–417. Washington, DC: US Government Printing Office. Accessed on 21 September 2013, http://www.gpo.gov/fdsys/granule/WCPD-2007-04-09/WCPD-2007-04-09-Pg409-2/content-detail.html.

Bush, George W. 2007c. "The President's News Conference." In *Public Papers of the Presidents of the United States: George W. Bush* 20 December 2007, vol. (Book) II, pp. 1573–1585. Washington, DC: US Government Printing Office. Accessed on 10 August 2013, http://www.gpo.gov/fdsys/granule/PPP-2007-book2/PPP-2007-book2-doc-pg1573-2/content-detail.html.

Bush, George W. 2007d. "Remarks during a meeting on Energy Security and Climate Change." In *Compilation of Presidential Documents* 1 October 2007, vol. 43, no. 39, pp. 1261–1265. Washington, DC: US Government Printing Office. Accessed on 8 October 2013, http://www.gpo.gov/fdsys/granule/WCPD-2007-10-01/WCPD-2007-10-01-Pg1261/content-detail.html.

Bush, George W. 2007e. "Remarks on Energy in Athens, Alabama." In *Compilation of Presidential Documents* 25 June 2007, vol. 43, no. 25, pp. 837–842. Washington, DC: US Government Printing Office Accessed on 23 August 2013, http://www.gpo.gov/fdsys/granule/WCPD-2007-06-25/WCPD-2007-06-25-Pg837/content-detail.html.

Bush, George W. 2008a. "Address before a Joint Session of the Congress on the State of the Union." In *Public Papers of the Presidents of the United States: George W. Bush*, 28 January 2008, vol. (Book) I, pp. 116–125. Washington, DC: US Government Printing Office. Accessed on 12 August 2013, http://www.gpo.gov/fdsys/granule/PPP-2008-book1/PPP-2008-book1-doc-pg116/content-detail.html.

Bush, George W. 2008b. "The President's News Conference." In *Compilation of Presidential Documents* 7 July 2008, vol. 44, no. 26, pp. 939–943. Washington, DC: US Government Printing Office. Accessed on 12 August 2013, http://www.gpo.gov/fdsys/granule/WCPD-2008-07-07/WCPD-2008-07-07-Pg939/content-detail.html.

Bush, George W. 2008c. "Remarks at the Washington International Renewable Energy Conference." In *Compilation of Presidential Documents* 10 March 2008, vol. 44, no. 9, pp. 324–330. Washington, DC: US Government Printing Office. Accessed on 2 October 2013, http://www.gpo.gov/fdsys/granule/WCPD-2008-03-10/WCPD-2008-03-10-Pg324-2/content-detail.html.

Bush, George W. 2008d. "Remarks on Energy and Climate Change." In *Compilation of Presidential Documents* 21 April 2008, vol. 44, no. 15, pp. 524–527. Washington, DC: US Government Printing Office. Accessed on 12 September 2013, http://www.gpo.gov/fdsys/granule/WCPD-2008-04-21/WCPD-2008-04-21-Pg524/content-detail.html.

Bush, George W. 2009. "Directive on Arctic Region Policy." In *Administration of George W. Bush*, 12 January 2009, pp. 47–54. Washington, DC: US Government Printing Office. Accessed on 5 September 2013, http://www.gpo.gov/fdsys/pkg/WCPD-2009-01-19/pdf/WCPD-2009-01-19-Pg47-2.pdf.

Butler, Josh. 2016. "It Took Just 35 Hours For A Massive Backflip On Emissions Policy." *The Guardian*, The Guardian, 7 December 2016. Accessed on 7 January 2017, http://www.huffingtonpost.com.au/2016/12/06/it-took-just-35-hours-for-a-massive-backflip-on-emissions-policy/.

Buzan, Barry. 1991. *People, states and fear: an agenda for international security studies in the post-Cold War era*. London: Harvester Wheatsheaf.

Buzan, Barry, and Lene Hansen. 2009. *The Evolution of International Security Studies*. Cambridge: Cambridge University Press.

Buzan, Barry, Ole Wæver, and Jaap de-Wilde. 1998. *Security: A New Framework for Analysis*. Boulder: Lynne Rienner Publishers.

CAHA. 2010. "Climate change is a health issue." Published online: Climate and Health Alliance. Accessed on 10 March 2015, http://caha.org.au/wp-content/uploads/2011/03/CAHA-Briefing-Paper-1_Climate-change-is-a-health-issue1.pdf.

Campbell, Angus. 2016. "Chief of Army opening address to the 2016 Chief of Army's Exercise." In *Chief of Army: Speeches and Transcripts*. Canberra: Commonwealth of Australia. Accessed on 5 May 2017, https://www.army.gov.au/our-work/speeches-and-transcripts/chief-of-army-opening-address-to-the-2016-chief-of-armys-exercise.

Campbell, Kurt M., Jay Gulledge, J.R. McNeill, John Podesta, Peter Ogden, Leon Fuerth, R. James Woolsey, Alexander T.J. Lennon, Julianne Smith, Richard Weitz, and Derek Mix. 2007. *The Age of Consequences: The Foreign Policy and National Security Implications of Global Climate Change*. Washington, DC: Center for a New American Century.

Caney, Simon. 2010. "Cosmopolitan Justice, Responsibility, and Global Climate Change." In *Climate Ethics: Essential Readings*, edited by Stephen M. Gardiner, Simon Caney, Dale Jamieson and Henry Shue. Oxford: Oxford University Press.

Carmen, Herbert E., Christine Parthemore, and Will Rogers. 2010. *Broadening Horizons: Climate Change and the U.S. Armed Forces*. Washington, DC: Center for a New American Century. Accessed on 15 May 2017, https://s3.amazonaws.com/files.cnas.org/documents/CNAS_Broadening-Horizons_Carmen-Parthemore-Rogers.pdf.

Carson, Rachel. 1962. *Silent Spring*. Boston: Mariner.

CC&S. 2015. *The U.S. Asia-Pacific Rebalance, National Security and Climate Change*. A Climate and Security Correlations Series: Washington, DC. Accessed on 10 November 2016, https://climateandsecurity.files.wordpress.com/2015/11/ccs_us_asia_pacific-rebalance_national-security-and-climate-change.pdf.

CC&S. 2016. *Briefing Book for a New Administration: Recommended Policies and Practicies for Addressing the Security Risks of a Changing Climate*. Washington, DC. Accessed on 10 November 2016, https://climateandsecurity.files.wordpress.com/2016/09/climate-and-security-advisory-group_briefing-book-for-a-new-administration_2016_091.pdf.

CDIAC. 2015. "Global, Regional, and National Fossil-Fuel CO2 Emissions." T. A. Boden, G. Marland and R. J. Andres (Eds.). Oak Ridge National Laboratory, Tenn., USA: U.S. Department of Energy, Carbon Dioxide Information Analysis Center. Accessed on 9 November 2016, http://cdiac.ornl.gov/trends/emis/meth_reg.html.

Chaturvedi, Sanjay, and Timothy Doyle. 2010. "Geopolitics of Climate Change and Australia's 'Re-engagement' with Asia: Discourses of Fear and Cartographic Anxieties." *Australian Journal of Political Science*, vol. 45, no. 1:95–115. doi:10.1080/10361140903517734.

Christoff, Peter, and Robyn Eckersley. 2014. "No island is an island: Security in a Four Degree World." In *Four Degrees of Global Warming: Australia in a hot world*, edited by Peter Christoff, pp. 190–204. London: Routledge.

Chubb, Philip. 2014a. "The day the Rudd Government lost its way on climate change." *The Age*, Fairfax Media, 9 May 2014. Accessed on 26 May 2014, http://www.theage.com.au/insight/the-day-the-rudd-government-lost-its-way-on-climate-change-20140509-zr7fm.html.

Chubb, Philip. 2014b. *Power Failure: The Inside Story of Climate Politics Under Rudd and Gillard*. Melbourne: Black Inc. Agenda.

Clapper, James. 2015. "Statement for the Record Worldwide Threat Assessment of the US Intelligence Community." In *Senate Armed Services Committee*. Accessed on 20 November 2016, https://www.dni.gov/files/documents/Unclassified_2015_ATA_SFR_-_SASC_FINAL.pdf.

Clapper, James. 2016. "U.S. Intelligence as a Pillar of Stability during Transition." In *DNI Clapper's As Delivered Remarks at the 2016 INSA & AFCEA Intelligence & National Security Summit*. Washington DC: Office of the Director of National Intelligence. Accessed on 15 May 2017, https://www.dni.gov/index.php/newsroom/speeches-interviews/speeches-interviews-2016/item/1627-dni-clapper-s-as-delivered-remarks-at-the-2016-insa-afcea-intelligence-national-security-summit.

Clapper, James R. 2013. "Worldwide Threat Assessment to the Senate Select Committee on Intelligence." Washington DC: US Public Affairs Office. Accessed on 14 November 2016, https://www.dni.gov/files/documents/Intelligence%20Reports/WWTA%20Remarks%20as%20delivered%2012%20Mar%202013.pdf.

A Bill to address security risks posed by global climate change, and for other purposes. Global Climate Change Security Oversight Act (110th Congress, First session, H.R. 1961, (Bill was not enacted)). Accessed on 3 March 2013, http://www.gpo.gov/fdsys/pkg/BILLS-110hr1961ih/pdf/BILLS-110hr1961ih.pdf.

CNA. 2007. *National Security and The Threat of Climate Change*. Virginia: The CNA Corporation. Accessed on 12 April 2013, https://www.cna.org/cna_files/pdf/National%20Security%20and%20the%20Threat%20of%20Climate%20Change.pdf.

CNA. 2014. *National Security and the Threat of Climate Change*. Virginia: The CNA Corporation. Accessed 17 May 2017, https://www.cna.org/cna_files/pdf/MAB_5-8-14.pdf

Cole, Steve. 2010. "Climate Change Implications for the Royal Australian Navy." Climate and Energy Symposium 2010: Proceedings on Climate and Energy, Imperatives for Future Naval Forces, Baltimore, 23 & 24 March 2010. Accessed on 18 October 2012, http://www.jhuapl.edu/ClimateAndEnergy/Presentations/2010/RT5_Cole.pdf.

Commonwealth Bank. 2008. *Commonwealth Bank of Australia Environment Policy*. Published online: Commonwealth Bank. Accessed on 12 Ocober 2015, https://www.commbank.com.au/about-us/download-printed-forms/EnvironmentalPolicy_0808.pdf.

Commonwealth of Australia. 2004. *Securing Australia's Energy Future*. Department of the Prime Minister and Cabinet. Canberra: Commonwealth of Australia. Accessed on 15 March 2011, http://www.efa.com.au/Library/cthEnergyWhitePaper.pdf.

Commonwealth of Australia. 2006. "Supplementary Budget Estimates." In *Standing Committee on Foreign Affairs, Defence and Trade Official Committee Hansard*, 1 November 2006. Canberra: Commonwealth of Australia.

Commonwealth of Australia. 2007a. "Budget Estimates." In *Standing Committee on Foreign Affairs, Defence and Trade Hansard*, 31 May 2007. Canberra: Commonwealth of Australia. Accessed on 13 February 2011, http://www.aph.gov.au/~/media/Estimates/Live/commttee/S10269.ashx.

Commonwealth of Australia. 2007b. *National Climate Change Adaptation Framework*. Plan of Collaborative Action on Climate Change: Canberra: Commonwealth of Australia. Accessed on 2 October 2012, http://www.environment.gov.au/system/files/resources/eaaf0350-9781-4006-957c-a5801fadc466/files/nccaf.pdf.

Commonwealth of Australia. 2008a. "Budget Estimates." In *Standing Committee on Foreign Affairs, Defence and Trade Official Committee Hansard*, 4 June 2008. Canberra: Commonwealth of Australia. Accessed on 12 June 2012, http://www.aph.gov.au/Parliamentary_Business/Senate_Estimates/fadtctte/estimates/bud0809/index.

Commonwealth of Australia. 2008b. *Energy Use in the Australian Government's Operations 2006–2007*. Water Department of the Environment, Heritage and the Arts. Canberra: Department of the Environment, Water, Heritage and the Arts. Accessed on 4 June 2011, http://industry.gov.au/Energy/EnergyEfficiency/Non-residentialBuildings/GovernmentBuildings/Documents/report-2006.pdf.

Commonwealth of Australia. 2009a. *Climate Change Risks to Australia's Coast: A First Pass National Assessment*. Department of Climate Change. Canberra: Commonwealth of Australia. Accessed on 4 July 2012, https://www.environment.gov.au/system/files/resources/fa553e97-2ead-47bb-ac80-c12adffea944/files/cc-risks-full-report.pdf.

Commonwealth of Australia. 2009b. "Estimates—Additional Estimates." In *Standing Committee on Foreign Affairs, Defence and Trade Hansard*, 25 February 2009. Canberra: Commonwealth of Australia. Accessed on 12 May 2011, http://www.aph.gov.au/Parliamentary_Business/Senate_Estimates/fadtctte/estimates/add0809/index.

Commonwealth of Australia. 2010. *Adapting to Climate Change in Australia An Australian Government Position Paper*. Department of Climate Change. Canberra: Commonwealth of Australia. Accessed on 5 October 2012, http://www.environment.gov.au/system/files/resources/93e900d7-3307-41c8-88e1-e0aa314c4cb3/files/adapt-climate-change-position-paper.pdf.

Commonwealth of Australia. 2011. *Climate Change Risks to Coastal Buildings and Infrastructure: A supplement to the First-Pass National Assessment*. Department of Climate Change and Energy Efficiency. Canberra: Commonwealth of Australia. Accessed on 12 July 2012, https://www.environment.gov.au/system/files/resources/0f56e5e6-e25e-4183-bbef-ca61e56777ef/files/risks-coastal-buildings.pdf.

Commonwealth of Australia. 2012a. "Senate Budget Estimates Questions On Notice—Committees." In *Senate Standing Committee on Foreign Affairs, Defence and Trade Hansard*, 28 & 29 May 2012. Canberra: Commonwealth of Australia. Accessed on 17 July 2013, http://www.aph.gov.au/Parliamentary_Business/Senate_Estimates/fadtctte/estimates/bud1213/def/Department-of-Defence/defenceqonsindex.

Commonwealth of Australia. 2012b. "Senate Supplementary Budget Estimates Hearing." In *Senate Standing Committee on Foreign Affairs, Defence and Trade Hansard*, 17 October 2012. Canberra: Commonwealth of Australia.

Commonwealth of Australia. 2013a. *Energy Use in the Australian Government's Operations 2010–11*. Energy and Tourism Department of Resources. Canberra: Commonwealth of Australia. Accessed 20 May 2017, http://webarchive.nla.gov.au/gov/20160615065736/, http://www.industry.gov.au/Energy/EnergyEfficiency/Non-residentialBuildings/GovernmentBuildings/EnergyUseOperations/Pages/PreviousEditions.aspx).

Commonwealth of Australia. 2013b. "Energy Use in the Australian Government's Operations 2011–12: Annex C." Energy and Tourism Department of Resources. Canberra: Commonwealth of Australia. [Website archived, report available at: http://webarchive.nla.gov.au/gov/20160615065736/, http://www.industry.gov.au/Energy/EnergyEfficiency/Non-esidentialBuildings/GovernmentBuildings/EnergyUseOperations/Pages/PreviousEditions.aspx].

Commonwealth of Australia. 2013c. "Energy Use in the Australian Government's Operations (Reports 2001–2011)." Energy and Tourism Department of Resources. Canberra: Commonwealth of Australia. [Website archived, report available at: http://webarchive.nla.gov.au/gov/20160615065736/, http://www.industry.gov.au/Energy/EnergyEfficiency/Non-tesidentialBuildings/GovernmentBuildings/EnergyUseOperations/Pages/PreviousEditions.aspx].

Commonwealth of Australia. 2013d. *Strong and Secure: A Strategy for Australia's National Security* Canberra: Commonwealth of Australia.

Commonwealth of Australia. 2015a. *Foreign Affairs, Defence and Trade Legislation Committee: Estimates*. Canberra: Commonwealth of Australia. Accessed on 20 May 2017, http://parlinfo.aph.gov.au/parlInfo/search/display/display.w3p;query=Id%3A%22committees%2Festimate%2F6662254f-7408-419c-b81e-4a2e7981ccea%2F0001%22

Commonwealth of Australia. 2015b. "Supplementary Budget Estimates 2015–2016." In *Foreign Affairs, Defence and Trade Legislation Committee Hansard*. Canberra: Commonwealth of Australia. Accessed on 2 November 2015, http://www.aph.gov.au/Parliamentary_Business/Senate_Estimates/fadtctte/estimates/sup1516/index.

Commonwealth of Australia. 2016. "National Greenhouse Gas Inventory—Kyoto Protocol classifications." Department of the Environment and Energy. Canberra: Commonwealth of Australia. Accessed on 9 November 2016, http://ageis.climatechange.gov.au/.

Corn, David. 2006. "Gore Warms Up." *The Nation*, Katrina vanden Heuvel, 25 May 2006. Accessed on 5 June 2014, http://www.thenation.com/article/gore-warms/.

Corry, Olaf. 2011. "Securitisation and 'Riskification': Second-order Security and the Politics of Climate Change." *Millenium: Journal of International Studies*, vol. 40, no. 2:235–258. doi:10.1177/0305829811419444.

Cotton, William R., and Roger A. Peilke. 2007. *Human Impacts on Weather*. Fort Collins: Cambridge University Press.

Council of the European Union. 2008. *Climate Change and International Security: Paper from the High Representative and the European Commission to the European Council*. Brussels: Council of the European Union. Accessed on 12 March 2012, http://www.consilium.europa.eu/uedocs/cms_data/docs/pressdata/en/reports/99387.pdf.

Council on Foreign Relations. 2007. "Political Interference with Climate Change Science Under the Bush Administration, December 2007." Online: Council on Foreign Relations. Accessed on 25 January 2017, http://www.cfr.org/climate-change/political-interference-climate-change-science-under-bush-administration-december-2007/p15079.

Dabelko, Geoffrey. 1999. "The Environmental Factor." *Wilson Quarterly*, vol. 23, no. 4:14–19.

Dalby, Simon. 1992. "Security, modernity, ecology: the dilemmas of post-Cold War security discourse." *Alternatives*, vol. 17, no.:95–134.

Dalby, Simon, Hans Günter Brauch, and Úrsula Oswald Spring. 2009. "Environmental Security Concepts Revisited During the First Three Phases (1983–2006)." In *Facing Global Environmental Change Environmental, Human, Energy, Food, Health and Water Security*

Concepts, edited by Hans Günter Brauch, Úrsula Oswald Spring, John Grin, Czeslaw Mesjasz, Patricia Kameri-Mbote, Navnita Chadha Behera, Béchir Chourou and Heinz Krummenacher, vol. 4, pp. 781–790. Berlin: Springer.

Defence, Department of. 2005a. *Australia's National Security: A Defence Update 2005*. Canberra: Commonwealth of Australia.

Defence, Department of. 2005b. *Defence Energy Management Strategy*. Canberra: Department of Defence. Accessed on 20 May 2017, http://www.uat.defence.gov.au/ips/parliament/qons/41st/slc/0506_budget_may05/responses/Q09_ems.pdf.

Defence, Department of. 2007a. *Defence Energy Policy*. Defence Support Group. Canberra: Commonwealth of Australia.

Defence, Department of. 2007b. *Joint Operations for the 21st Century*. Canberra: Department of Defence. Accessed on 11 September 2011, http://www.defence.gov.au/publications/docs/FJOC.pdf.

Defence, Department of. 2008a. *Combat Climate Change*. Canberra: Department of Defence. Accessed on 7 July 2011, http://www.defence.gov.au/environment/climate_change/. [Author note: Website archived, document in author's possession].

Defence, Department of. 2008b. *Looking over the Horizon: Australians Consider Defence*. Canberra: Department of Defence. Accessed on 7 August 2011, http://www.defence.gov.au/whitepaper/2009/docs/Looking_Over_The_Horizon.pdf.

Defence, Department of. 2009a. *Defence Annual Report 2008–09*. Department of Defence. Canberra: Department of Defence. Accessed on 10 July 2012, http://www.defence.gov.au/AnnualReports/08-09/.

Defence, Department of. 2009b. *Defending Australia in the Asia Pacific Century: Force 2030*. Canberra: Commonwealth of Australia. Accessed on 5 May 2017, http://www.defence.gov.au/whitepaper/2009/docs/defence_white_paper_2009.pdf.

Defence, Department of. 2010a. *Defence Annual Report 2009–10*. Department of Defence. Canberra: Department of Defence. Accessed on 1 March 2011, http://www.defence.gov.au/AnnualReports/09-10/.

Defence, Department of. 2010b. *Defence Environmental Strategic Plan 2010–2014*. Canberra: Defence Support Group for Department of Defence. Accessed on 16 June 2011, http://www.defence.gov.au/environment/strat_plan.pdf. [Author note: Hyperlink inoperative, author retains original. Superseded reference accessible at: http://www.defence.gov.au/estatemanagement/governance/Policy/Environment/Policy/EnvironmentStrategy2016.PDF].

Defence, Department of. 2010c. *The Strategy Framework 2010*. Canberra: Department of Defence. Accessed on 14 March 2011, http://www.defence.gov.au/environment/strat_plan. [Author note: Hyperlink inoperative, author retains original. Superseded reference accessible at: http://www.defence.gov.au/estatemanagement/governance/Policy/Environment/Policy/EnvironmentStrategy2016.PDF].

Defence, Department of. 2011. *Global Change and Energy Sustainability Initiative*. Canberra: Defence Preparedness Branch for Department of Defence. [Website archived. Contents available on request to author].

Defence, Department of. 2012. "Answers to Questions on Notice—Committees." In *Senate Standing Committee on Foreign Affairs, Defence and Trade Senate Budget Estimates Hansard*, 29 May 2012. Canberra: Commonwealth of Australia. Accessed on 10 May 2013, http://www.aph.gov.au/~/media/Estimates/Live/fadt_ctte/estimates/bud_1213/def/Department-of-Defence/Defence_QONs.ashx.

Defence, Department of. 2013a. *Defence Energy Integration Framework*. Department of Defence. Canberra: Commonwealth of Australia. Accessed 5 May 2017, http://www.natolibguides.info/ld.php?content_id=1675730.

Defence, Department of. 2013b. *Defence White Paper*. Canberra: Commonwealth of Australia. Accessed on 10 January 2014, http://www.defence.gov.au/whitepaper/2013/docs/WP_2013_web.pdf.

Defence, Department of. 2014. *Defence Procurement Policy Manual*. Canberra: Department of Defence. Accessed on 5 June 2015, http://www.defence.gov.au/dmo/Multimedia/DPPM-9-5247.pdf. [Website archived. Document available on request to author].

Defence, Department of. 2016a. *Additional Budget Estimates 2015–2016: Responses to Questions on Notice*. Canberra: Commonwealth of Australia. Accessed on 8 January 2017, http://www.aph.gov.au/Parliamentary_Business/Senate_Estimates/fadtctte/estimates/add1516/def/defenceqonsindex.

Defence, Department of. 2016b. *Defence Annual Report 2014–15*. Department of Defence. Canberra: Commonwealth of Australia. Accessed on 1 November 2016, http://www.defence.gov.au/annualreports/14-15/downloads/DAR_2014-15_Vol_1.pdf.

Defence, Department of. 2016c. *Defence White Paper*. Canberra: Commonwealth of Australia. Accessed on 31 October 2016, http://www.defence.gov.au/WhitePaper/Docs/2016-Defence-White-Paper.pdf.

Defence, Department of. 2016d. *Environmental Strategy 2016–2036*. Canberra: Defence Estate and Infrastructure Group (E&IG). Accessed on 16 October 2016, http://www.defence.gov.au/estatemanagement/governance/Policy/Environment/Policy/EnvironmentStrategy2016.PDF.

Defence, Department of. 2016e. "Estate and Infrastructure Group." Canberra: Commonwealth of Australia. Accessed on 1 November 2016, http://www.defence.gov.au/EIG/.

Defense Science Board Task Force. 2008. *Report of the Defense Science Board Task Force on DoD Energy Strategy: More Fight—Less Fuel*. Washington, DC: Department of Defense.

Dempsey, Martin. 2011. "General Dempsey at the Military Reporters and Editors Conference." In *Joint Chiefs of Staff Speeches*, 18 November 2008, vol. 2013. Arlington: Department of Defense. Accessed on 12 July 2012, http://www.jcs.mil/Media/Speeches/tabid/3890/Article/571837/gen-dempseys-remarks-and-qa-at-at-the-military-reporters-and-editors-conference.aspx.

Dempsey, Martin. 2012. "Gen. Dempsey's Remarks at the Ambassador S. Davis Phillips Family International Lecture Series." In *Joint Chiefs of Staff Speeches*, 12 January 2012. Duke University: Department of Defense. Accessed on 12 July 2013, http://www.jcs.mil/Media/Speeches/tabid/3890/Article/571848/gen-dempseys-remarks-and-qa-at-the-ambassador-s-davis-phillips-family-internati.aspx.

Department of Industry. 2011. *Low Emission Technology Demonstration Fund (LETDF): Funded Projects*. Canberra: Department of Industry for Commonwealth of Australia.

Deudney, Daniel. 1990. "The Case Against Linking Environmental Degradation and National Security." *Millenium: Journal of International Studies*, vol. 19, no. 3:461–476. doi:10.1177/03058298900190031001.

DiPeso, Jim. 2010. "Can DOD lead the way to a better energy future?" *Environmental Quality Management*, vol. 20, no. 2:95–100. doi:10.1002/tqem.20283.

DSTO. 2008. "Bubbles to bust a bio-fouling problem." *Australian Defence Science*, vol. 16, no. 4:4–5.

Dupont, Alan, and Graeme Pearman. 2006. *Heating up the Planet: Climate Change and Security*. Lowy Institute Paper 12. Sydney: Lowy Institute. Accessed on 10 March 2010, http://www.lowyinstitute.org/publications/heating-planet-climate-change-and-security.

Dyer, Gwynne. 2008. *Climate Wars: The fight for survival as the world overheats*. Melbourne: Scribe Publications.

Edwards, Michael J. 1999. "Security Implications of a Worst-case Scenario of Climate Change in the South-west Pacific." *Australian Geographer*, vol. 30, no. 3:311–330.

Ehrlich, Paul R. 1968. *The Population Bomb*. New York: Ballantine Books.

ESSP. 2012. "Global Carbon Budget 2012." Corinne Le Quéré, Glen Peters, Robbie Andrew, Bob Andres, Tom Boden, Thomas Conway, Skee Houghton, Jo House, Gregg Marland, Guido van der Werf, Anders Ahlström, Laurent Bopp, Pep Canadell, Philippe Ciais, Scott Doney, Clare Enright, Pierre Friedlingstein, Chris Huntingford, Atul Jain, Charlotte Jourdain, Etsushi Kato, Ralph Keeling, Kees Klein Goldewijk, Samuel Levis, Peter Levy, Mark Lomas, Ben Poulter, Mike Raupach, Jörg Schwinger, Stephen Sitch, Benjamin Stocker, Nicolas Viovy, Charlie

Wilson, Soenke Zaehle and Ning Zeng: Earth System Science Partnership. Accessed on 15 May 2017, http://www.globalcarbonproject.org/carbonbudget/archive/2012/CarbonBudget_2012.pdf.

EU. 2003. *European Security Strategy: A Secure Europe in a Better World*. Brussels: European Union. Accessed on 23 September 2014, http://www.consilium.europa.eu/uedocs/cmsUpload/78367.pdf.

European Commission. 2015. *The European Agenda on Security* Strasbourg: European Commission. Accessed on 6 January 2017, https://ec.europa.eu/home-affairs/sites/homeaffairs/files/e-library/documents/basic-documents/docs/eu_agenda_on_security_en.pdf.

European Council. 2016a. "The European Union in a changing global environment: A more connected, contested and complex world." Brussels: European Union. Accessed on 15 May 2017, http://www.europarl.europa.eu/sides/getDoc.do?pubRef=-//EP//NONSGML+REPORT+A8-2016-0069+0+DOC+PDF+V0//EN.

European Council. 2016b. "Implementation Plan on Security and Defence." In *EU global strategy on foreign and security policy*. Brussels: European Union. Accessed on 20 November 2016, http://www.consilium.europa.eu/en/press/press-releases/2016/11/14-conclusions-eu-global-strategy-security-defence/.

Evans, Ann. 2012. *Australian Survey of Social Attitudes, 2011*. The Australian Survey of Social Attitudes: Canberra: Australian Data Archive.

Evans, Michael. 2005. *The Tyranny of Dissonance: Australia's Strategic Culture and Way of War 1901–2005*. Land Warfare Studies Centre Study Paper Number 306: Canberra: Land Warfare Studies Centre.

Executive Order 13514. 2009. "Executive Order 13514 Federal Leadership in Environmental, Energy, and Economic Performance." In *Presidential Documents, Federal Register* 5 October 2009, vol. 74, no. 194, pp. 52117–52127. Washington, DC: US Government Printing Office Accessed on 10 October 2013, http://www.gpo.gov/fdsys/granule/FR-2009-10-08/E9-24518/content-detail.html.

Field, Chris. 2011. "Operation Queenslander: Ten Ideas for Australian Defence Force Support to Disaster Relief Operations." *Australian Army Journal For The Profession of Arms*, vol. 8, no. 3:121–140.

Fitzgibbon, Joel. 2008a. "Defence Supports Earth Hour 2008." In *Transcripts of the Minister for Defence*, 29 March 2008. Canberra: Department of Defence. [Site archived. Document in author possession and available on request].

Fitzgibbon, Joel. 2008b. "Labor's Defence Reform Project—Meeting the Strategic Challenges of the 21st Century." In *Transcripts of the Minister for Defence*, 30 July 2008. Canberra: Commonwealth of Australia. [Site archived. Document in author possession and available on request].

Fitzgibbon, Joel. 2008c. "Speech by the Minister for Defence to the Brookings Institution, Washington DC." In *Transcripts of the Minister for Defence*, 15 July 2008. Canberra: Commonwealth of Australia. [Site archived. Document in author possession and available on request].

Fitzgibbon, Joel. 2008d. "Speech by the Minister for Defence to the Shangri-La Conference Singapore, 31 May 2008, 'Making Defence Policy in Uncertain Times'." In *Transcripts of the Minister for Defence*, 31 May 2008. Canberra: Commonwealth of Australia. [Site archived. Document in author possession and available on request].

Fitzgibbon, Joel. 2008e. "Speech to the National Resilience Institute (Lemhannas), Jakarta, Indonesia." In *Transcripts of the Minister for Defence*, 18 September 2008. Canberra: Department of Defence. [Site archived. Document in author possession and available on request].

Fitzgibbon, Joel. 2014. "Hunter Electorate." Federal Member for Hunter—Joel Fitzgibbon MP. Accessed on 20 February 2014, http://www.joelfitzgibbon.com/about.

Flounders, Sara. 2014. "The Pentagon, The Climate Elephant. The US Military Machine is the World's Worst Polluter of Greenhouse Gas Emissions." Online: Global Research: Centre for Research on Globalization. Accessed on 10 November 2016, http://www.globalresearch.ca/the-pentagon-the-climate-elephant-2/5402505.

Floyd, Rita. 2010. *Security and the Environment: Securitisation Theory and US Environmental Security Policy*. Cambridge: Cambridge University Press.

Foley, Catherine, and Andrew Holland. 2012. *Climate Security Report*. Washington, DC. Accessed on 23 November 2016, http://www.americansecurityproject.org/climate-security-report/.

Foucault, Michel. 1982. "Afterword: The Subject and Power." In *Michel Foucault: Beyond structuralism and hermeneutics*, edited by Hubert L. Dreyfus and Paul Rabinow, pp. 208–225. Chicago: University of Chicago Press.

Friedrich, Tobias, Axel Timmermann, Michelle Tigchelaar, Oliver Elison Timm, and Andrey Ganopolski. 2016. "Nonlinear climate sensitivity and its implications for future greenhouse warming." *Science Advances*, vol. 2, no. 11. doi:10.1126/sciadv.1501923.

Gardiner, Stephen M. 2010a. *Climate Ethics: Essential Readings*. Oxford: Oxford University Press.

Gardiner, Stephen M. 2010b. "Ethics and Global Climate Change." In *Climate Ethics: Essential Readings*, edited by Stephen M. Gardiner, Simon Caney, Dale Jamieson and Henry Shue. Oxford: Oxford University Press.

Garnaut, Ross. 2008. *The Garnaut Climate Change Review: Final Report*. Melbourne: Cambridge University Press.

Gates, Robert M. 2014. *Duty: Memoirs of a Secretary at War*. New York: Alfred A. Knopf.

German Government. 2016. *White Paper: On German Security Policy and The Future of the Bundeswehr*. Ministry of Defence. Berlin: Federal Ministry of Defence. Accessed on 22 November 2016, http://www.new-york-un.diplo.de/contentblob/4847754/Daten/6718448/160713weibuchEN.pdf.

Gibbs, Meredith, and Tony Hill. 2011. *Coastal Climate Change Risk: Legal and Policy Responses in Australia*. Department of Climate Change and Energy Efficiency. Canberra: Commonwealth of Australia. Accessed on 12 March 2012, https://www.environment.gov.au/climate-change/adaptation/publications/coastal-climate-change-risk.

Gillard, Julia E. 2011a. "Address to the Carbon Expo Australasia 2011, Melbourne." In *Transcripts of the Prime Ministers of Australia*, 09 November 2011. Canberra: Commonwealth of Australia. Accessed on 12 March 2012, http://pmtranscripts.dpmc.gov.au/release/transcript-18253.

Gillard, Julia E. 2011b. "A great clean energy future, a great reform agenda: Address to the National Press Club, Canberra." In *Transcripts from the Prime Ministers of Australia*, 14 July 2011. Canberra: Commonwealth of Australia. Accessed on 12 March 2012, http://pmtranscripts.dpmc.gov.au/release/transcript-18004.

Gillespie, Ken J. 2008. "Chief of Army Speech." In *Address to Australian Command and Staff College*. Canberra. (Website archived. Document in transcript format in possession of author).

Gillespie, Ken J. 2009. "The Adaptive Army Initiative." *Australian Army Journal For The Profession of Arms*, vol. 6, no. 3:7–19.

Goldenberg, Suzanne. 2009. "The worst of times: Bush's environmental legacy examined." *The Guardian*, Guardian News and Media, 17 January 2009. Accessed on 25 January 2017, https://www.theguardian.com/politics/2009/jan/16/greenpolitics-georgebush.

Greenert, Jonathon W. 2011. "Adapting to Climate and Energy Challenges: Options for U.S. Maritime Forces." Adapting to Climate and Energy Challenges Symposium, Laurel, 29 & 30 March 2011. Accessed on 15 June 2013, http://www.jhuapl.edu/ClimateAndEnergy/Book2011/2011CEBook-LR.pdf.

Guggenheim, Davis. 2006. An Inconvenient Truth. Paramount Classics.

Gurran, Nicole, Elisabeth Hamin, and Barbara Norman. 2008. *Planning for climate change: Leading Practice Principles and Models for Sea Change Communities in Coastal Australia.* National Sea Change Taskforce: Sydney: The University of Sydney. Accessed on 19 April 2012, http://stokes2013.files.wordpress.com/2014/10/planningforclimatechange.pdf.

Hagel, Chuck. 2013. "Halifax International Security Forum (DoD Arctic Strategy)." Online: US Department of Defense. Accessed on 15 January 2017, http://archive.defense.gov/speeches/speech.aspx?speechid=1821.

Hagel, Chuck. 2014a. "The Department of Defense Must Plan for the National Security Implications of Climate Change." Online: US Department of Defense. Accessed on 18 January 2018, https://obamawhitehouse.archives.gov/blog/2014/10/13/defense-department-must-plan-national-security-implications-climate-change.

Hagel, Chuck. 2014b. "Realizing the Asia-Pacific Rebalance." Online: Defense One. Accessed on 15 January 2017, http://www.defenseone.com/ideas/2014/04/realizing-asia-pacific-rebalance/81730/.

Halden, Peter. 2007. *The Geopolitics of Climate Change.* Stockholm: Swedish Defence Research Agency.

Hansen, J., D. Johnson, A. Lacis, S. Lebedeff, P. Lee, D. Rind, and G. Russell. 1981. "Climate Impact of Increasing Atmospheric Carbon Dioxode." *Science*, vol. 213, no. 4511:958–966.

Hansen, James, M. Sato, P. Kharecha, and K. von Scuckmann. 2011. "Earth's energy imbalance and implications." *Atmospheric Chemistry and Physics*, no. 11:13421–13449. doi:10.5194/acp-11-13421-2011.

Hardin, Garrett. 1968. "The Tragedy of the Commons." *Science*, vol. 162, no. 13:1243–1248.

Hartcher, Peter. 2014. "Tony Abbott must shift with the tide on climate change." *The Age*, Fairfax Media, 9 December 2014, http://www.theage.com.au/federal-politics/political-opinion/tony-abbott-must-shift-with-the-tide-on-climate-change-20141208-122ie7.html.

Hartmann, Betsy. 2010. "Rethinking Climate Refugees and Climate Conflict: Rhetoric, Reality and the Politics of Policy Discourse." *Journal of International Development*, vol. 22, no. 2:233–246. doi:10.1002/jid.1676.

Hawke, Allan, and Ric Smith. 2012. *Australian Defence Force Posture Review.* Canberra: Commonwealth of Australia.

Hayes, Jarrod, and Janelle Knox-Hayes. 2014. "Security in Climate Change Discourse: Analysing the Divergence between US and EU Approaches to Policy." *Global Environmental Politics*, vol. 14, no. 2:82–101. doi:10.1162/GLEP_a_00230.

HM Government. 2010. *Foreign and Commonwealth Office Annual Report 2008–09: Fifth Report of Session 2009–10.* London: Foreign and Commonwealth Office. Accessed on 14 April 2013, http://www.publications.parliament.uk/pa/cm200910/cmselect/cmfaff/145/145.pdf.

HM Government. 2015. *National Security Strategic and Strategic Defence and Security Review 2015.* London: HM Government. Accessed on 22 November 2016, https://www.gov.uk/government/uploads/system/uploads/attachment_data/file/478936/52309_Cm_9161_NSS_SD_Review_PRINT_only.pdf.

Homer-Dixon, Thomas. 1996. "Strategies for Studying Causation in Complex Ecological-Political Systems." *The Journal of Environment and Development*, vol. 5, no. 2:132–148. doi:10.1177/107049659600500202.

Homer-Dixon, Thomas F. 1991. "On the Threshold: Environmental Changes as Causes of Acute Conflict." *International Security*, vol. 16, no. 2:76–116. doi:10.2307/2539061.

Homer-Dixon, Thomas F. 1994. "Environmental scarcities and violent conflict: Evidence from cases." *International Security*, vol. 19, no. 1:5–40. doi:10.2307/2539147.

Homer-Dixon, Thomas F. 1999. *Environment, Scarcity and Violence.* Princeton: Princeton University Press.

Homer-Dixon, Thomas F. 2007. "Terror in the Weather Forecast." *New York Times*, Arthur Ochs Sulzberger Jr, 24 April 2007.

Houston, Angus. 2007. "The ADF of the Future." *Australian Defence Force Journal: Journal of the Profession of Arms*, no. 173.

Houston, Angus. 2010. "Future security prospects, future capability plans and how the ADF makes a contribution to a 'whole of government' approach to national security." In *Address to National Security Lecture Series*, 29 October 2010. Canberra: University of Canberra. [Site archived. Document in author possession and available on request].

Howard, John. 2010. *Lazarus Rising: A Personal and Political Autobiogrpahy*. Sydney: Harper Collins.

Howard, John W. 2004. "Address to National Press Club Canberra, Securing Australia's Energy Future." In *Transcripts from the Prime Ministers of Australia*, 15 June 2004. Canberra: Department of the Prime Minister and Cabinet. Accessed on 13 March 2012, http://pandora.nla.gov.au/pan/10052/20040821-0000/www.pm.gov.au/news/speeches/speech912.html.

Howard, John W. 2006a. "Address at the Business Council of Australia Annual Dinner Sofitel Wentworth, Sydney." In *Transcripts from the Prime Ministers of Australia* 13 November 2006. Canberra: Department of Prime Minister and Cabinet. Accessed on 13 March 2013, http://pmtranscripts.dpmc.gov.au/release/transcript-22579.

Howard, John W. 2006b. "Address at the Launch of the Asia-Pacific Partnership on Clean Development and Climate 'Partnership for Action 2006' CSIRO, Yarralumla." In *Transcripts from the Prime Ministers of Australia*, 1 November 2006. Department of Prime Minister and Cabinet. Accessed on 11 March 2013, http://pmtranscripts.dpmc.gov.au/release/transcript-22552.

Howard, John W. 2006c. "Address at the Liberal Party of Australia (Queensland Division) Annual State Convention." In *Transcripts from the Prime Ministers of Australia*, 4 November 2006. Canberra: Department of the Prime Minister and Cabinet. Accessed on 13 March 2013, http://pmtranscripts.dpmc.gov.au/release/transcript-22559.

Howard, John W. 2006d. "Address to the APEC CEO Summit National Convention Centre, Hanoi." In *Transcripts of the Prime Ministers of Australia*, 18 November 2006. Canberra: Department of Prime Minister and Cabinet. Accessed on 15 March 2013, http://pmtranscripts.dpmc.gov.au/release/transcript-22590.

Howard, John W. 2006e. "Address to the Asia-Pacific Partnership on Clean Development and Climate Inaugural Ministerial Meeting Government House, Sydney." In *Transcripts from the Prime Ministers of Australia*, 12 January 2006. Canberra: Department of Prime Minister and Cabinet. Accessed on 8 March 2013, http://pmtranscripts.dpmc.gov.au/release/transcript-22099
.

Howard, John W. 2006f. "Address to the Australia/UK Leadership Forum Parliament House, Canberra." In *PM Transcripts: Transcripts from the Prime Ministers of Australia*, 27 March 2006. Canberra: Department of the Prime Minister and Cabinet. Accessed on 10 March 2013, http://pmtranscripts.dpmc.gov.au/release/transcript-22190.

Howard, John W. 2007a. "Address to the National Press Club Great Hall, Parliament House." In *Transcripts from the Prime Ministers of Australia*, 25 January 2007. Canberra: Department of Prime Minister and Cabinet. Accessed on 13 March 2013, http://pmtranscripts.dpmc.gov.au/release/transcript-15149.

Howard, John W. 2007b. "APEC Leaders' Sydney Declaration on Climate Change, Energy Security and Clean Development." In *Transcripts from the Prime Ministers of Australia*, 8 September 2007. Canberra: Department of Prime Minister and Cabinet. Accessed on 5 April 2013, http://pmtranscripts.dpmc.gov.au/release/transcript-15401.

Howard, John W. 2007c. "Climate Change Announcement." In *Transcripts from the Prime Ministers of Australia*, 17 July 2007. Canberra: Department of Prime Minister and Cabinet. Accessed on 12 April 2013, http://pmtranscripts.dpmc.gov.au/release/transcript-15296.

Howard, John W. 2007d. "Interview with Simon Beaumont, Radio 6PR, Perth." In *Transcripts from the Prime Ministers of Australia*, 19 February 2007. Canberra: Department of Prime Minister and Cabinet. Accessed on 19 April 2013, http://pmtranscripts.dpmc.gov.au/release/transcript-15212.

Howard, John W. 2007e. "Meeting of Major Economies of Energy Security and Climate Change." In *Transcripts from the Prime Ministers of Australia*, 4 August 2007. Canberra: Department of Prime Minister and Cabinet. Accessed on 29 March 2013, http://pmtranscripts.dpmc.gov.au/release/transcript-15377.

Howard, John W. 2007f. "National Plan for Water Security; Qantas; Australia-Japan relations." In *Transcripts from the Prime Ministers of Australia*, 5 February 2007. Canberra: Department of Prime Minister and Cabinet. Accessed on 3 March 2013, http://pmtranscripts.dpmc.gov.au/release/transcript-15193.

Howard, John W. 2007g. "Nuclear Power Station Plebiscites." In *Transcripts of the Prime Ministers of Australia*, 23 August 2007. Canberra: Department of Prime Minister and Cabinet. Accessed on 8 April 2013, http://pmtranscripts.dpmc.gov.au/release/transcript-15491.

Howard, John W. 2007h. "Weekly Radio Message—Climate Change." In *Transcripts from the Prime Ministers of Australia*, 16 July 2007. Canberra: Department of Prime Minister and Cabinet. Accessed on 18 April 2013, http://pmtranscripts.dpmc.gov.au/release/transcript-15292.

Howard, John W. 2013. "One Religion is Enough." London: The Global Warming Policy Foundation. Accessed on 7 February 2014, http://www.thegwpf.org/john-howard-religion/.

Huebert, Rob, Heather Exner-Pirot, Adam Lajeunesse, and Jay Gulledge. 2012. *Climate Change & International Security: The Arctic as a Bellweather*. Arlington, Virginia. Accessed on 23 November 2016, http://www.c2es.org/publications/climate-change-international-arctic-security.

Hulme, Mike. 2009. *Why we disagree about climate change: Understanding controversy, inaction and opportunity*. Cambridge: Cambridge Univesity Press.

Hurley, David J. 2011. "Chief of the Defence Force—Address to the Australian Strategic Policy Institute." In *Defence News and Media*, 18 August 2011. Canberra: Department of Defence.

Huysmans, Jef. 2004. "A Foucaultian view on spill-over: freedom and security in the EU." *Journal of International Relations and Development*, vol. 7, no.:294–318. doi:10.1057/palgrave.jird.1800018.

Huysmans, Jef. 2006. *The Politics of Insecurity: Fear, Migration and Asylum in the EU*. London: Routledge.

ICAO. 2015. "Uniting Aviation on Climate Change." International Civil Aviation Organization. Accessed on 23 September 2015, http://www.icao.int/environmental-protection/Pages/act-global.aspx.

IEA. 2011. "Executive Summary." In *2011 World Energy Outlook*. Paris: International Energy Agency. Accessed on 10 September 2012, http://www.iea.org/publications/freepublications/publication/English.pdf.

Inhofe, James M. 2012a. "Inhofe Again Requests Sec. Mabus to disclose full cost of greening fleet demonstration." Washington, DC: US Senate. Published online 2 August 2012, http://www.inhofe.senate.gov/newsroom/press-releases/inhofe-again-requests-sec-mabus-to-disclose-full-cost-of-green-fleet-demonstration.

Inhofe, James M. 2012b. "Inhofe: Jolly "Green Fleet" Throwing Dollars Out to Sea." Washington, DC: US Senate. Published online 19 July 2012, http://www.epw.senate.gov/public/index.cfm/2012/7/post-a145fe86-802a-23ad-4b07-45be3935642e.

Inhofe, James M. 2012c. "Press Release: Inhofe Questions Excessive Cost of 'Greening' Military (25 July)." Washington, DC: US Senate. Accessed on 8 October 2013, http://www.epw.senate.gov/public/index.cfm/press-releases-republican?ID=BFB1C8CE-802A-23AD-4849-5001DD46198D.

Insurance Council of Australia. 2008. *Improving Community Resilience to Extreme Weather Events*. Published online: Insurance Council of Australia. Accessed on 10 October 2015, http://www.insurancecouncil.com.au/assets/files/community%20resilience%20policy%20150408.pdf.

International Federation of Red Cross. 2015. "Aggravating factors: climate change." Published online: International Federation of Red Cross and Red Crescent Societies. Accessed on 1 October 2015, http://www.ifrc.org/en/what-we-do/disaster-management/about-disasters/aggravating-factors/climate-change/.

IPCC. 2007. "Summary for Policymakers." In *Climate Change 2007: The Physical Science Basis. Contribution of Working Group I to the Fourth Assessment Report of the Intergovernmental Panel on Climate Change*, edited by S. Solomon, D. Qin, M. Manning, Z. Chen, M. Marquis, K. B. Averyt, M. Tignor and H. L. Miller. Cambridge: Cambridge University Press.

IPCC. 2015. "Climate Change 2014 Synthesis Report." In *Climate Change 2014: Synthesis Report. Contribution of Working Group I to the Fifth Assessment Report of the Intergovernmental Panel on Climate Change*, edited by Core Writing Team, Rajendra K. Pachauri and Leo Meyer. Geneva: Intergovernmental Panel on Climate Change. Accessed on 14 March 2015, http://www.ipcc.ch/pdf/assessment-report/ar5/syr/SYR_AR5_FINAL_full.pdf.

Jamieson, Dale. 2010. "Ethics, Public Policy, and Global Warming." In *Climate Ethics: Essential Readings*, edited by Stephen M. Gardiner, Simon Caney, Dale Jamieson and Henry Shue. Oxford: Oxford University Press.

Jotzo, Frank. 2013. "Australia needs climate institutions, whoever is in power." *The Conversation*, no.

Kaplan, Robert D. 1994. "The Coming Anarchy." *World Policy Journal*, vol. 17, no. 2:95–96.

Kelly, Mike. 2008. "Regional Security and Middle Power Diplomacy." In *John Gee Memorial Lecture* 29 August 2008. Canberra: Department of Defence. Accessed on 20 May 2017, https://www.lowyinstitute.org/sites/default/files/pubfiles/Kelly%2C_Regional_security_and_middle_power_diplomacy_1.pdf.

Kelly, Mike. 2012. "PNG: Securing a Prosperous Future, Domestic and Regional Security (Prosperous Future Conference, Deakin University Geelong)." In *Transcripts Parliamentary Secretary for Defence*, 13 April 2012. Canberra: Department of Defence. [Site archived. Document in author possession and available on request].

Keohane, Robert O., and David G. Victor. 2011. "The Regime Complex for Climate Change." *Perspectives on Politics*, vol. 9, no. 1:7–23. doi:10.1017/S1537592710004068.

Klare, Michael T. 2007. "Global Warming Battelfields: How Climate Change Threatens Security." *Current History*, vol. 106, no. 703:355–361.

Klötzli, Stefan. 1994. "The Water and Soil Crisis in Central Asia: A Source for Future Conflicts?" In *Environment and Conflicts Project ENCOP Occasional Papers*, edited by Günter Bächler and Kurt R. Spillmann, vol. 11, no. 1. Zurich: Center for Security Studies.

Knieling, Jörg, and Walter Leal Filho. 2013. "Climate Change Governance: The Challenge for Politics and Public Administration, Enterprises and Civil Society." In *Climate Change Governance*. Heidelberg: Springer.

Lang, Christoph I. 1995. "Environmental Degradation in Kenya as a Cause of Political Conflict, Social Stress, and Ethnic Tensions." In *CSS Environment and Conflicts Project (ENCOP)*, edited by Günter Bächler and Kurt R. Spillmann, vol. 12, no. 1. Zurich: Centre for Security Studies.

Lawson, Robert. 2007. "Climate Change in the Asia-Pacific Region: Security Implications for Australia." In *Centre for Defence and Strategic Studies Australian Defence College Shedden Papers*. Canberra: Department of Defence. Accessed on 5 May 2012, http://www.defence.gov.au/ADC/Publications/Shedden/2008/Publctns_ShedPaper_050310_ClimateChangeintheAsiaPacificRegion.pdf.

Layton, Peter. 2014. "What strategy for the new Defence White Paper?" In *The Strategist*, 13 March 2014. Canberra: Australian Strategic Policy Institute. Accessed on 23 March 2014, http://www.aspistrategist.org.au/long-live-the-force-structure/.

Leahy, Peter. 2004. "The Australian Army Reserve: Relevant and Ready." *Australian Army Journal For the Profession of Arms*, vol. 2, no. 1.

Leahy, Peter. 2008. "Introduction By The Chief of Army." *Australian Army Journal For The Profession of Arms*, vol. 5, no. 2.

Leahy, Peter. 2010. "How do we know when we are at war?" In *Lowy Institute Perspectives: Occasional Papers and Speeches*, 22 July 2010. Sydney: Lowy Institute. Accessed on 12 May 2013, http://www.lowyinstitute.org/publications/how-do-we-know-when-we-are-war.

Leal-Arcas, Rafael. 2011. "Top-Down Versus Bottom-Up Approaches for Climate Change Negotiations: An Analysis." *The IUP Journal of Governance and Public Policy* vol. 6, no. 4:7–52.

LeQuéré, C., R. Moriarty, R. M. Andrew, J. G. Canadell, S. Sitch, J. I. Korsbakken, P. Friedlingstein, G. P. Peters, R. J. Andres, T. A. Boden, R. A. Houghton, J. I. House, R. F. Keeling, P. Tans, A. Arneth, D. C. E. Bakker, L. Barbero, L. Bopp, J. Chang, F. Chevallier, L. P. Chini, P. Ciais, M. Fader, R. Feely, T. Gkritzalis, I. Harris, J. Hauck, T. Ilyina, A. K. Jain, E. Kato, V. Kitidis, K. Klein Goldewijk, C. Koven, P. Landschützer, S. K. Lauvset, N. Lefèvre, A. Lenton, I. D Lima, N. Metzl, F. Millero, D. R. Munro, A. Murata, J. E. M. S. Nabel, S. Nakaoka, Y. Nojiri, K. O'Brien, A. Olsen, T. Ono, F. F. Pérez, B. Pfeil, D. Pierrot, B. Poulter, G. Rehder, C. Rödenbeck, S. Saito, U. Schuster, J. Schwinger, R. Séférian, T. Steinhoff, B. D. Stocker, A. J. Sutton, T. Takahashi, B. Tilbrook, I. T. van der Laan-Luijkx, G. R. van der Werf, S. van Heuven, D. Vandemark, N. Viovy, A. Wiltshire, S. Zaehle, and N. Zeng. 2015. "Global Carbon Budget 2015." *Earth System Science Data Discussions*, vol. 7, no. 2:349–396. doi:10.5194/essd-7-349-2015.

LeQuéré, Corinne, Robbie M. Andrew, Josep G. Canadell, Stephen Sitch, Jan Ivar Korsbakken, Glen P. Peters, Andrew C. Manning, Thomas A. Boden, Pieter P. Tans, Richard A. Houghton, Ralph F. Keeling, Simone Alin, Oliver D. Andrews, Peter Anthoni, Leticia Barbero, Laurent Bopp, Frédéric Chevallier, Louise P. Chini, Philippe Ciais, Kim Currie, Christine Delire, Scott C. Doney, Pierre Friedlingstein, Thanos Gkritzalis, Ian Harris, Judith Hauck, Vanessa Haverd, Mario Hoppema, Kees Klein Goldewijk, Atul K. Jain, Etsushi Kato, Arne Körtzinger, Peter Landschützer, Nathalie Lefèvre, Andrew Lenton, Sebastian Lienert, Danica Lombardozzi, Joe R. Melton, Nicolas Metzl, Frank Millero, Pedro M. S. Monteiro, David R. Munro, Julia E. M. S. Nabel, Shin-ichiro Nakaoka, Kevin O'Brien, Are Olsen, Abdirahman M. Omar, Tsuneo Ono, Denis Pierrot, Benjamin Poulter, Christian Rödenbeck, Joe Salisbury, Ute Schuster, Jörg Schwinger, Roland Séférian, Ingunn Skjelvan, Benjamin D. Stocker, Adrienne J. Sutton, Taro Takahashi, Hanqin Tian, Bronte Tilbrook, Ingrid T. van der Laan-Luijkx, Guido R. van der Werf, Nicolas Viovy, Anthony P. Walker, Andrew J. Wiltshire, and Sönke Zaehle. 2016. "Global Carbon Budget 2016." *Earth System Science Data*, vol. 8, no. 2:605–649. doi:10.5194/essd-8-605-2016.

LeTreut, H., R. Somerville, U. Cubasch, Y. Ding, C. Mauritzen, A. Mokssit, T. Peterson, and M. Prather. 2007. "Historical Overview of Climate Change." In *Climate Change 2007: The Physical Science Basis. Contribution of Working Group I to the Fourth Assessment Report of the Intergovernmental Panel on Climate Change*, edited by S Solomon, D. Qin, M. Manning, Z. Chen, M. Marquis, K.B. Averyt, M. Tignor and H.L. Miller, pp. 93–128. Cambridge: Cambridge University Press. Accessed on 15 June 2012, http://www.ipcc.ch/pdf/assessment-report/ar4/wg1/ar4-wg1-chapter1.pdf.

Levy, Marc A. 1995. "Is the Environment a National Security Issue?" *International Security*, vol. 20, no. 2:35–62.

Libiszewski, Stephan. 1995. "Water Disputes in the Jordan Basin Region and their Role in the Resolution of the Arab-Israeli Conflict." In *CSS Environment and Conflicts Project (ENCOP)*, edited by Günter Bächler and Kurt R. Spillmann, vol. 13, no. 1. Zurich: Center for Security Studies. Accessed on 3 May 2017, http://www.mideastweb.org/Mew_water95.pdf.

Lizza, Ryan. 2010. "As The World Burns." *The New Yorker*, Condé Nast, 11 October 2010. Accessed on 10 November 2014, http://www.newyorker.com/magazine/2010/10/11/as-the-world-burns.

Lucke, Franziskus von, Thomas Diez, and Zehra Wellmann. 2013. "What's at Stake in Securitising Climate Change? Towards a Differentiated Approach." European Conference on International Relations, Warsaw, September 2013.

Lynas, Mark. 2009. "How do I know China wrecked the Copenhagen deal? I was in the room." *The Guardian*, The Guardian, 23 December 2009. Accessed on 10 May 2013, http://www.theguardian.com/environment/2009/dec/22/copenhagen-climate-change-mark-lynas.

Lynn, William J. 2009. "Remarks at the Connect U.S. Fund." In *Deputy Secretary of Defense Speech*, 14 December 2009. Washington, DC: US Department of Defense. [Site archived. Document in author possession and available on request].

Mabus, Ray. 2009a. "Remarks by the Honourable Ray Mabus Secretary of the Navy at Current Strategy Forum Naval War College, Newport, Rhode Island." In *Transcripts of the Secretary of Navy*, 16 June 2009. Washington, DC: United States Navy. Accessed on 12 June 2013, http://www.navy.mil/navydata/people/secnav/Mabus/Speech/Current%20Strategy%20Forum.pdf.

Mabus, Ray. 2009b. "Remarks by the Honourable Ray Mabus Secretary of the Navy at International Seapower Symposium Naval War College, Newport, Rhode Island." In *Transcripts of the Secretary of Navy*, 7 October 2009. Washington, DC: United States Navy. Accessed on 8 August 2013, http://www.navy.mil/navydata/people/secnav/Mabus/Speech/International%20Seapower%20Symposium%207%20Oct%2009%20Rel.pdf.

Mabus, Ray. 2009c. "Remarks by the Honourable Ray Mabus Secretary of the Navy, Defense Attache Association, Washington DC." In *Transcripts of the Secretary of Navy*, 18 November 2009. Washington, DC: United States Navy. Accessed on 6 September 2013, http://www.navy.mil/navydata/people/secnav/Mabus/Speech/Defense%20Attache%20Association%20paginated%20-%2018%20Nov%2009.pdf.

Mabus, Ray. 2009d. "Remarks by the Honourable Ray Mabus Secretary of the Navy, Naval Energy Forum, Hilton McLean, Tysons Corner, McLean, Virginia." In *Transcripts of the Secretary of Navy*, 14 October 2009. Washington, DC: United States Navy. Accessed on 8 September 2013, http://www.navy.mil/navydata/people/secnav/Mabus/Speech/SECNAV%20Energy%20Forum%2014%20Oct%2009%20Rel1.pdf.

Mabus, Ray. 2009e. "Remarks by the Honourable Ray Mabus Secretary of the Navy, Hamer Institute Speech, Jackson State University, Jackson, Mississippi." In *Transcripts of the Secretary of Navy*, 9 October 2009. Washingston, DC: United States Navy. Accessed on 7 September 2013, http://www.navy.mil/navydata/people/secnav/Mabus/Speech/Hamer%20Institute%20Naming%20of%20Medgar%20Evers%209%20Oct%2009%20Rel.pdf.

Mabus, Ray. 2010a. "Remarks by the Honourable Ray Mabus Secretary of the Navy, Center for Naval Analysis Earth Day Luncheon, Alexandria, VA (29 April)." In *Transcripts of the Secretary of Navy*, 29 April 2010. Washington, DC: United States Navy. Accessed on 8 September 2013, http://www.navy.mil/navydata/people/secnav/Mabus/Speech/Center%20for%20Naval%20Analysis%20Earth%20Day%20Luncheon%2029%20Apr%2010.pdf.

Mabus, Ray. 2010b. "Remarks by the Honourable Ray Mabus Secretary of the Navy, The Commonwealth Club, San Francisco, CA." In *Transcripts of the Secretary of Navy*, 16 August 2010. Washington, DC: United States Navy. Accessed on 9 October 2013, http://www.navy.mil/navydata/people/secnav/Mabus/Speech/commonwealthclub.pdf.

Mabus, Ray. 2012a. "Remarks by the Honourable Ray Mabus Secretary of the Navy, Global Conference on Oceans, Climate and Security: Making the Connections, Boston, Mass." In *Transcripts of the Secretary of Navy*, 22 May 2012. Washington, DC: United States Navy. Accessed on 8 October 2013, http://www.navy.mil/navydata/people/secnav/Mabus/Speech/OceansConferenceMay12v2.pdf.

Mabus, Ray. 2012b. "Remarks by the Honourable Ray Mabus Secretary of the Navy, Naval Energy Forum, Washington DC." In *Transcripts of the Secretary of Navy*, 17 October 2012. Washington, DC: United States Navy. Accessed on 1 October 2013, http://www.navy.mil/navydata/people/secnav/Mabus/Speech/navalenergyforumtranscriptfinal.pdf.

Mabus, Ray. 2012c. "Remarks by the Honourable Ray Mabus Secretary of the Navy, Senate Committee on Energy and Natural Resources, Norfolk, VA." In *Transcripts of the Secretary of Navy*, 12 March 2012. Washington, DC: United States Navy. Accessed on 4 October 2013, http://www.navy.mil/navydata/people/secnav/Mabus/Speech/Senate%20Committee%20on%20Energy%20and%20Natural%20Resources%20oral%20testimonyAsPrepared.pdf.

Macintosh, Andrew, Deb Wilkinson, and Richard Denniss. 2010. "Climate Change." In *The Rudd Government: Australian Commonwealth Administration 2007–2010*, edited by Chris Aulich and Mark Evans, pp. 596–652. Canberra: The Australian National University E Press.

Accessed on 3 March 2014, http://press.anu.edu.au/apps/bookworm/view/The+Rudd+Government%3A+Australian+Commonwealth+Administration+%092007+-+2010/5091/ch11.xhtml#toc-anchor.

Mackellar, Dorothea. 1908. "My Country." Pippa Masson. Published online: The Estate of Dorothea Mackellar. Published online 5 September 1908, http://www.dorotheamackellar.com.au/archive/mycountry.htm.

Major Economies Forum Declaration. 2009. "Declaration by the Leaders of the Major Economies Forum on Energy and Climate." Washington, DC: Office of the Federal Register, National Archives and Records Administration. Accessed on 11 October 2013, http://www.gpo.gov/fdsys/granule/PPP-2009-book2/PPP-2009-book2-Doc-pg1072/content-detail.html.

Mathews, Jessican Tuchman. 1989. "Redefining Security." *Foreign Affairs*, vol. 68, no. 2:162–177.

McCain, John. 2012. "Letter from Senator McCain to Navy Secretary Ray Mabus Regarding Biofuels." In *John McCain Press Releases*, 27 July 2012. Published online: John McCain Accessed on 12 March 2013, http://www.mccain.senate.gov/public/index.cfm/press-releases?ID=ca4414f1-e686-7cb6-820e-995f3289f68a.

McDonald, Matt. 2012. "The Failed Securitization of Climate Change in Australia." *Australian Journal of Political Science*, vol. 47, no. 4:579–592. doi:10.1080/10361146.2012.731487.

Meadows, Donella H., Dennis L Meadows, Jorgen Randers, and William W. Behrens. 1972. *The Limits to Growth*. New York: Universe Books.

Michaelowa, Axel, and Tobias Kock. 2001. "Military Emissions, Armed Conflicts, Border Changes and the Kyoto Protocol." *Climatic Change*, vol. 50, no. 4:383–394.

Ministry of Defence. 2013. *White Paper on Defence and National Security*. Ministry of Defence. France: French Government,. Accessed on 22 November 2016, http://www.rpfrance-otan.org/White-Paper-on-defence-and.

Ministry of Defense. 2008. *National Strategy of Defense*. Ministry of Defense. Brazil: Ministry of Defense. Accessed on 18 November 2016, http://www.defesa.gov.br/projetosweb/estrategia/arquivos/estrategia_defesa_nacional_ingles.pdf.

Ministry of Ecology. 2009. *French Climate Plan*. Sustainable Development and Energy Ministry of Ecology. Paris: Government of France. Accessed on 7 January 2017, http://ec.europa.eu/environment/archives/networks/greenspider/doc/climate_change_campaigns/ccc_france.pdf.

Ministry of Ecology. 2011. *National Plan: Climate Change Adaptation*. Sustainable Development and Energy Ministry of Ecology. Paris: Government of France. Accessed on 7 January 2017, http://www.developpement-durable.gouv.fr/IMG/pdf/ONERC_PNACC_Eng_part_1.pdf.

Ministry of Interior. 2014. *National Internal Security Policy 2014–2018*. Ministry of Interior. Government of Pakistan. http://nacta.gov.pk/policies/NISP.pdf.

Morris, Sophie. 2015. "Tony Abbott's affinity with ADF reflected in parliament." *The Saturday Paper*, Morry Schwartz. Accessed on 5 May 2017, https://www.thesaturdaypaper.com.au/news/politics/2015/03/14/tony-abbotts-affinity-with-adf-reflected-parliament/14262516001605.

Morrison, David. 2012. "Speech to Australian Strategic Policy Institute." In *Chief of Army Speeches and Transcripts*, 11 April 2012. Canberra: Department of Defence. Accessed on 12 April 2014, http://www.army.gov.au/Our-work/Speeches-and-transcripts/Australian-Strategic-Policy-Institute-2012.

Morrison, David. 2013. "Speech to Royal United Services Institute, Sydney." In *Chief of Army Speeches and Transcripts*, 26 March 2013. Canberra: Department of Defence. Accessed on 5 May 2017, http://103.11.78.168/Our-work/Speeches-and-transcripts/Royal-United-Services-Institute.

Mullen, Michael G. 2011. "From the Chairman." *Joint Force Quarterly*, no. 60:2–3.

Myers, Norman. 1994. *Ultimate Security: the environmental basis of political stability*. New York: W. W. Norton.

NAB. 2015. "Climate Change." Published online: National Australia Bank. Accessed on 12 October 2015, https://www.nab.com.au/about-us/corporate-responsibility/responsibility-management-of-our-business/environment/climate-change.

NASA Earth Observatory. 2016. "The Atmosphere's Energy Budget." Goddard Space Flight Center, Washington DC: NASA Earth Observatory. Accessed on 2 November 2016, http://earthobservatory.nasa.gov/Features/EnergyBalance/page6.php.

An Act to establish a comprehensive and coordinated national climate policy and program, and for other purposes. National Climate Program Act (95th Congress, USC 2901, Public Law 95-367, 17 September 1978. Accessed on 21 September 2015, http://www.gpo.gov/fdsys/pkg/STATUTE-92/pdf/STATUTE-92-Pg601.pdf.

National Energy Policy Development Group. 2001. *National Energy Policy: Report of the National Energy Policy Development Group* Washington, DC: US Government Printing Office. Accessed on 13 January 2013, http://www.wtrg.com/EnergyReport/National-Energy-Policy.pdf.

NCCARF. 2013. *Emergency management and climate change adaptation*. Policy Guidance Briefs: Brisbane: National Climate Change Adaptation Research Facility. Accessed on 5 October 2015, https://www.nccarf.edu.au/sites/default/files/attached_files_publications/EM-A4-Webview.pdf.

National Defense Authorization Act for Fiscal Year 2008. National Defense Authorization Act for Fiscal Year 2008 (110th US Congress, Public Law 110-181, 28 January 2008. Accessed on 21 September 2013, http://www.gpo.gov/fdsys/pkg/PLAW-110publ181/html/PLAW-110publ181.htm.

Nelson, Brendan. 2007a. "Address to the CEDA State of the Nation Conference 2007, 'Constructing a Regional Security Community'." In *Transcripts of the Minister for Defence* 14 June 2007. Canberra: Department of Defence. [Site archived. Document in author possession and available on request].

Nelson, Brendan. 2007b. "Opening Address to D+I Conference Adelaide." In *Transcripts of the Minister for Defence* 22 August 2007. Canberra: Department of Defence. [Site archived. Document in author possession and available on request].

Neslen, Arthur. 2015. "Pentagon to lose emissions exemption under Paris climate dea." Online: The Guardian. Accessed on 10 November 2016, https://www.theguardian.com/environment/2015/dec/14/pentagon-to-lose-emissions-exemption-under-paris-climate-deal.

Newell, Peter, Harriet Bulkeley, Karen Turner, Christopher Shaw, Simon Caney, Elizabeth Shove, and Nicholas Pidgeon. 2015. "Governance traps in climate change politics: re-framing the debate in terms of responsibilities and rights." *WIREs Climat Change*, vol. 6, no. Nov/Dec:535–540. doi:10.1002/wcc.356.

NIC. 2008. "National Intelligence Assessment on the National Security Implications of Global Climate Change to 2030", Thomas Fingar. Washington D.C.: National Intelligence Council. Accessed on 17 February 2013, http://online.wsj.com/public/resources/documents/FingarSFR.GCC06252008.pdf.

NOAA ESRL. 2016. "Full Mauna Loa CO_2 record." P. Tans and R. Keeling (Eds.). Boulder: National Oceanic and Atmospheric Administration Earth System Research Laboratory. Accessed on 2 November 2016, http://www.esrl.noaa.gov/gmd/ccgg/trends/.

NRC. 2013. *Climate and Social Stress: Implications for Security Analysis*. Committee on Assessing the Impacts of Climate Change on Social and Political Stresses: Washington, DC: The National Academies Press. Accessed on 23 November 2016, https://www.nap.edu/catalog/14682/climate-and-social-stress-implications-for-security-analysis.

O'Lear, Shannon, and Simon Dalby. 2016. *Reframing Climate Change: Constructing Ecological Geopolitics*. Edited by Shannon O'Lear and Simon Dalby. London: Routledge.

Obama, Barack H. 2008. "Videotaped Remarks to the Bi-Partisan Governors Global Climate Summit November 18, 2008." In *Obama-Biden Transition Project Press Releases*. Washington, DC: The Office of the President Elect. Accessed on 20 May 2017, http://www.presidency.ucsb.edu/ws/index.php?pid=84875.

Obama, Barack H. 2009a. *Executive Order 13514: Federal Leadership in Environmental, Energy, and Economic Performance*. Washington, DC: White House. Accessed on 20 May 2017, https://www.gpo.gov/fdsys/pkg/FR-2009-10-08/pdf/E9-24518.pdf.

Obama, Barack H. 2009b. "A Just and Lasting Peace." In *Nobel Lecture*, 10 December 2009. Oslo: Nobel Media. Accessed on 13 October 2013, http://www.nobelprize.org/nobel_prizes/peace/laureates/2009/obama-lecture_en.html.

Obama, Barack H. 2009c. "The President's News Conference in Copenhagen." In *Public Papers of the Presidents of the United States: Barack Obama* 18 December 2009, vol. (Book) II, pp. 1823–1828. Washington, DC: US Government Printing Office. Accessed on 20 May 2017, https://www.gpo.gov/fdsys/pkg/PPP-2009-book2/pdf/PPP-2009-book2-Doc-pg1823-2.pdf.

Obama, Barack H. 2009d. "President Obama on Historic Energy Bill." In *White House Briefing Room*, 24 June 2009. Washington, DC: White House. [Site archived. Document in author possession and available on request].

Obama, Barack H. 2009e. "Remarks at a Plenary Session of the United Nations Climate Change Conference in Copenhagen, Denmark." In *Public Papers of the Presidents of the United States: Barack Obama* 18 December 2009, vol. (Book) II, pp. 1820–1822. Washington, DC: US Government Printing Office. Accessed on 13 October 2013, http://www.gpo.gov/fdsys/granule/PPP-2009-book2/PPP-2009-book2-Doc-pg1820/content-detail.html.

Obama, Barack H. 2009f. "Remarks at the United Nations Climate Change Summit in New York City." In *The American Presidency Project*, edited by Gerhard Peters and John T. Woolley, 22 September 2009. Published online: The American Presidency Project. Accessed on 3 March 2014, http://www.presidency.ucsb.edu/ws/index.php?pid=86657&st=&st1=.

Obama, Barack H. 2009g. "Remarks by the President at United Nations Secretary General Ban Ki-Moon's Climate Change Summit (United Nations Headquarters, New York)." In *President Barack Obama, Speeches and Remarks*, 22 September 2009. Washington, DC: White House. [Site archived. Document in author possession and available on request].

Obama, Barack H. 2009h. "Remarks by the President to the United Nations General Assembly (United Nations Headquarters, New York)." In *President Obama, Speches and Remarks*, 23 September 2009. Washington, DC: White House. [Site archived. Document in author possession and available on request].

Obama, Barack H. 2009i. "Remarks on Energy." In *Compilation of Presidential Documents*, (January, 26), 26 January 2009. Washington, DC: US Government Printing Office. Accessed on 3 September 2013, http://www.gpo.gov/fdsys/pkg/DCPD-200900019/content-detail.html.

Obama, Barack H. 2009j. "Remarks on the Declaration by the Major Economies Forum on Energy and Climate Change in L'Aquila, Italy." In *Public Papers of the Presidents of the United States: Barack Obama* 9 July 2009, vol. (Book) II, pp. 1072–1075. Washington, DC: US Government Printing Office. Accessed on 11 October 2013, http://www.gpo.gov/fdsys/granule/PPP-2009-book2/PPP-2009-book2-Doc-pg1072/content-detail.html.

Obama, Barack H. 2009k. "Remarks to the Ghanian Parliament in Accra." In *Public Papers of the Presidents of the United States: Barack Obama*, 11 July 2009, vol. (Book) II, pp. 1087–1092. Washington, DC: US Government Printing Office. Accessed on 8 November 2013, http://www.gpo.gov/fdsys/granule/PPP-2009-book2/PPP-2009-book2-Doc-pg1087/content-detail.html.

Obama, Barack H. 2009l. "Remarks to the United Nations General Assembly in New York City." In *Public Papers of the Presidents of the United States: Barack Obama* 23 September 2009, vol. (Book) II, pp. 1440–1447. Washington, DC: US Government Printing Office. Accessed on 4 May 2013, http://www.gpo.gov/fdsys/granule/PPP-2009-book2/PPP-2009-book2-Doc-pg1440/content-detail.html.

Obama, Barack H. 2010. "Remarks by The President on Energy Security at Andrews Air Force Base." In *President Barack Obama, Speeches and Remarks*, 31 March 2010. Washington, DC: White House. Accessed on 3 March 2014, http://www.presidency.ucsb.edu/ws/?pid=107665.

Obama, Barack H. 2011. "The President's news conference in Kapolei, Hawaii." In *Compilation of Presidential Documents*, 13 November 2011. Washington, DC: US Government Printing Office. Accessed on 13 October 2013, http://www.gpo.gov/fdsys/pkg/DCPD-201100864/content-detail.html.

Obama, Barack H. 2012a. "Address before a Joint Session of Congress on the State of the Union." In *The American Presidency Project*, edited by Gerhard Peters and John T. Woolley. Published online: The American Presidency Project. Accessed on 18 August 2013, http://www.presidency.ucsb.edu/ws/?pid=99000.

Obama, Barack H. 2012b. "Remarks by the President in a News Conference." In *President Barack Obama, Speeches and Remarks*, 14 November 2012. Washington, DC: White House. Accessed on 7 October 2014, http://www.whitehouse.gov//the-press-office/2012/11/14/remarks-president-news-conference.

Obama, Barack H. 2013. *Executive Order 13653: Preparing the United States for the Impacts of Climate Change*. Washington, DC: White House. Accessed on 6 January 2017, https://www.gpo.gov/fdsys/pkg/FR-2013-11-06/pdf/2013-26785.pdf.

Obama, Barack H. 2014. "Remarks by the President at U.N. Climate Change Summit." In *White House Briefing Room*, 23 September 2014. Washington, DC: White House. Accessed on 12 January 2015, http://www.whitehouse.gov/the-press-office/2014/09/23/remarks-president-un-climate-change-summit.

Obama, Barack H. 2015. *Executive Order 13693: Planning for Federal Sustainability in the Next Decade*. Washington, DC: White House. Accessed on 6 January 2017, https://www.whitehouse.gov/the-press-office/2015/03/19/executive-order-planning-federal-sustainability-next-decade.

Obama, Barack H. 2016. *Presidential Memorandum: Climate Change and National Security*. Washington, DC: White House. Accessed on 23 December 2016, https://www.whitehouse.gov/the-press-office/2016/09/21/presidential-memorandum-climate-change-and-national-security.

Oels, Angela. 2012. "From 'Securitization' of Climate Change to 'Climatization' of the Security Field: Comparing Three Theoretical Perspectives." In *Climate Change, Human Security and Violent Conflict: Challenges for Societal Stability*, edited by Jürgen Scheffran, Michael Brzoska, Hans Günter Brauch, Peter Michael Link and Janpeter Schilling, vol. 8, pp. 185–206. Berlin: Springer.

Oliver, Alex. 2015. *Lowy Institute Poll 2015*. Sydney: The Lowy Institute. Accessed on 12 September 2015, http://www.lowyinstitute.org/files/final_2015_lowy_institute_poll.pdf.

Olson, Wyatt. 2014. "PACOM not waiting on politics to plan for climate change challenges." Hawaii: Stars and Stripes. Accessed on 14 December 2014, http://www.stripes.com/news/pacom-not-waiting-on-politics-to-plan-for-climate-change-challenges-1.297433.

Oswald Spring, Úrsula, and Hans Günter Brauch. 2011. "Coping with Global Environmental Change—Sustainability Revolution and Sustainable Peace." In *Coping with Global Environmental Change, Disasters and Security: Threats, Challenges, Vulnerabilities and Risks*, edited by Hans Gunter Brauch, Ursula Oswald Spring, Czeslaw Mesjasz, John Grin, Patricia Kameri-Mbote, Bechir Chourou, Pal Dunay and Jorn Birkmann, vol. 5, pp. 1487–1503. Berlin: Springer.

Palazzo, Albert. 2012. "The Future of War Debate in Australia: Why has there not been one? Has the need for one now arrived?" In *Land Warfare Studies Centre Working Paper Series*, Working Paper 140. Canberra: Land Warfare Studies Centre. Accessed on 20 May 2017, https://www.army.gov.au/sites/g/files/net1846/f/wp140-the-future-of-war-debate-in-australia.pdf.

Palmer, Dr Jennifer L. 2009. "Addressing Energy as a Military Cost." *Australian Army Journal For The Profession of Arms*, no. 178:6–31.

Panetta, Leon E. 2012a. "Environmental Defense Fund Reception (Energy, Security, and the Environment)." In *Secretary of Defense Speech*, 2 May 2012. Washington, DC: Department of Defense. [Site archived. Document in author possession and available on request].

Panetta, Leon E. 2012b. "Shangri-La Security Dialogue (Singapore)." In *The IISS Shangri-La Dialogue 14th Asia Security Summit*, 2 June 2012. Washington, DC: International Institute for Strategic Studies. [Site archived. Document in author possession and available on request].

Paul Ehrlich and Anne Ehrlich. 1988. "The Environmental Dimensions of National Security." In *Global Problems and Common Security: Annals of Pugwash 1988*, edited by J. Rotblat and V. I. Goldanskii, pp. 180–190. Heidelberg: Springer.

Pearce, Alix. 2013. "The New Security Climate: Rethinking Australian Defence Policy in an Age of Climate Change." Honours in Government and International Relations, School of Social and Political Sciences, University of Sydney.

Pecl, Gretta, Stewart Frusher, Caleb Gardner, Marcus Haward, Alistair Hobday, Sarah Jennings, Melissa Nursey-Bray, André Punt, Hilary Revill, and Ingrid van Putten. 2009. *The east coast Tasmanian rock lobster fishery—vulnerability to climate change impacts and adaptation response options*. Department of Climate Change. Canberra: Commonwealth of Australia. Accessed on 3 October 2014, http://www.environment.gov.au/system/files/resources/48b0b14e-9234-480f-9a6b-602274396991/files/rock-lobser-report.pdf.

Penny, Christopher K. 2007. *Climate Change and the Security Council: A Preliminary Framework for Implementing Remedial Measures through Chapter VII of the UN Charter*. CISDL Legal Working Paper Series on Climate Change Law and Policy: Montreal: Centre for International Sustainable Development Law. Accessed on 12 June 2013, http://cisdl.org/public/docs/legal/Penny%20-%20Climate%20Change%20and%20the%20Security%20Council.pdf.

Perlo-Freeman, Sam, Aude Fleurant, Pieter Wezeman, and Siemon Wezeman. 2016. *Trends in World Military Expenditure, 2015*. SIPRI Military Expenditure Database: Stockholm: SIPRI. Accessed on 1 November 2016, http://books.sipri.org/files/FS/SIPRIFS1604.pdf.

Personal Communication. 2013. "Email correspondence." Unpublished. 17 July 2013.

PEW. 2009. Economy, Jobs Trump All Other Policy Priorities In 2009: Environmental, Immigration, Health Care Slip Down the List. edited by Andrew Kohut. Washington, DC: The PEW Research Center.

Pittock, A. Barrie. 2009. *Climate Change: The Science, Impacts and Solutions*. Second ed. Melbourne: CSIRO Publishing

PLA. 2015. *China's Military Strategy*. Beijing: The State Council Information Office of the People's Republic of China. Accessed on 25 July 2015, http://news.xinhuanet.com/english/china/2015-05/26/c_134271001.htm.

Pope Francis. 2015. *On care for our common home, Encyclical Letter Laudato Si' Of the Holy Father Francis*. Vatican: Vatican Press.

Press, Anthony, Anthony Bergin, and Eliza Garnsey. 2013. *Heavy Weather: Climate and the Australian Defence Force*. Special Report 49. Canberra: Australian Strategic Policy Institute. Accessed on 12 March 2014, https://www.aspi.org.au/publications/special-report-issue-49-heavy-weather-climate-and-the-australian-defence-force.

Reich, Jenny, John Hearps, Andrew Cohn, Jeromy Temple, and Peter McDonald. 2006. *Defence Personnel Environment Scan 2025*. Canberra: Director Strategic Personnel Planning and Research for Department of Defence. Accessed on 13 February 2012, http://trove.nla.gov.au/work/26715104?selectedversion=NBD41326420.

Revelle, Roger R., Robert B. Bacastow, James Baker, Charles D. Keeling, Hans H. Landsberg, Ralph A. Llewellyn, Murray Mitchell, Walter H. Munk, Harry Perry, George D. Robinson, Joseph Smagorinsky, Verner E. Suomi, and Warren M. Washington. 1977. *Studies in Geophysics: Energy and Climate*. Washington, DC: National Academy of Sciences. Accessed on 20 May 2017, https://www.nap.edu/read/12024/chapter/1.

Revkin, Andrew C. 2005. "Bush Aide Softened Greenhouse Gas Links to Global Warming." *The New York Times*, Arthur Ochs Sulzberger, 8 June 2005. Accessed on 10 January 2017, http://www.nytimes.com/2005/06/08/politics/bush-aide-softened-greenhouse-gas-links-to-global-warming.html.

Rintoul, Stuart. 2009. "Town of Beaufort changed Tony Abbott's view on climate change." The Australian. Accessed on 31 October 2016, http://www.theaustralian.com.au/archive/politics/the-town-that-turned-up-the-temperature/story-e6frgczf-1225809567009.

Roughead, Gary. 2009. "Chief of Naval Operations, ADM Gary Roughead delivers remarks at the Current Strategy Forum, Naval War College, Newport, RI." In *2010 Speeches: ADM Gary Roughead*, 16 June 2009. Washington, DC: US Navy. [Site archived. Document in author possession and available on request].

Roughead, Gary. 2010a. "Chief of Naval Operations Adm. Gary Roughead deliver remarks at Government Executive Magazine at the National Press Club." In *2010 Speeches: ADM Gary Roughead*, 10 November 2010. Washington, DC: US Navy. http://www.navy.mil/navydata/people/cno/Roughead/Speech/101111-govexec-roughead.pdf.

Roughead, Gary. 2010b. "Chief of Naval Operations Adm. Gary Roughead Delivers Remarks during the Current Strategy Forum Newport, Rhode Island." In *2010 Speeches: ADM Gary Roughead*, 8 June 2013. Washington, DC: US Navy. Accessed on 6 August 2013, http://www.navy.mil/navydata/leadership/mist.asp?x=S&yr=2010.

Roughead, Gary. 2010c. "Chief of Naval Operations Admiral Gary Roughead Remarks as delivered World Affairs Council of Northern California." In *2010 Speeches: ADM Gary Roughead*, 5 August 2010. Washington, DC: US Navy. Accessed on 6 September 2013, http://www.navy.mil/navydata/leadership/mist.asp?x=S&yr=2010.

Roughead, Gary. 2010d. "Chief of Naval Operations, ADM Gary Roughead delivers remarks at the Navy Energy Forum." In *2010 Speeches: ADM Gary Roughead*, 12 October 2010. Washington, DC: US Navy. Accessed on 12 October 2013, http://www.navy.mil/navydata/leadership/mist.asp?x=S&yr=2010.

Roughead, Gary. 2011a. "Chief of Naval Operations (CNO) Adm. Gary Roughead speaks at Active in the Arctic Seminar June 16, 2011." In *2011 Speeches: ADM Gary Roughead*, 16 June 2011. Washington, DC: US Navy. Accessed on 8 June 2013, http://www.navy.mil/navydata/people/cno/Roughead/Speech/110616%20Arctic%20Capitol%20Hill.pdf.

Roughead, Gary. 2011b. "Chief of Naval Operations Adm. Gary Roughead delivers remarks at the 4th Symposium on the Impacts of an Ice-Diminishing Arctic on Naval & Maritime Operations." In *2011 Speeches: ADM Gary Roughead*, 20 June 2011. Washington, DC: US Navy. Accessed on 7 September 2013, http://www.navy.mil/navydata/people/cno/Roughead/Speech/110620%20Arctic%20Symposium.pdf.

Royal Australian Navy. 2007. *Future Maritime Operating Concept—2025: Maritime Force Projection and Control*. Canberra: Department of Defence. Accessed on 12 June 2013, http://www.navy.gov.au/media-room/publications/future-maritime-operating-concept-2025.

Rudd, Kevin. 2010a. "Prime Minister Speech at the opening of the National Security College Australian National University, Canberra: The establishment of the National Security College of Australia." In *PM Transcripts: Transcripts from the Prime Ministers of Australia*, 24 April 2010. Canberra: Department of the Prime Minister and Cabinet. Accessed on 3 March 2014, http://pmtranscripts.pmc.gov.au/release/transcript-17245.

Rudd, Kevin M. 2008a. "Address to the East Asia Forum in conjunction with the Australian National University, Advancing Australia's Global and Regional Economic Interests." In *PM Transcripts: Transcripts by the Prime Ministers of Australia*, 26 March 2008. Canberra: Department of Prime Minister and Cabinet. Accessed on 11 February 2014, http://pmtranscripts.pmc.gov.au/release/transcript-15823.

Rudd, Kevin M. 2008b. "Address to the London School of Economics: Australia and the UK—Global Partners in shaping the future Global Order, London." In *PM Transcripts: Transcripts from the Prime Ministers of Australia*, 7 April 2008. Canberra: Department of Prime Minister and Cabinet. Accessed on 18 February 2014, http://pmtranscripts.pmc.gov.au/release/transcript-15847.

Rudd, Kevin M. 2008c. "Australia 2020—Setting our Nation's Sights for the Future, Address to the Sydney Institute Annual Dinner 2008, Grand Habour Ballroom, Star City, Sydney." In *PM Transcripts: Transcripts from the Prime Ministers of Australia*, 16 April 2008. Canberra: Department of Prime Minister and Cabinet. Accessed on 5 March 2014, http://pmtranscripts.pmc.gov.au/release/transcript-15873.

Rudd, Kevin M. 2008d. "Building a Better World Together, Speech at Kyoto University." In *PM Transcripts: Transcripts from the Prime Ministers of Australia*, 09 June 2008. Canberra: Department of Prime Minister and Cabinet. Accessed on 2 April 2013, http://pmtranscripts.pmc.gov.au/release/transcript-15955.

Rudd, Kevin M. 2008e. *Building on ASEANs success: towards an Asia-Pacific Community, Singapore Lecture Series*. Singapore: Institute of South East Asian Studies Publishing.

Rudd, Kevin M. 2008f. "Climate Change and Water." Canberra: National Library of Australia Accessed on 10 February 2014, http://pandora.nla.gov.au/pan/79983/20080812-0001/www.pm.gov.au/topics/climate.html.

Rudd, Kevin M. 2008g. "The First National Security Statement to the Parliament Address by the Prime Minister of Australia The Hon. Kevin Rudd MP." In *PM Transcripts: Transcripts of the Prime Ministers of Australia*, 4 December 2008. Canberra: Department of Prime Minister and Cabinet. Accessed on 5 March 2013, http://pmtranscripts.pmc.gov.au/release/transcript-16289.

Rudd, Kevin M. 2009a. "Address to the 64th session of the United Nations General Assembly, New York." In *PM Transcripts: Transcripts from the Prime Ministers of Australia*, 23 September 2009. Canberra: Department of Prime Minister and Cabinet. Accessed on 12 March 2014, https://pmtranscripts.pmc.gov.au/release/transcript-16828.

Rudd, Kevin M. 2009b. "Address to the Australia-Israel Chamber of Commerce and General Sir John Monash Foundation Leadership Luncheon Docklands, Melbourne." In *PM Transcripts: Transcripts by the Prime Ministers of Australia*, 24 April 2009. Canberra: Department of Prime Minister and Cabinet. Accessed on 12 March 2014, https://pmtranscripts.pmc.gov.au/release/transcript-16515.

Rudd, Kevin M. 2009c. "Address to the Lowy Institute." In *PM Transcripts: Transcripts from the Prime Ministers of Australia*, 06 November 2009. Canberra: Department of Prime Minister and Cabinet. Accessed on 12 March 2014, http://pmtranscripts.pmc.gov.au/release/transcript-16904.

Rudd, Kevin M. 2009d. "Focus on Climate Change—Prime Minister Blog." Canberra: National Library of Australia. Accessed on 11 February 2014, http://pandora.nla.gov.au/pan/79983/20091030-1529/www.pm.gov.au/PM_Connect/PMs_Blog/Climate_Change_Blog.html.

Rudd, Kevin M. 2009e. "Partners in the Global Economy—Australia and the United States, Address to the US Chamber of Commerce, Washington." In *PM Transcripts: Transcripts from the Prime Ministers of Australia*, 31 March 2009. Canberra: Department of Prime Minister and Cabinet. Accessed on 3 March 2014, https://pmtranscripts.pmc.gov.au/release/transcript-15832.

Rudd, Kevin M. 2009f. "Transcript of address at the close of the Pacific Islands Forum Cairns." In *PM Transcripts: Transcripts from the Prime Ministers of Australia*, 7 August 2009. Canberra: Department of Prime Minister and Cabinet. Accessed on 4 March 2013, https://pmtranscripts.pmc.gov.au/release/transcript-16737.

Rudd, Kevin M. 2009g. "Transcript of joint doorstop at the United Nations with Minister Wong New York." In *PM Transcripts: Transcripts by the Prime Ministers of Australia*, 23 September 2009. Canberra: Department of Prime Minister and Cabinet. Accessed on 2 March 2014, https://pmtranscripts.pmc.gov.au/release/transcript-16829.

Rudd, Kevin M. 2010b. "Prime Minister Transcript of interview with Kerry O'Brien The 7.30 Report 12 May 2010." In *PM Transcripts: Transcripts from the Prime Ministers of Australia*, 12 May 2010. Canberra: Department of Prime Minister and Cabinet. Accessed on 4 March 2014, https://pmtranscripts.pmc.gov.au/release/transcript-17292.

Rumsfeld, Donald H. 2003. "FY2004 Defense Budget Testimony (As Prepared)." In *Transcripts of the Secretary of Defense*, 5 February 2003. Washington, DC: Department of Defense. [Site archived. Document in author possession and available on request].

Russian Federation. 2009. *Russia's National Security Strategy to 2020*. Moscow. Accessed on 22 November 2016, http://rustrans.wikidot.com/russia-s-national-security-strategy-to-2020.

Russian Federation. 2015. *Russian National Security Strategy*. Moscow. Accessed on 22 November 2016, http://www.ieee.es/Galerias/fichero/OtrasPublicaciones/Internacional/2016/Russian-National-Security-Strategy-31Dec2015.pdf.

Sales, Leigh. 2013. "The toxic political debate of climate change." Canberra: Australian Broadcasting Corporation. Accessed on 11 February 2014, http://www.abc.net.au/7.30/content/2013/s3831850.htm.

Savage, Lisa. 2015. "Elephant in the Room: The Pentagon's Massive Carbon Footprint." Online: Counterpunch. Accessed on 10 November 2016, http://www.counterpunch.org/2015/07/23/72279/.

Schubert, Renate, Hans Joachim Schellnhuber, Nina Buchmann, Astrid Epiney, Rainer Grießhammer, Margareta E. Kulessa, Dirk Messner, Stefan Rahmstorf, and Jürgen Schmid. 2008. *Climate Change as a Security Risk*. Berlin: Earthscan. Accessed on 20 May 2017, http://ccsl.iccip.net/wbgu_jg2007_engl.pdf.

Schwartz, Peter, and Doug Randall. 2003. *An Abrupt Climate Change Scenario and Its Implications for United States National Security*. Washington, DC: Department of Defense. Accessed on 23 October 2010, http://stephenschneider.stanford.edu/Publications/PDF_Papers/SchwartzRandall2004.pdf.

Scott, Steven, and Sarah Vogler. 2014. "G20 Brisbane: Tony Abbott clashes with Barack Obama over Climate Change." *The Courier Mail*, News Corp, 17 November 2014. Accessed on 10 March 2015, http://www.couriermail.com.au/news/queensland/g20-brisbane-tony-abbott-clashes-with-barack-obama-over-climate-change/story-fnmd7bxx-1227125019686.

Shabecoff, Philip. 1988. "Global Warming Has Begun, Expert Tells Senate." *The New York Times*, The New York Times, 24 June 1988. Accessed on 20 May 2017, http://www.nytimes.com/1988/06/24/us/global-warming-has-begun-expert-tells-senate.html?pagewanted=all.

Shanahan, Rodger. 2013. "Editorial." *Australian Army Journal for The Profession of Arms*, vol. 10, no. 4:5–7.

Sindico, Francesco. 2007. "Climate Change: A Security (Council) Issue?" *Carbon and Climate Law Review*, vol. 1, no.:26–31.

Slaughter, Anne-Marie, Troy Sternberg, Sarah Johnstone, Jeffrey Mazo, Francesco Femia, Caitlin Werrell, Michael Werz, Max Hoffman, David Michel, and Mona Yacoubian. 2013. *The Arab Spring and Climate Change*. A Climate and Security Correlations Series: Washington, DC: The Center for American Progress.

Smith, Stephen. 2012. "Minister for Defence and Minister for Defence Materiel—Joint Media Release—Ocean Shield arrives in Australia." In *Transcripts of the Minister for Defence* 30 June 2012. Canberra: Department of Defence. [Site archived. Document in author possession and available on request].

Snowden, Warren. 2008. "Australia's Strategic Imperatives Presented to the ASPI Global Forces Conference." In *Defence Minister: Ministerial Releases*, 2 July 2008. Canberra: Department of Defence. [Site archived. Document in author possession and available on request].

Spencer, Tom, Nick Mabey, Chad Briggs, Elena Belluci, and Geraud de Ville. 2009. *Climate Change & The Military: The State of the Debate*. Institute for Environmental Security Military Advisory Council: Belgium: Institute for Environmental Security. Accessed on 4 April 2013, http://www.tomspencer.info/articles/StateoftheDebate.pdf.

Spratt, David. 2012. "Scientists call for war on climate change, but who on earth is listening." Published online: Climate Code Red. Accessed on 6 October 2015, http://www.climatecodered.org/2012/12/scientists-call-for-war-on-climate.html.

Stang, Gerald. 2014. "Climate change and EU Security: When and how they intersect." Online: European Union Institute for Security Studies. Accessed on 20 November 2016, http://www.iss.europa.eu/uploads/media/Brief_32_climate_change.pdf.

Steffen, Will, Andrew A. Burbidge, Lesley Hughes, Roger Kitching, David Lindenmayer, Warren Musgrave, Mark Stafford Smith, and Patricia A. Werner. 2009. *Australia's Biodiversity and Climate Change: A Strategic Assessment of the vulnerability of Australia's biodiversity to climate change*. Canberra: Commonwealth of Australia. http://www.environment.gov.au/system/files/resources/eab369d6-76f9-46c8-beb4-aaae8ece112e/files/biodiversity-vulnerability-assessment.pdf.

Steffen, Will, John Hunter, and Lesley Hughes. 2014. *Counting the Coast: Climate Change and Coastal Flooding*. Sydney: Climate Council. Accessed on 5 April 2014, http://www.climatecouncil.org.au/uploads/56812f1261b168e02032126342619dad.pdf.

Stern, Nicholas. 2007. *Stern Review on the Economics of Climate Change*. Cambridge: Cambridge University Press.

Sturrock, Robert, and Peter Ferguson. 2015. *The Longest Conflict: Australia's Climate Security Challenge*. Melbourne: Centre for Policy Development.

Tanzler, Dennis, Alexander Carius, and Sebastian Oberthur. 2002. *Climate Change and Conflict: Can Climate Impacts Increase Conflict Potentials? What is the Relevance of this Issue for the International Process on Climate Change?* Nature Conservation and Nuclear Safety Federal Ministry for the Environment. Berlin: Federal Ministry for the Environment, Nature Conservation and Nuclear Safety. Accessed on 10 February 2013, http://www.afes-press.de/pdf/Brauch_ClimateChange_BMU.pdf.

Tauberer, Josh. 2015. "H.R. 1961 (100th): Global Climate Change Security Oversight Act." Ben Hammer. Published online: Civic Impulse, LLC. Accessed on 10 November 2015, https://www.govtrack.us/congress/bills/110/hr1961.

Terreu, Julia. 2014. "Can the Australian Military Think Out Loud?" In *The Diplomat*, edited by James Pach, 1 April 2014. Tokyo: Trans-Asia Inc. Accessed on 12 March 2015, http://thediplomat.com/2014/04/can-the-australian-military-think-out-loud/.

The New Daily. 2014. "Runaway climate debate keeps Abbott hopping." In *The New Daily*, 16 November 2014. Published online: The New Daily Pty. Ltd. Accessed on 12 January 2015, http://thenewdaily.com.au/news/2014/11/16/climate-debate-tony-abbott-g20/.

Thomas, Michael. 2011. "Climate Change and the ADF." *Journal of the Australian Profession of Arms*, no. 185:34–43.

Thomas, Michael. 2012. "Assessing the Risk of Global Climate Change on the Australian Defence Force." e-IR. Accessed on 10 October 2012, http://www.e-ir.info/author/michael-thomas/.

Thomas, Michael. 2013. "The Securitisation of Climate Change: A Military Perspective." *Journal of the Australian Profession of Arms* vol. Nov/Dec, no. 192:7–18.

Thomas, Michael. 2015. "Climate securitization in the Australian political-military establishment." *Global Change Peace and Security*, vol. 27, no. 1:97–118. doi:10.1080/14781158.2015.990879.

Thomson, Mark. 2013. *The Cost of Defence: ASPI Defence Budget Brief 2013–14*. Canberra: Australian Strategic Policy Institute. Accessed on 15 May 2014, https://www.aspi.org.au/publications/the-cost-of-defence-aspi-defence-budget-brief-2013-2014/ASPI-CostDefence2013.pdf.

Titley, David. 2015. "The Australian Climate Security Summit." Canberra. Personal communication. 29 October 2015.

Titley, David W. 2010a. "Climate Change Finance: Providing assistance for vulnerable countries." In *Hearing Before the Subcommittee on Asia, the Pacific and the Global Environment of the Committee on Foreign Affairs House of Representatives One Hundred Eleventh Congress Second Session*, 111th Congress, Serial No. 111–120, 27 July 2010. Washington, DC: US Government Printing Office. http://www.gpo.gov/fdsys/pkg/CHRG-111hhrg57687/html/CHRG-111hhrg57687.htm.

Titley, David W. 2010b. "Climate Change Investments in a Fiscally Restrained Environment Time." In *Department of Defense Bloggers Roundtable*, edited by John Ohab, 18 June 2010. Washington, DC: Federal News Service. Accessed on 10 May 2013, http://www.dodlive.mil/files/2010/06/0618_Rear-Adm_Titley.pdf.

Titley, David W. 2010c. "Global Climate Change." Climate and Energy Symposium 2010: Proceedings on Climate and Energy, Imperatives for Future Naval Forces, Baltimore. Accessed on 18 October 2012, http://www.jhuapl.edu/ClimateAndEnergy/Book/2010CEBook.pdf.

Titley, David W. 2010d. "A Rational Discussion of Climate Change: The Science, the Evidence, the Response." In *Hearing before the Sub Committee on Energy and Environment Committee on Science and Technology House of Representatives One Hundred Eleventh Congress Second*

Session, 111th Congress, Serial No. 111–114, 17 November 2010. Washington, DC: US Government Printing Office. Accessed on 10 April 2013, http://www.gpo.gov/fdsys/pkg/CHRG-111hhrg62618/pdf/CHRG-111hhrg62618.pdf.

Torney, Diarmuid. 2015. *European Climate Leadership in Question: Policies toward China and India*. Online: MIT Press.

Townsend, Mark, and Paul Harris. 2004. "Now the Pentagon tell Bush: climate change will destroy us." *The Observer*, Guardian News. Accessed on 20 May 2017, https://www.theguardian.com/environment/2004/feb/22/usnews.theobserver.

Tremblay, Pascal, Anna Boustead, Stephen Turton, Robyn Wilson, Tracey Dickson, Bruce Hayllar, Rosallie Chapple, Sue Uzabeaga, Kate Hammill, Beatrice Pegard, Mario Rimini, Phoenix Lawhon Isler, John Merson, Bradley Jorgensen, Maureen Rogers, Peter Hayman, Graham Brown, Jenny Davies, Amanda Lynch, Carolina Roman, Lee Tryhorn, Tien Duc Pham, Peter Forsyth, and Ray Spurr. 2008. *The Impact of Climate Change on Australian Tourism Destinations*. Published online: Commonwealth of Australia. Accessed on 10 October 2015, http://www.bmwhi.org.au/docs/The%20Impacts%20of%20Climate%20Change%20Summary.pdf.

Trenberth, Kevin E., J. T. Fasullo, and J. Kiehl. 2009. "Earth's Global Energy Budget." *Bulletin of American Meteorological Society*, vol. 90, no. 3:311–324. doi:10.1175/2008BAMS2634.1.

Trombetta, Maria J. 2008. "Environmental Security and Climate Change: Analysing the discourse." *Cambridge Review of International Affairs*, vol. 21, no. 4:585–602. doi:10.1080/09557570802452920.

Trombetta, Maria J. 2012. "Climate Change and the Environmental Conflict Discourse." In *Climate Change, Human Security and Violent Conflict: Challenges for Societal Stability*, edited by Jürgen Scheffran, Michael Brzoska, Hans Günter Brauch, Peter Michael Link and Janpeter Schilling, vol. 8, pp. 151–164. Berlin: Springer.

Tschakert, Petra. 2015. "1.5°C or 2°C: a conduit's view from the science-policy interface at COP20 in Lima, Peru." *Climate Change Responses*, vol. 2, no. 3. doi:10.1186/s40665-015-0010-z.

Turnbull, Malcolm. 2016. "Remarks at the Pacific Island Forum." In *Media Transcripts*, 9 September 2016. Canberra: Commonwealth of Australia. Accessed on 31 October 2016, https://www.pm.gov.au/media/2016-09-09/remarks-pacific-island-forum.

UK. 2007. "Letter dated 5 April 2007 from the Permanent Representative of the United Kingdom of Great Britain and Northern Ireland to the United Nations addressed to the President of the Security Council." Online: United Nations. Accessed on 20 May 2017, http://www.securitycouncilreport.org/atf/cf/%7B65BFCF9B-6D27-4E9C-8CD3-CF6E4FF96FF9%7D/CC%20S2007%20186.pdf.

UK DECC. 2016. *Final UK Greenhouse gas emissions national statistics: 1990–2014 (2014 UK ghg: final figures—data tables)*. Department of Energy and Climate Change. London: UK Government. Accessed on 12 November 2016, https://www.gov.uk/government/statistics/final-uk-greenhouse-gas-emissions-national-statistics-1990-2014.

UK MoD. 2010a. *MOD Climate Change Delivery Plan*. Ministry of Defence. London: Ministry of Defence. Accessed on 10 April 2014, https://www.gov.uk/government/publications/mods-climate-change-strategy. (Document archived, see website. In possession of author).

UK MoD. 2010b. *MoD Climate Change Strategy 2010*. Ministry of Defence. London: Ministry of Defence UK. Accessed on 10 April 2014, https://www.gov.uk/government/publications/mods-climate-change-strategy. (Document archived, see website. In possession of author).

UK MoD. 2013. *Annex D to Sustainability Report—MOD Annual Report and Accounts 2011–2012*. Ministry of Defence. London: UK Ministry of Defence. Accessed on 12 November 2016, https://www.gov.uk/government/uploads/system/uploads/attachment_data/file/35010/20121205_mod_annualreport_Annex_D.pdf.

UK MoD. 2015. *Sustainable MOD: Annual Report 2014/15: Sustainability in the Ministry of Defence*. UK Ministry of Defence. London: UK Government. Accessed on 12 November 2016, https://www.gov.uk/government/uploads/system/uploads/attachment_data/file/447951/20150723-Sustainable_MOD_Annual_Report-internet-ver.pdf.

Ullman, Richard H. 1983. "Redefining Security." *International Security*, vol. 8, no. 1:129–153. doi:10.2307/2538489.

UN. 1988. "Our Common Future: Report of the World Commission on Environment and Development." In *Secretary General A/42/427—Development and International Co-operation: Environment*, 4 August 1987. New York: United Nations. Accessed on 10 May 2013, http://www.un-documents.net/a42-427.htm.

UN. 2002. "Background Note on the "Arria-Formula" Meetings of the Security Council Members." United Nations Secretariat. New York: United Nations. Accessed on 19 January 2017, http://www.un.org/en/sc/about/methods/bgarriaformula.shtml.

UN. 2016. "Leading on climate change." Online: United Nations. Accessed on 17 January 2017, http://www.un.org/sustainabledevelopment/ban-ki-moon-climate-change/.

UNFCCC. 1997. *Kyoto Protocol to the United Nations Framework Convention on Climate Change*. Kyoto: United Nations. Accessed on 12 June 2013, http://unfccc.int/essential_background/kyoto_protocol/items/1678.php.

UNFCCC. 2012. "Kyoto Protocol: Doha Amendment." New York: United Nations. Accessed on 6 February 2013, http://unfccc.int/kyoto_protocol/items/2830.php.

UNFCCC. 2015. "National Inventory Submissions." New York: United Nations. Accessed on 9 November 2016, http://unfccc.int/national_reports/annex_i_ghg_inventories/national_inventories_submissions/items/8108.php.

UNGA. 2008. "Battle against climate change calls for 'war footing', Deputy Secretary-general says as General Assembly holds follow-up to February thematic debate." In *General Assembly Plenary Press Release*, GA/10725, 8 July 2008. New York: United Nations. Accessed on 20 May 2017, http://www.securitycouncilreport.org/atf/cf/%7B65BFCF9B-6D27-4E9C-8CD3-CF6E4FF96FF9%7D/res%2063%20281.pdf.

UNGA. 2009. "Resolution 63/281: Climate change and its possible security implications." In *United Nations General Assembly 63rd Session* Agenda item 107, A/RES/63/291, 11 June 2009. New York: United Nations. Accessed on 28 January 2013, http://www.securitycouncilreport.org/atf/cf/%7B65BFCF9B-6D27-4E9C-8CD3-CF6E4FF96FF9%7D/res%2063%20281.pdf.

UNSC. 2007. "Letter dated 5 April 2007 from the Permanent Representative of the United Kingdom of Great Britain and Northern Ireland to the United Nations addressed to the President of the Security Council (S/2007/186)." In *United Nations Security Council 5663rd Meeting*, S/PV.5663, 17 April 2007. New York: United Nations.

UNSC. 2011. "Maintenance of international peace and security: Impact of climate change." In *United Nations Security Council 6587th meeting*, S/PV.6587, 20 July 2011. New York: United Nations. Accessed on 20 May 2017, http://www.circleofblue.org/wp-content/uploads/2012/09/UN-Security-Council_climate-change-July-2011.pdf.

UNSG. 2008. "We have moved climate change up to top of the agenda." In *Secretary-General Statements and Messages, Press Release*, vol. SGM/SM/11408. New York: United Nations. Accessed on 18 April 2012, http://www.un.org/News/Press/docs/2008/sgsm11408.doc.htm.

UNSG. 2009. "Report of the Secretary-General: Climate change and its possible security implications (A/64/350)." In *United Nations General Assembly 64th session*, Item 114, A/64/350, 11 September 2009. New York: United Nations. Accessed on 15 May 2017, http://www.un.org/ga/search/view_doc.asp?symbol=A/64/350.

UNSG. 2011. "Possible Security Implications of Climate Change." New York: United Nations. Accessed on 1 March 2013, http://www.un.org/News/Press/docs/2011/sc10332.doc.htm.

UNSG. 2012. "Secretary-General's video message to Planet Under Pressure Conference." In *Statement by the Secretary-General*, 29 March 2012. New York: United Nations. Accessed on 16 March 2013, http://www.un.org/sg/statements/index.asp?nid=5960.

US DoD. 2005. *The National Defense Strategy of the United States of America*. Washington, DC: Department of Defense. Accessed on 6 August 2013, http://www.au.af.mil/au/awc/awcgate/nds/nds2005.pdf.

US DoD. 2006. *Quadrennial Defense Review Report*. Washington, DC: Department of Defense. Accessed on 18 August 2013, http://archive.defense.gov/pubs/pdfs/QDR20060203.pdf.

US DoD. 2008. *National Defense Strategy*. Washington, DC: Department of Defense. Accessed on 2 May 2013, http://www.defense.gov/Portals/1/Documents/pubs/2008NationalDefenseStrategy.pdf.

US DoD. 2010. *Quadrennial Defense Review*. Washington, DC: Department of Defense. Accessed on 20 May 2017, https://www.defense.gov/Portals/1/features/defenseReviews/QDR/QDR_as_of_29JAN10_1600.pdf.

US DoD. 2011. *The National Military Strategy of the United States of America 2011: Redefining America's Military Leadership*. Washington, DC: Department of Defense. Accessed on 5 March 2013, http://www.defense.gov/Portals/1/Documents/pubs/2011-National-Military-Strategy.pdf.

US DoD. 2012a. "Department of Defense FY-2012 Climate Change Adaptation Roadmap." In *Department of Defense Strategic Sustainability Performance Plan FY-2012*. Washington, DC: Department of Defence. Accessed on 20 May 2017, http://www.denix.osd.mil/sustainability/dod-sspp/unassigned/dod-sspp-fy-2012/. (FY-2012 SSPP document archived. Document in possession of author. See FY-2014 SSPP weblink herein for nearest version).

US DoD. 2012b. *Department of Defense Strategic Sustainability Performance Plan FY-2012*. Washington, DC: Department of Defense. Accessed on 9 November 2016, http://www.denix.osd.mil/sustainability/dod-sspp/unassigned/dod-sspp-fy-2012/.

US DoD. 2013a. *Department of Defense Strategic Sustainability Performance Plan FY-2013*. Washington, DC: Department of Defense. Accessed on 5 March 2014, http://www.denix.osd.mil/sustainability/loader.cfm?csModule=security/getfile&pageid=35931

US DoD. 2013b. *National Defense Budget Estimates for FY-2014*. Washington, DC: Office of the Under Secretary of Defense. Accessed on 20 May 2017, http://comptroller.defense.gov/Portals/45/Documents/defbudget/fy2014/FY14_Green_Book.pdf.

US DoD. 2014a. "2014 Climate Change Adaptation Roadmap." In *Department of Defense Strategic Sustainability Performance Plan FY-2014*. Washington, DC: Department of Defense. Accessed on 20 May 2017, http://www.denix.osd.mil/sustainability/dod-sspp/unassigned/dod-sspp-fy-2014/.

US DoD. 2014b. *Department of Defense Strategic Sustainability Performance Plan FY 2014*. Defense. Alexandria: Office of the Deputy Secretary of Defense for Installations and Environment.

US DoD. 2014c. *Quadrennial Defense Review 2014*. Washington, DC: Department of Defense. Accessed on 20 May 2017, http://archive.defense.gov/pubs/2014_Quadrennial_Defense_Review.pdf.

US DoD. 2015. *National Security Implications of Climate-Related Risks and a Changing Climate*. US DoD. Washington, DC: US Government. Accessed on 11 February 2016, http://archive.defense.gov/pubs/150724-congressional-report-on-national-implications-of-climate-change.pdf?source=govdelivery.

US DoD. 2016a. *Department of Defense Strategic Sustainability Performance Plan FY 2016*. Washington, DC: Department of Defense. Accessed on 9 November 2016, http://www.denix.osd.mil/sustainability/dod-sspp/unassigned/department-of-defense-strategic-sustainability-performance-plan-fy-2016/ and http://www.denix.osd.mil/sustainability/dod-sspp/.

US DoD. 2016b. *DoD Directive 4715.21: Climate Change Adaptation and Resilience*. US DoD. Washington, DC: Office of the Under Secretary of Defence for Acquisition, Technology, and Logistics. Accessed on 11 February 2016, http://www.dtic.mil/whs/directives/corres/pdf/471521p.pdf.

US DoE. 2016. "Greenhouse Gas Emissions." Washington, DC: US Department of Energy. Accessed on 10 November 2016, http://catalog.data.gov/dataset/federal-greenhouse-gas-inventories-and-performance/resource/347d9ac8-0388-4b41-b064-41404b9915ec.

US EPA. 2016. "Annex 3 (Part A) Methodological Descriptions for Additional Source or Sink Categories." United States: US Environmental Protection Agency. Accessed on 10 November 2016, https://www.epa.gov/ghgemissions/inventory-us-greenhouse-gas-emissions-and-sinks-1990-2014 and https://www.epa.gov/sites/production/files/2016-04/documents/us-ghg-inventory-2016-annex-3-additional-source-or-sink-categories-part-a.pdf.

US GAO. 2014. *Climate Change Adaptation: DOD Can Improve Infrastructure Planning and Processes to Better Account for Potential Impacts*. US Government Accountability Office. Washington, DC: US Government. Accessed on 7 January 2017, http://www.gao.gov/assets/670/663734.pdf.

US GAO. 2015. "Limiting the Federal Government's Fiscal Exposure by Better Managing Climate Change Risks." Washington DC: US Government Accountability Office. Accessed on 23 July 2015, http://www.gao.gov/highrisk/limiting_federal_government_fiscal_exposure/why_did_study#t=1.

US House of Representatives. 2006. "Department of Energy's Plan for Climate Change Technology Programmes." In *Hearing Before the Subcommittee on Energy and Committee on Science*, 109th Congress, Second session, Serial No. 109-62, 20 September 2006. Washington, DC: US Government Printing Office. Accessed on 16 March 2013, http://www.gpo.gov/fdsys/pkg/CHRG-109hhrg29851/html/CHRG-109hhrg29851.htm.

US House of Representatives. 2007a. "Global Climate Change Security Oversight Act." In, H.R. 1961 (110th), 19 April 2007. Washington, DC: US Congress. Accessed on 23 December 2016, https://www.govtrack.us/congress/bills/110/hr1961.

US House of Representatives. 2007b. "The National Security Implications of Climate Change." In *Hearing Before the Subcommittee on Investigations and Oversight Committee on Science and Technology*, 110th Congress, First session, Serial No. 110-58, 27 September 2007. Washington, DC: US Government Printing Office. Accessed on 16 March 2013, http://www.gpo.gov/fdsys/pkg/CHRG-110hhrg34720/pdf/CHRG-110hhrg34720.pdf.

US House of Representatives. 2008a. "National Intelligence Assessment on the National Security Implications of Global Climate Change." In *Joint Hearing Before the Select Committee on Energy Independence and Global Warming and Subcommittee on Intelligence, Community Management, Permanent Select Committee on Intelligence House of Representatives*, 110th Congress, Serial No. 110-41, 25 June 2008. Washington, DC: US Government Printing Office. Accessed on 1 September 2013, https://www.fas.org/irp/congress/2008_hr/climate.pdf.

US House of Representatives. 2008b. "National Security Implications of Global Climate Change." In *Joint Hearing before the Select Committee on Energy Independence and Global Warming and Subcommittee on Intelligence, Community Management, Permanent Select Committee on Intelligence*, 110th Congress, Second session, Serial No. 110-41, 25 June 2008. Washington, DC: US Government Printing Office. Accessed on 6 March 2014, http://www.gpo.gov/fdsys/pkg/CHRG-110hhrg62523/html/CHRG-110hhrg62523.htm.

US House of Representatives. 2013. "President Obama's Fiscal 2014 Budget Proposal for the U.S. Navy and Marine Corps." In *House Appropriations Subcommittee on Defense Hearing*, 7 May 2013. Washington, DC: US Government Printing Office. Accessed on 6 March 2014, http://www.navy.mil/navydata/people/cno/Greenert/Testimony/130507%20HACD%20Hearing%20on%202014%20Navy%20and%20Marine%20Corps%20Budget.pdf.

US Navy. 2007. *A Cooperative Strategy for 21st Century Seapower*. Washington, DC: US Navy. Accessed on 20 May 2017, http://www.navy.mil/local/maritime/150227-CS21R-Final.pdf.

US Navy. 2009a. *Task Force Climate Change Charter*. Washington, DC: US Navy. Accessed on 6 March 2013, http://greenfleet.dodlive.mil/files/2010/09/Task-Force-Climate-Change-Charter.pdf.

US Navy. 2009b. *US Navy Arctic Roadmap*. Washington, DC: US Navy. Accessed on 3 August 2013, http://www.navy.mil/navydata/documents/USN_artic_roadmap.pdf.

US Navy. 2010a. *Energy Programme for Security and Independence*. Washington, DC: US Navy. Accessed on 6 August 2013, http://greenfleet.dodlive.mil/files/2010/04/Naval_Energy_Strategic_Roadmap_100710.pdf.

US Navy. 2010b. *Navy Climate Change Roadmap*. Washington, DC: US Navy. Accessed on 6 March 2012, http://www.navy.mil/navydata/documents/CCR.pdf.

US Navy. 2010c. *A Navy Energy Vision for the 21st Century*. Washington, DC: US Navy. Accessed on 6 March 2013, http://greenfleet.dodlive.mil/files/2010/10/Navy-Energy-Vision-Oct-2010.pdf.

US Navy. 2011. *Arctic Environmental Assessment and Outlook Report In support of The Navy Arctic Roadmap Action Item 5.7*. Washington, DC: US Navy. Accessed on 7 June 2013, http://greenfleet.dodlive.mil/files/2011/08/U.S.-Navy-Arctic-Environmental-Assessment.pdf.

US Senate. 2006. "The Lugar-Biden Climate Change Resolution." In *U.S. Senate Foreign Relations Committee*, S. Res. 312, 23 May 2006. Washington, DC: US Senate Accessed on 14 November 2016, http://www.c2es.org/docUploads/S%20Res%20312%20as%20passed%20by%20SFRC%20May%2023%2006.pdfhttp://www.c2es.org/federal/congress/109/summary-lugar-biden-climate-change-senate-resolution-312.

US Senate. 2007. "Climate Change: National Security Threats." In *Hearing Before the Committee on Foreign Relations*, 110th Congress, Hearing No. 110-402, 9 May 2007. Washington, DC: US Government Printing Office. Accessed on 14 November 2016, https://www.gpo.gov/fdsys/pkg/CHRG-110shrg42725/pdf/CHRG-110shrg42725.pdf.

US Senate. 2009. "Climate Change and Global Security: Challenges, Threats and Diplomatic Opportunities." In *Hearing Before the Committee on Foreign Relations*, 111th Congress, Senate Hearing No. 111-207, 21 July 2009. Washington, DC: US Government Printing Office. Accessed on 14 November 2016, https://www.gpo.gov/fdsys/pkg/CHRG-111shrg54243/html/CHRG-111shrg54243.htm.

USACE. 2014. Adaption Policy Statement. In *USACE Climate Preparedness and Resilience Policy Statement*, edited by Jo-Ellen Darcy: US Army Corps of Engineers.

Victorian Government. 2011. *Review of the Climate Change Act 2010*. Department of Sustainability and Environment. Melbourne: Department of Premier and Cabinet. Accessed on 5 May 2017, http://www.delwp.vic.gov.au/__data/assets/pdf_file/0008/323936/Independent-Review-of-the-Climate-Change-Act-2010.pdf.

Vis, Peter, and Jos Delbeke. 2015. *EU Climate Policy Explained*. Abingdon: Routledge.

Warren, Dane. 2015. *Climate change and International Peace and Security: Possible Roles for the U.N. Security Council in Addressing Climate Change*. New York: Columbia Law School. Accessed on 18 January 2017, https://web.law.columbia.edu/sites/default/files/microsites/climate-change/warren_-_cc_and_international_peace_and_security_-_roles_for_the_un_security_council.pdf.

Watt, David. 2012. "Budget 2011–12: Australian Defence Force (ADF) operations." In *Parliamentary Library Research Publications, Australian Defence Force (ADF) Operations*. Canberra: Commonwealth of Australia. Accessed on 10 May 2013, http://www.aph.gov.au/About_Parliament/Parliamentary_Departments/Parliamentary_Library/pubs/rp/BudgetReview201112/ADF.

Watts, Jonathon, and John Vidal. 2009. "China blamed as anger mounts over climate deal." *The Guardian*, The Guardian, 20 December 2009. Accessed on 5 June 2014, http://www.theguardian.com/environment/2009/dec/20/china-blamed-copenhagen-climate-failure?guni=Article:in%20body%20link.

Weart, Spencer R. 2008. *The Discovery of Global Warming*. Second ed. Cambridge: Harvard University Press.

Werrell, Caitlin, and Francesco Femia. 2014. "Must Watch: PACOM Commander on the Climate Change Threat to the Asia-Pacific." Online: The Center for Climate and Security. Accessed on 10 October 2015, http://climateandsecurity.org/2014/03/07/must-watch-pacom-commander-on-the-climate-change-threat-to-the-asia-pacific/.

Werz, Michael, and Laura Conley. 2012. *Climate Change, Migration, and Conflict in Northwest Africa: Rising Dangers and Policy Options Across the Arc of Tension*. Climate Migration Series: Washington, DC. Accessed on 20 November 2016, https://cdn.americanprogress.org/wp-content/uploads/issues/2012/04/pdf/climate_migration_nwafrica.pdf.

Wesley, Michael. 2012. "Climate views have moved on." *The Australian*, Lowy Institute for International Policy, 2 July 2012. Accessed on 23 March 2013, http://www.lowyinstitute.org/publications/climate-views-have-moved.

White, Alexander. 2014. "The ten worst environment decisions in Abbott's first year." *The Guardian*, The Guardian, 6 September 2014. Accessed on 9 December 2016, https://www.theguardian.com/environment/southern-crossroads/2014/sep/05/abbott-first-year-environment-climate-ten-worst-decisions.

White, Ben. 2010. "Sustainable Defence Capability: Australia's national security and the role of defence industry." *Journal of the Australian Profession of Arms*, no. 183:87–96.

White House. 2002. *The National Security Strategy of the United States of America* White House. Washington, DC: Government of the United States of America. Accessed on 4 June 2013, http://www.state.gov/documents/organization/63562.pdf.

White House. 2006. *The National Security Strategy of the United States of America*. White House. Washington, DC: Government of the United States of America. Accessed on 15 September 2013, http://georgewbush-whitehouse.archives.gov/nsc/nss/2006/index.html.

White House. 2010. *The National Security Strategy of the United States of America*. White House. Washington, DC: Government of the United States of America. Accessed on 20 May 2017, http://nssarchive.us/NSSR/2010.pdf.

White House. 2015. *National Security Strategy*. Washington, DC: Government of the United States of America. Accessed on 20 May 2017, http://nssarchive.us/wp-content/uploads/2015/02/2015.pdf.

White, Martin. 2008. "The Compelling Requirement to Energy-Proof the Australian Defence Force." *Journal of the Australian Profession of Arms*, no. 175:12–21.

Wikipedia. 2017a. "Chairman of the Joint Chiefs of Staff." Online: Wikipedia. Accessed on 15 January 2017, https://en.wikipedia.org/wiki/Chairman_of_the_Joint_Chiefs_of_Staff.

Wikipedia. 2017b. "Chief of Naval Operations." Online: Wikipedia. Accessed on 15 January 2017, https://en.wikipedia.org/wiki/Chief_of_Naval_Operations.

Wikipedia. 2017c. "Chief of the Defence Force (Australia)." Online: Wikipedia. Accessed on 15 January 2017, https://en.wikipedia.org/wiki/Chief_of_the_Defence_Force_(Australia)).

Wikipedia. 2017d. "Department of Defence (Australia)." Online: Wikipedia. Accessed on 20 May 2017, https://en.wikipedia.org/wiki/Department_of_Defence_(Australia).).

Wikipedia. 2017e. "List of United States Secretaries of Defense by time in office." Online: Wikipedia. Accessed on 15 January 2017, https://en.wikipedia.org/wiki/List_of_United_States_Secretaries_of_Defense_by_time_in_office.

Wikipedia. 2017f. "Minister for Defence (Australia)." Online: Wikipedia. Accessed on 20 May 2017, https://en.wikipedia.org/wiki/Minister_for_Defence_(Australia)).

Wikipedia. 2017g. "United States Secretary of the Navy." Online: Wikipedia. Accessed on 15 January 2017, https://en.wikipedia.org/wiki/United_States_Secretary_of_the_Navy.

Williams, Richard J., Ross A. Bradstock, Geoffrey J. Cary, Neal J. Enright, A. Malcolm Gill, Adam C. Liedloff, Christopher J. Whelan, Alan N. Andersen, David J. M. S. Bowman, Peter J. Clarke, Garry D. Cook, Kevin J. Hennessy, and Alan York. 2009. *Interactions between climate change, fire regimes and biodiversity in Australia: A Preliminary Assessment*. Water Department of Climate Change and Department of the Environment, Heritage and the Arts. Canberra: Commonwealth of Australia. Accessed on 5 October 2014, http://www.environment.gov.au/system/files/resources/ab711274-004c-46cb-a9f1-09dba68d98c3/files/20100630-climate-fire-biodiversity-pdf.pdf.

WMO. 1988. "The Changing Atmosphere: Implications for Global Security." In *Toronto Conference on the Changing Atmosphere*, Series No. 710, pp. 292–304. Toronto: Secretariat of the World Meteorological Organization.

Wolfers, Arnold. 1952. "'National Security' as an Ambiguous Symbol." *Political Science Quarterly*, vol. 67, no. 4:481–502.

Woodward, Peter. 2011. "The Reserve Response Force and Public Safety." *Australian Army Journal For The Profession of Arms*, vol. 8, no. 3:101–120.

World Bank. 2013. *Turn Down the Heat: Climate Extremes, Regional Impacts and the Case for Resilience*. Washington, DC: World Bank. Accessed on 5 May 2017, http://documents.worldbank.org/curated/en/975911468163736818/Turn-down-the-heat-climate-extremes-regional-impacts-and-the-case-for-resilience-full-report.

World Bank. 2016. "Total greenhouse gas emissions." Online: World Bank. Accessed on 12 November 2016, http://data.worldbank.org/indicator/EN.ATM.GHGT.KT.CE.

World Vision. 2015. "Climate change: The effects on children." Published online: World Vision. Accessed on 23 September 2015, http://www.worldvision.com.au/global-issues/work-we-do/climate-change/climate-change-the-effects-on-children.

Youngs, Richard. 2014. *Climate Change and EU Security Policy: An Unmet Challenge*. Washington, DC: Carnegie Endowment for International Peace.

Youngs, Richard. 2015. *Climate Change and European Security*. London: Taylor and Francis.

Zhang, Chun. 2009. "Regional Reactions to the Defence White Paper 2009: Rebuilding middle power leadership for Australia." *Australian Defence Force Journal: Journal of the Australian Profession of Arms*, no. 180:7–9.

About the Author

Dr. Michael Thomas is an Australian climate and water security research officer. He is a graduate of the Australian Defence Force Academy, the Royal Military College (Duntroon) and the Australian Capability and Technology Management College. He served for twenty years as an officer in the Australian Army, including stints in signals-intelligence, capability management and military instruction. His doctorate examined how military forces have responded to the threat posed by climate change, and he has been active in educating global military leaders for six years through his previous work with the Australian Defence Force Global Change and Energy Sustainability Initiative, as a Visiting Fellow at the Australian Defence College and as a visiting lecturer at the University of New South Wales and Deakin University. Dr. Thomas has completed climate resilience and capacity building studies at the United Nations University Tokyo and, together with his Ph.D. from the University of New South Wales, holds three post-graduate degrees in Oceanography, in International Relations and in Management (Defence Capability and Acquisition). He was a co-author on the influential 2015 Australian Climate Council publication *Be Prepared* that addressed climate security risks in the Australian Defence sector. Any views expressed in this book are his own. This is his first book.

Email: mickdthomas@gmail.com.

Additional information on this book is at: http://afes-press-books.de/html/APESS_10.htm.

M.D. Thomas, *The Securitization of Climate Change: Australian and United States' Military Responses (2003–2013)*, The Anthropocene: Politik—Economics—Society—Science 10, DOI 10.1007/978-3-319-49658-0

MIX
Papier aus verantwortungsvollen Quellen
Paper from responsible sources
FSC® C105338

If you have any concerns about our products,
you can contact us on
ProductSafety@springernature.com

In case Publisher is established outside the EU,
the EU authorized representative is:
Springer Nature Customer Service Center GmbH
Europaplatz 3, 69115 Heidelberg, Germany

Printed by Libri Plureos GmbH
in Hamburg, Germany